MICROBIAL
BIOMASS

ECONOMIC MICROBIOLOGY

Series Editor

A. H. ROSE

ECONOMIC MICROBIOLOGY
Volume 4

MICROBIAL BIOMASS

edited by

A. H. ROSE

*School of Biological Sciences
University of Bath,
Bath, England*

1979

ACADEMIC PRESS

LONDON NEW YORK TORONTO SYDNEY SAN FRANCISCO

A Subsidiary of Harcourt Brace Jovanovich, Publishers

ACADEMIC PRESS INC. (LONDON) LTD.
24/28 Oval Road
London NW1

United States edition published by
ACADEMIC PRESS INC.
111 Fifth Avenue
New York, New York 10003

British Library Cataloguing in Publication Data

Microbial biomass. - (Economic
  microbiology; vol. 4).
  7. Industrial microbiology 2. Microbial growth
  I. Rose, Anthony Harry II. Series
  660'.62   QR53   77-77361

  ISBN 0-12-596554-0

*Filmset and printed in Great Britain by
Willmer Brothers Limited Birkenhead*

CONTRIBUTORS

K. BAYER, Institute of Applied Microbiology, University of Agriculture, Vienna, Austria.

JOHN R. BENEMANN, Sanitary Engineering Research Laboratory, University of California, 1301 South 46 Street, Richmond, California 94804, U.S.A.

LEE A. BULLA, Jr., Science and Education Administration, U.S. Grain Marketing Research Laboratory, U.S. Department of Agriculture, Manhattan, Kansas 66502, U.S.A.

S. BURROWS, Distillers Company Limited, Glenochil Technical Centre, Menstrie, Clackmannanshire, Scotland.

JOE C. BURTON, Nitragin Co., 3101 West Custer Avenue, Milwaukee, Wisconsin 53209, U.S.A.

CLAYTON D. CALLIHAN, Department of Chemical Engineering, Louisiana State University, Baton Rouge, Louisiana 70803, U.S.A.

JAMES E. CLEMMER, Department of Chemical Engineering, Louisiana State University, Baton Rouge, Louisiana 70803, U.S.A. (Present address: Department of Chemical Engineering, University of Mississippi, Oxford, Mississippi, U.S.A.)

KO SWAN DJIEN, Department of Food Science, Agricultural University, De Dreyen, 12 Wageningen, The Netherlands.

G. P. EBBON, BP Research Centre, Chertsey Road, Sunbury-on-Thames, Middlesex TW16 7LN, England.

A. J. FORAGE, Philip Lyle Memorial Research Laboratory, P.O. Box 68, Reading RG6 2BX, England.

G. HAMER, Kuwait Institute for Scientific Research, P.O. Box 12009, Kuwait.

W. A. HAYES, Mushroom Science Unit, Department of Biological Sciences, University of Aston, Gosta Green, Birmingham B4 7ET, England.

C. W. HESSELTINE, Northern Regional Research Center, U.S. Department of Agriculture, Peoria, Illinois 61604, U.S.A.

J. D. LEVI, BP Research Centre, Chertsey Road, Sunbury-on-Thames, Middlesex TW16 7LN, England. (Present address: Conservation of Clean Air and Water in Europe, The Hague, The Netherlands.)

J. MEYRATH, Institute of Applied Microbiology, University of Agriculture, Vienna, Austria.

WILLIAM J. OSWALD, Sanitary Engineering Research Laboratory, University of California, 1301 South 46 Street, Richmond, California 94804, U.S.A.

R. C. RIGHELATO, Philip Lyle Memorial Research Laboratory, P.O. Box 68, Reading RG6 2BX, England.

ANTHONY H. ROSE, Zymology Laboratory, School of Biological Sciences, Bath University, Claverton Down, Bath BA2 7AY, England.

JEAN L. SHENNAN, BP Research Centre, Chertsey Road, Sunbury-on-Thames, Middlesex TW16 7LN, England.

JOSEPH C. WEISSMAN, Sanitary Engineering Research Laboratory, University of California, 1301 South 46 Street, Richmond, California 94804, U.S.A.

S. H. WRIGHT, Mushroom Science Unit, Department of Biological Sciences, University of Aston, Gosta Green, Birmingham B4 7ET, England.

ALLAN A. YOUSTEN, Department of Biology, Virginia Polytechnic Institute and State University, Blacksburg, Virginia 24061, U.S.A.

PREFACE

Previous volumes in this series described production by microbiological processes of chemical compounds, including ethanol and associated flavour compounds in alcoholic beverages, that are of established commercial importance. This volume deals with commercial production of bulk quantities of micro-organisms, otherwise known as microbial biomass. Since by definition biomass is formed during exponential growth, it can be considered as a primary product of metabolism. Moreover, almost all of the commercially viable processes for production of microbial biomass have as their objective the availability of microbial protein, be it enzymically active or not. As such, the aim of these processes can be considered production of microbial protein.

The processes described in this volume fall into two classes. Firstly, there are those which entail production of viable, and therefore enzymically active, biomass. These processes are few in number, but range from the large and well established manufacture of baker's yeast to such small, and still to be fully exploited, processes as production of bacteria for inoculation into soil to promote nitrogen fixation and for use as insecticides.

The second class of process is equally diverse, although all of the processes have as their aim the manufacture of bulk quantities of micro-organism for use as food or fodder for animals, and conceivably in some instances for human beings. The *raison d'etre* for these processes is that it is now firmly established that the major nutritional deficiency in diets of man and animals is protein, and that the most rapid means of manufacturing large quantities of protein is by growing microbes in bulk. Human and animal diets have, for centuries, been supplemented with valuable protein, usually inadvertently, in the form of traditional dishes whose preparation involves microbial growth. These dishes, particularly in the less well developed regions of the world, still make and will continue to make a valuable contribution to diets. At the other extreme, there are processes, mostly still at the developmental stage, which can produce large bulk quantities of micro-organism, with sophisticated technology, often from relatively expensive substrates. In many respects, this volume appears at a propitious time, for the rapid escalation in costs of energy world-wide has halted the development of most of these processes, and has forced operators to look in greater detail at the possibility of producing dietary microbial biomass from cheaper sources, including waste materials. The future of the more sophisticated process is very problematical indeed. Nevertheless, recent changes in the economic climate of the world have not altered the conviction of many nutritionists that microbial biomass can in the future make a valuable contribution to animal and human diets. We must wait and see.

August 1979   ANTHONY H. ROSE

CONTENTS

1. History and Scientific Basis of Large-Scale Production of Microbial Biomass

ANTHONY H. ROSE

2. Baker's Yeast

S. BURROWS

3. Bacteria for Azofication
JOE C. BURTON

4. Bacterial Insecticides
LEE A. BULLA,Jr. and ALLAN A. YOUSTEN

5. Tempe and Related Foods

KO SWAN DJIEN and C. W. HESSELTINE

6. Edible Mushrooms

W. A. HAYES and S. H. WRIGHT

7. Algal Biomass

JOHN R. BENEMANN, JOSEPH C. WEISSMAN
and WILLIAM J. OSWALD

8. Biomass from Whey

J. MEYRATH and K. BAYER

9. Biomass from Cellulosic Materials

CLAYTON D. CALLIHAN and JAMES E. CLEMMER

10. Biomass from Carbohydrates

A. J. FORAGE and R. C. RIGHELATO

11. Biomass from Natural Gas

G. HAMER

12. Biomass from Liquid *n*-Alkanes

J. D. LEVI, JEAN L. SHENNAN and G. P. EBBON

NOTES

Abbreviations

The abbreviations used for chemical and biochemical compounds in this book are those recommended by the International Union of Pure and Applied Chemistry—International Union of Biochemistry Commission on Biochemical Nomenclature, and summarized in the *Biochemical Journal* (1976; **153**, 1–24). S.C.P. indicates single-cell protein.

Names of Micro-Organisms

In general, the names of bacteria used are those recommended in *Bergey's Manual of Determinative Bacteriology* (8th edition, 1974, edited by R. E. Buchanan and N. E. Gibbons, and published by Williams and Wilkins Co. of Baltimore) and those of filamentous fungi which were adopted in the *Dictionary of the Fungi* (6th edition, written by G. C. Ainsworth and W. Bisby, and published in 1971 by the British Commonwealth Mycological Institute at Kew). Names of yeasts are those recommended in *The Yeasts, a Taxonomic Study* (2nd edition, 1970, edited by J. Lodder, and published by the North-Holland Publishing Co. of Amsterdam, Holland).

1. History and Scientific Basis of Large-Scale Production of Microbial Biomass

ANTHONY H. ROSE

Zymology Laboratory, School of Biological Sciences,
Bath University, Bath, Avon, England

I. HISTORY OF LARGE-SCALE MICROBE CULTIVATION

Plant and animal material which is acceptable and nutritionally valuable as a food for man is, by its very nature, susceptible to spoilage by micro-organisms. Where microbial spoilage of a foodstuff leads to the formation

1

of toxic substances in that food, Man has over the centuries learned to avoid eating the food, and more recently to prevent microbial spoilage of that type of food. On the other hand, microbial spoilage of a food may alter the appearance rather than the safety of a food. A case in point is spoilage of fermented beverages by bacteria and wild yeasts: a type of spoilage that is still encountered, but which must have been commonplace when these beverages were drunk from pewter vessels. Other types of microbial spoilage of foodstuffs are not only accepted but are actually encouraged. These include growth of moulds in cheeses and in the various tempe-type foods, in all of which microbial activity adds to the flavour and acceptability of the food. Encouraging growth of microbes in foods also adds to the nutritional value of the food, and it is interesting to note the comment made by Thaysen (1957) that an average daily diet of a fermented food could furnish between 6 and 10 g of microbial matter, and therefore 2–4 g of protein.

In a historical context, the development of processes for large-scale cultivation of micro-organisms is best considered under the following three headings.

A. Edible Fungi

When early man looked round for food to eat, he was quickly attracted by the fruiting bodies of several fungi—mushrooms—which have a pleasant and attractive taste. Early civilizations came to marvel at the ability of mushrooms to appear seemingly spontaneously overnight, and without seeds having to be sown, and as a consequence they acquired certain mystical qualities. Francis Bacon, in *Sylva Sylvarum* (1627), referred to mushrooms as a venerous feast. There is evidence, too, that they came to be associated with various forms of worship, and John Marco Allegro, in his book *The Sacred Mushroom and the Cross* (1970), has argued that the cult of the mushroom was a seminal influence in the rise of Christianity. Collecting mushrooms in the wild has continued over the centuries. It is today particularly popular in continental Europe where, sadly, several hundred deaths occur annually because some hunters are unable to recognize poisonous mushrooms.

The first organized efforts at cultivating mushrooms—species of *Agaricus*—apparently took place in France during the reign of Louis XIV (1639–1715). Underground caves in Paris provided a suitable environment and indeed still do. It had been discovered that horse manure was a good medium on which to grow mushrooms, and this was stacked in rows on the floors of the caves. The earliest description of mushroom growing was by N. de Bonnefons in an article entitled *Le jardinier françois*, and published in 1650. This was translated into English eight years later by John Evelyn under the title *The French Gardiner*. But the first authoritative account of the new art came from the eminent French botanist Tournefort in 1707. Mushroom growing was soon taken up in England and, in 1779, Abercrombie published his treatise on the garden mushroom and its cultivation. Thereafter, mushroom growing gradually became a stable industry. Today, world-wide production is approaching half a million metric tonnes annually. More detailed accounts of its development can be found in Ramsbottom's (1953) book and in the chapter by Hayes and Nair (1975).

Regular attempts have been made over the years to cultivate other fungi, often without success. Particular attention has been given to truffles, *Tuber melanosporum*, which grow underground, and which are traditionally sniffed out using trained pigs or occasionally goats. They grow beneath oak trees, and have for centuries been prized for their culinary virtues. The *département* of Dordogne in France is a well known truffle region, and at the turn of the present century sold through its town markets well over 100 tons each year. Schemes have been started in that *département* to cultivate truffles in a more organized fashion, and more than a 100 hectares of new oaks have recently been planted and treated to induce truffle growth (Scargill, 1974).

B. Baker's Yeast

Bread, which has been prepared from time immemorial, consists basically of a slurry of ground cereal, known as dough, that has been baked. The Egyptians are credited with the discovery, undoubtedly made quite unwittingly that, if the dough was left to develop a microbial flora, gases produced by fermentation of carbohydrates by the micro-organisms gave rise to bubbles in the dough which, when baked, gave a lighter and much more acceptable bread. This is the process of leavening, one which figured prominently in the history of the Jewish race. The original leavening agents were probably a mixture of yeasts and lactobacilli, which was maintained by retaining a portion of each dough before baking, and using it to start the next batch (Burrows, 1970). This practice continued for many centuries but, during the Middle Ages, use gradually was made of the surplus yeast produced during beer brewing and wine making; this was referred to as *barm*. However, using barm for leavening bread is fundamentally unsatisfactory, since the quality and activity of barm, even from a single brewery, are very variable.

Around the turn of the nineteenth century, increasing attention was given to processes that involved propagation of strains of *Saccharomyces cerevisiae* under conditions that gave a yeast with consistently good baking qualities. In the intervening years, the manufacture of baker's yeast has evolved into a highly specialized and sophisticated large-scale process (Frey, 1930). The first major advance came with the advent of the Vienna process around 1860 (Frey, 1930) which recognized that, when growing strains of *Sacch. cerevisiae* largely anaerobically in a grain mash, the yield of yeast was improved by passing a gentle stream of air through the mash. Then came Pasteur's researches, which stressed the need for a more intense aeration in order to develop high yields. A shortage of grain in most developed countries of the world during the 1914–1918 war led to molasses, a by-product in sugar manufacture, being substituted for grain. Finally, as a result of work patented almost simultaneously in Denmark (Sak, 1919) and Germany (Verein der Spiritusfabrikanten in Deutschland, 1919), and usually associated with the name of Soren Sak, there was introduced the fed batch or *Zulauf* process. In this, the carbohydrate supply (molasses solution) is fed incrementally during the growth phase, so that there is never an appreciable concentration of sugar in the culture in which, as a result, almost none of the sugar is converted into ethanol. Ethanol production represents a loss of carbohydrate which ideally should be converted into yeast cell material.

As a result of further development of the manufacturing process and genetic selection of strains of *Sacch. cerevisiae* with excellent leavening properties, many countries in the world now have highly efficient industries for manufacture of baker's yeast. Not enough credit is given to this industry which is able to produce, daily, huge quantities of a highly perishable commodity, namely pressed yeast, which must have a guaranteed fermenting capacity. Recent years have seen some attempts to foster production of active dried yeast, a form of baker's yeast which contains less water (about 8%, w/w) than pressed yeast (70%, w/w), and therefore has a much longer shelf life, and can safely be shipped to and used in hot climates.

C. Single-Cell Protein

I have already stressed that consumption of microbes in food, inadvertent or not, has long been recognized as improving the nutritional quality of foods. During the present century, ever increasing interest has been shown in the growth of micro-organisms, either in a food or in a fermenter, as a source of nutrients for domestic animals and man.

1. Mass Cultivation of Microbial Biomass

a. *Developments up to 1954.* Mass cultivated microbial biomass was often used for nutritional purposes in the period before the end of the Second World War. Nevertheless, it involved cultivation of a relatively small number of yeasts, was carried out with a limited understanding of the nutritional contributions and problems involved, and was frequently fostered by actual or impending national economic disasters rather than by a properly conceived scheme to exploit the nutritional qualities of micro-organisms.

The Germans, largely from research carried out by Max Delbrück and his colleagues at the Institut für Gärungsgewerbe in Berlin (Delbrück, 1910), were the first to appreciate the value of surplus brewer's yeast as a feeding supplement for animals. They termed this 'fodder yeast'. This experience proved more than useful in the First World War, when Germany was subjected to an economic blockade and ultimately managed to replace as much as 60% of its imported protein sources by dried or wet yeast. Braude (1942), who has reviewed early work on single-cell protein, recounts that yeast was used in wartime Germany to feed cows and poultry. Some of this was surplus brewer's yeast, but some also was specially cultivated 'mineral yeast', the yeast being grown in a semi-defined medium containing ammonium salts as the nitrogen source (Hayduck, 1913).

After the end of World War I, German interest in fodder and mineral yeast waned, but was revived around 1936 by the 'Heeresverwaltung', when both brewer's yeast, and a variety of yeasts specially mass cultured, were used to supplement human and animal diets. Around this time, two books were published in Germany describing the nutritive value of yeast (Weitzel and Winchel, 1932; Schülein, 1937). Yeasts were incorporated first into army diets, and later into civilian diets, mainly in soups and sausages. They were used too, in the earlier years, in the form of briquettes for horse fodder. Although plans were laid for production of well over 100 000 tons per annum, the figure in all likelihood never reached higher than 15 000, probably because of the extensive disruption of the German economy during World War II. Nevertheless, in technological terms, and bearing in mind the industrial difficulties that prevailed in Germany at that time, several important advances were made in yeast production. Extensive use was made of sulphite waste liquor, a by-product of the sulphite process in paper manufacture. The report of the British Intelligence Objectives Sub-Committee, which visited Germany towards the end of hostilities, states that, with one exception (*Oidium lactis*), strains of *Candida arborea* and *Torulopsis* (= *Candida*) *utilis* were employed, although most of the product consumed was probably a mixture of strains. There were at least eight producing centres, the most important being that run by I.G. Farbenindustries at Wolfen, near Leipzig.

The sustained interest in fodder yeast showed by the German workers in the inter-war years had not gone unnoticed elsewhere in the World. As part of a larger programme for utilizing natural sources, the Forest Products Laboratory of the United States Department of Agriculture, in Madison, Wisconsin, had examined mass cultivation of yeast on sulphite waste liquor, the strain used being *C. utilis* which can, unlike strains of *Sacch. cerevisiae*, utilize the pentose sugars present in these liquors. Production of fodder yeast in the mid-western states of the U.S.A. expanded steadily, and this development was reviewed by Peterson *et al.* (1945) and later by Harris (1949).

Workers in Britain also developed an interest in fodder yeast. This had begun even before the Second World War, but grew considerably when the Scientific Committee on Food Policy of the Royal Society undertook a detailed exploration of the feasibility of manufacturing fodder yeast, and of its value in animal and human nutrition. They took the view that, although it had been successful in Germany and in the United States, there was little prospect in Britain for mass cultivation of yeast on sulphite waste liquor, since that raw material was not available in bulk in the United Kingdom. They saw, however, a potentially rich source of raw material in several colonial countries, including fresh pulp of the coffee cherry and the seed pulp of *Palmyra* palm (Thaysen, 1957). However, these are available only at certain times of the year. Not so, crude cane sugar molasses. It was therefore decided to set up a pilot-plant production using molasses, under the direction of A. C. Thaysen at Teddington in

Middlesex, and transfer the process when developed to Jamaica. The developmental programme included selection of a mutant of *C. utilis* (var. *major*) which was larger in size than the wild type and therefore more easily separated from large cultures. Thaysen (1957) has provided a most valuable account of this development. The process went to Jamaica in 1944, the production plant being situated next to a sugar refinery at Frome, Westmorland, near the western tip of the island. However, the factory closed down just a few years later. This development did, however, lead to the installation of a similar process under the auspices of the Industrial Development Corporation of South Africa Ltd. The factory was at Merebank, and was technologically based on that developed in Jamaica (Thaysen, 1957). Wishing to differentiate the product from brewer's, baker's, fodder and mineral yeast, the product manufactured in Jamaica and South Africa was referred to as 'food' yeast.

b. *Acceptability of single-cell protein.* In the decade following the end of World War II, interest in production of microbial biomass for animal and human consumption continued albeit in a rather desultory manner. Most countries in the World were too pre-occupied with ensuring that the traditional sources of animal and human food were brought back to at least the standard of production that obtained in the late 1930s. Serious interest in using single-cell protein for nutritional purposes began to reappear in the mid-to-late 1950s, but this new era of interest, which is described in Section I.C.1 c (p. 14), took place in a climate that was different in two important respects from that in the years before the Second World War. The first of these was the considerably greater knowledge available on animal and human nutrition; the second was the very much more widespread awareness of the extent of food shortage in the World and the realization, later in the early 1970s, that the problem was rapidly worsening.

Hardly a day passes without the publication of yet another newspaper article, review or book on the world food crisis. The basic problem is that the expanding world population, which is currently about 4400 million and is expected to rise to about 7000 million by the year 1990, brings with it an enormous increase in the demand for food which, with present day science and technology coupled with political manoeuvring and national taboos, cannot conceivably be met. Pirie's paperback (1976) describes the situation succinctly, while Borgstrom (1973) has produced a more authoritative account. A selective bibliography of reviews on the subject has been complied by Rechcigl (1975). Brief perusal of these texts quickly reveals that scientists and politicians are very far from being in agreement as to the magnitude of the world food crisis, and of the best way to tackle it. However, these considerations are not the concern of this chapter which is rather to assess briefly the extent to which microbial biomass, produced as cheaply as possible, has, and can in the future, help to solve problems of food shortage in various countries of the world.

Any microbe that is to be grown as a source of animal or human food must have certain basic properties. The more important of these are discussed in the following paragraphs.

(i) Nutritional quality. When discussing problems of food shortage in the world, it has been customary in the past to dwell on those classes of nutrient—principally protein, vitamins and minerals—that are thought to be deficient in the diets of people in various countries of the world. These nutrient classes still deserve prime consideration, but nutritionists are increasingly coming round to the opinion that the basic need is for a balanced diet, with adequate calorie intake.

Classical research on the nutritional requirements of man and of domestic animals has shown that they are unable to synthesize for themselves certain of the twenty amino acids that are incorporated into proteins. These are the essential amino acids, and they vary in nature with each animal. Man requires eight such amino acids. Full information on these essential amino acids can be found in any of the basic texts on human and animal nutrition. It has been shown, moreover, that some of the protein sources consumed by man and domestic animals, principally in the underdeveloped countries of the world, are deficient in one or more essential amino acids. Interest in single-cell protein for animal and human consumption stems from the possibility that it can provide, when incorporated into diets, essential amino acids that are deficient in the traditionally consumed foods. Essentially the same desiderata apply when contemplating the use of single-cell protein as a source of essential vitamins and minerals. A further consideration is the daily requirement for each essential nutrient. These requirements have been determined with some accuracy for man and many domestic animals, and they dictate the amounts of single-cell protein that need to be incorporated into a traditional diet in order to make it nutritionally acceptable.

(ii) Palatability and digestibility. Although microbial biomass has seriously been considered for human consumption—and as we have noted has, in the form mainly of food yeast, been so used—it is as an animal fodder that in the foreseeable future it is most likely to be used.

The palatability of microbial biomass to domestic animals is extremely difficult to assess. Nevertheless, there is no reason to believe that, for the majority of domestic animals, microbial cell material consumed as a sizeable proportion of the diet is not perfectly acceptable. Some forms of microbial biomass, such as bacteria and algae, may however present problems. Shacklady (1975), for example, reviewed the value to animals of alkane-grown yeast biomass, and reported successful results with poultry and pigs; he also described results of trials with cattle, young stock animals, fish, rabbits and mink.

The palatability of microbial biomass for human beings has been gauged in a large number of trials, and with some disappointing results. There is no doubt that strains of *Sacch. cerevisiae* and *C. utilis* are, more often than not, quite acceptable to human beings. Nevertheless, as Scrimshaw (1975) reported in his most valuable review, there have been many reports of adverse reactions in human beings following consumption of microbial biomass. As long ago as 1947, Goyco and Asenjo (1947) reported gastro-intestinal disorders in subjects receiving as little as 15 g each day of *C. utilis* in Puerto Rico. Similar reports have followed at regular intervals, and these have been thoroughly documented by Scrimshaw (1975). Particular problems have been encountered with bacterial biomass. For example, amounts ranging from 12 to 25 g of *Aerobacter aerogenes* or *Hydrogenomonas eutropha* per day caused nausea, vomiting and diarrhoea when fed to young male volunteers, although these bacteria had been fed to experimental animals without any evidence of distress (Waslien *et al.*, 1969, 1970). Hardly anything is known of the physiological basis of the effect of some types of microbial biomass on human subjects, and it is a topic which deserves closer attention from nutritionists.

Microbial biomass is nutritionally valuable to man and animals only if proteins and vitamins contained in the biomass are released and, with proteins, digested by intestinal enzymes. Frequently, this consideration does not constitute a major obstacle to the use of microbial biomass as a food supplement. Nevertheless, it may on occasion be advisable to feed microbial biomass that has been subjected to a treatment that releases protein from the cells. Such processes have been considered most appropriate when protein is to be extracted and retextured to provide, for example, a meat analogue.

Several methods for releasing proteins from microbial biomass have been investigated over the years, and these have been reviewed by

Wimpenny (1967), Tannenbaum (1968) and Dunnill and Lilly (1975). There are two classes of method, namely mechanical and non-mechanical. Several high-pressure homogenizers have been designed to disrupt bulk quantities of microbial biomass; an example is that described by Hetherington *et al.* (1971). A more recently described process, developed by workers at Anheuser-Busch in St. Louis, Missouri, U.S.A., disrupts baker's yeast to release protein, and separates the wall fragments from the homogenate to give an additional commercial product which has been recommended as a fat analogue (Sucher *et al.*, 1973). Other mechanical methods for disrupting bulk microbial biomass involve use of high-speed ball mills. Earlier work used a commercial paint mill (Currie *et al.*, 1972) although more recently specially designed mills have been described (Dunnill and Lilly, 1975).

Of the non-mechanical methods that have been evaluated, enzymic digestion of microbial wall material is the most practicable. These methods are slow, however, and even more costly than use of homogenizers and ball mills, and they are unlikely to find a large-scale use in any cost-effective programme for producing microbial biomass for human and animal consumption.

(iii) Freedom from toxic compounds. Governments and international agencies are paying ever increasing attention to the possible toxicity of foodstuffs in general, and in particular the so-called novel food materials. Over the past quarter of a century, novel foods such as soybean concentrates manufactured by solvent extraction, peanut and cottonseed oil flours and fish protein concentrates, have been very closely scrutinized. So too has microbial biomass for possible human and animal consumption. Some years ago, the United Nations set up a Protein Advisory Group (PAG) to oversee all problems relating to novel sources of protein. The PAG in turn spawned a working party to consider microbial biomass under the chairmanship of the Russian A. A. Pokrovsky. It held its first meeting in Marseilles in 1970. The main functions of the working party have been to prepare and publish guidelines on various aspects of production of microbial biomass for food and fodder. Prominent among these aspects has been the problem of possible toxicity of some of these novel foodstuffs.

One of the first problems encountered with mass-produced microbial cell material was its content of nucleic acid. This problem was encountered, moreover, in micro-organisms, namely yeasts, that had over several decades been used widely as a human and animal food. The problem, which was first discussed at length at the conference on single-cell protein held at the Massachusetts Institute of Technology in 1967 (Mateles and Tannenbaum, 1968), arises from the fact that dietary nucleic acid is hydrolysed by nucleases in pancreatic juice, and the hydrolysis products are converted by intestinal enzymes to nucleosides which, after adsorption, give rise to free purines and pyrimidines (Kihlberg, 1972). Adenine and guanine are then converted into uric acid, but man has lost the ability to synthesize the enzyme uricase which oxidizes uric acid to the more water-soluble allantoin. Consumption of protein sources which contain high levels of nucleic acids, principally RNA, results, in man, in high blood levels of uric acid and increased excretion of this acid in the urine. Because of its low solubility in water, high levels of uric acid in the blood may lead to precipitation of uriates in tissues and joints, giving rise to symptoms that are similar to those of gout. Clearly, the problem would be of greatest concern with subjects who are genetically prone to develop gout. The estimated safe intake of nucleic acids for a healthy adult person is about 2 g per day (Scrimshaw, 1975). Since this limits the possible usefulness and application of microbial biomass for human consumption, several methods have been considered for lowering the nucleic acid content of bulk microbial biomass (Sinskey and Tannenbaum, 1975).

It is well established that the content of RNA in a micro-organism—RNA being quantitatively the more important of the two cellular nucleic acids—is proportional to the rate at which it is grown (Rose, 1976). The suggestion has been made, therefore, that the content of RNA in a feed microbe could be lowered by growing cultures at slower rates. It is all too obvious, however, that lowering the growth rate must correspondingly decrease productivity, and this means of minimizing the RNA content of bulk-produced microbial biomass has not engendered very much enthusiasm.

Methods which seem more promising are those which involve chemical or enzymic breakdown, extraction or hydrolysis of cellular RNA. Ribonucleic acid is very susceptible to base-catalysed hydrolysis, and the RNA content of biomass can be lowered by treating it with 0.3 N potassium hydroxide (Kihlberg, 1972). Alternatively, cellular RNA can be extracted with 10% hot sodium chloride or 85% phenol. Disruption of the cellular biomass may be advisable before extraction. Enzymic hydrolysis, although expensive, is in many ways to be preferred. The RNA content of yeast can be lowered considerably by treatment with pancreatic ribonuclease (Castro *et al.*, 1971). Penetration of the enzyme into the cells may be a problem, and it has been discovered that heating *C. utilis* at 80°C for 30 minutes aids entry of bovine pancreatic ribonuclease into cells (Sinskey and Tannenbaum, 1975). It has also been shown that the RNA content of biomass could be lowered by activating endogenous ribonucleases.

Toxic compounds, collectively known as mycotoxins, are produced by many types of filamentous fungi. These compounds have been described in detail by Mirocha and his colleagues in this series (Mirocha *et al.*, 1979). The extreme toxicity, and even carcinogenicity, possessed by some of these compounds has had a salutary effect on food toxologists, although in many Oriental countries of the world foods containing a lot of fungal mycelium have been the staple diet for centuries. Any filamentous fungus which is proposed as a microbial food or feed is, in consequence, subjected to very close scrutiny for possible mycotoxin production.

Recent years have seen the examination of several strains of Gram-negative bacteria as possible sources of bulk-produced microbial biomass for animal consumption. Many Gram-negative bacteria produce an endotoxin, associated with the outermost lipopolysaccharide layer of the bacterial envelope, and there is always the possibility that this type of toxin may render mass-produced bacterial biomass unacceptable for animal consumption. Little if anything has been published on this subject, probably because of the silence that accompanies any prolonged toxicity tests on a novel foodstuff.

Toxicity in a microbial foodstuff may arise not from a microbial component but from chemicals used to treat the biomass. Reference has already been made to the scrutiny to which solvent-extracted soybeans were subjected. A similarly rigorous examination has been made of microbial biomass grown on alkanes in view of the toxic nature, when present above certain concentrations, of these substrates. As described in Section III.B (p. 26), this aspect of possible toxicity in microbial biomass has had a traumatic impact on the entire industry.

(iv) Low production costs. In the foreseeable future and probably also the long term, microbial biomass will be used very largely as a source of protein in animal feed and fodder. If large-scale production of microbial biomass is to compete effectively with the alternative sources of cheap protein that are used in animal feeds, it must be produced at a cost that competes with those for production of alternatives. The two principal sources of cheap protein with which microbial biomass competes are fish meal and soybean meal, the latter being the product which remains after oil has been extracted from soybeans. Here, the picture is a complex and changing one. Availability of cheap fish meal depends to a considerable extent on the annual catch of anchovies off the coasts of Chile and Peru. Fisheries are in serious trouble in many parts of the world due to a variety of factors, including principally overfishing and to a lesser extent a lack of enforcement of international regulations concerning the size of catches, and pollution of some coastal waters. These factors have operated in the Chilean and Peruvian waters. At first it was thought that the rapid decrease in the catch off these coasts was caused by a temporary shift in the direction of the Humboldt Current, but conservationists now believe that this was not the major factor. Shortage of cheap fishmeal has caused a rise in the costs of both commodities (Fig. 1), and an increased world production of, and demand for, soybeans. Brazil, for example, increased

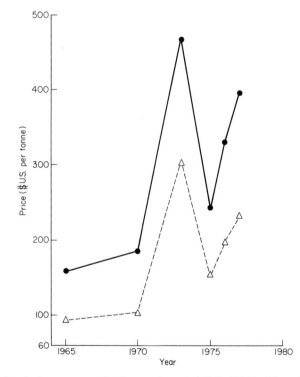

Fig. 1. Variations in the average retail price over the period 1965–1977 for fishmeal, shipped from New York (●), and soybean meal shipped from Rotterdam (△). From Done (1978) who also quotes similar data for variations in the price of fishmeal shipped from Hamburg, and for soybean meal dispatched from Chicago, over the same time period.

its production of soybeans from 523 000 tonnes in 1965 to 9 892 000 tonnes in 1975; during this period, the United States, for many years a major producer of soybeans, almost doubled its production. During the 1960s, the time therefore seemed propitious for the introduction of bulk-produced microbial biomass for animal consumption. For some time, it appeared that microbial biomass could be manufactured on a large scale at a price that would compete with fish meal and soybeans. However, problems, some of them financial arising from the four-fold increase in the world price for petroleum in the mid 1970s but also others of a non-financial nature, have since made the microbial route to cheap protein less attractive. These matters are discussed more fully in later sections of this chapter.

c. *Developments since 1945.* The proven success of food yeast before and during the Second World War led in the post-war years to the construction of biomass-production plants in several underdeveloped countries of the world, including Puerto Rico and Formosa. However, in the mid 1950s, world-wide interest in the use of microbial biomass as food and fodder rapidly intensified and, in the period up to 1977, many new processes were developed. The main features of this upsurge in interest were the use of a much wider range of raw materials, and of a greater variety of micro-organisms, including bacteria and filamentous fungi as well as yeasts other than strains of *Sacch. cerevisiae* and *C. utilis*. Numerous symposia and conferences have been held on the subject. Notable among these were the two meetings held, in 1967 and 1973, at the Massachusetts Institute of Technology in Cambridge, Massachusetts, U.S.A; the proceedings of these symposia (Mateles and Tannenbaum, 1968; Tannenbaum and Wang, 1975) remain definitive sources of information. It was at the 1967 meeting that the term 'single-cell protein' (S.C.P.) gained acceptance, although it had been coined a year earlier at M.I.T. While, microbiologically, it may often be inaccurate, the term S.C.P. is generally preferred to 'microbial protein', although many microbiologists have difficulty in understanding why. I venture to suggest, however, that they would prefer either term to 'petroprotein' which was for a time used to describe the product grown on petroleum as a substrate. Today, the literature on S.C.P. is voluminous, not least because the concept of using microbes as an animal feed, and a potential human food, has caught the imagination of the public. Particularly recommended sources of information on S.C.P. are the text edited by Davis (1974) and the excellent review by Laskin (1977).

Plants are still in operation growing yeasts on molasses for animal feed (Sobkowicz, 1976). However, the principal feature of research on S.C.P. during the past 20 years has been the exploration of a much wider range of natural materials as substrates, a development which inevitably was accompanied by the use of a greater variety of micro-organisms. In this context, the introduction of hydrocarbons as a substrate for S.C.P. manufacture was a major development, not least because it brought with it extensive publicity and a considerable amount of microbiological and biochemical research.

While most, if not all, of the major petroleum companies in the World have shown interest in producing S.C.P., the pioneering work in this area was done by a British Petroleum (B.P.) team led by Alfred Champagnat at their Cap Lavéra plant near Marseilles in France (Champagnat *et al.*, 1963). Originally, the aim was to use micro-organisms to remove higher *n*-alkanes (waxes) from petroleum fractions, but soon microbial protein became the goal. The B.P. interest went from strength to strength, with plants being installed at Cap Lavéra to grow *Candida tropicalis* on gas oil, and at Grangemouth in Scotland, and in collaboration with ANIC S.p.A. the chemical subsidiary of the Italian state oil company E.N.I. at Sarroch in Sardinia, to produce *Candida (Saccharomycopsis) lipolytica* from *n*-alkanes. Their product was to be marketed under the trade name *toprina*. Other petroleum companies followed in the wake of the B.P. development, although not all using alkanes as a raw material (Laskin, 1977). Earlier work on production of S.C.P. from alkanes was reviewed in the text edited by Gounelle de Pontanel (1973). Levi and his colleagues bring the story up to date in Chapter 12.

The Shell Oil Company, believing that B.P. had too great a head start in research on petroleum-based S.C.P., decided to use another raw material, namely natural gas, an energy source which consists largely of methane and which in Britain had come on stream following exploration, largely by oil companies, in the North Sea. Predictably, initial work by Shell showed that working with methane was, to say the least, hazardous; air–methane mixtures all too easily explode. Notwithstanding, they developed some interesting microbial technology for cultivating microbes on methane. Shunning the use of pure cultures of micro-organisms—they termed them monocultures—the Shell workers turned to mixed symbiotic bacterial cultures. They claim, with justification, that such cultures are less prone to contamination, because there is invariably present a member in the mixed population with the ability to utilize a compound excreted by another member, and which in a pure culture would be available to a chance contaminant. This exciting advance in S.C.P. production is further chronicled by Hamer in Chapter 11. Other companies, in Japan and Russia, also developed S.C.P.-producing technology based on methane (Laskin, 1977).

In order to circumvent the problem of exploding air–methane mixtures, other companies, notably Imperial Chemical Industries (I.C.I.) in Britain, turned to methanol as a raw material. Methanol can be obtained from methane, by chemical means, or from petroleum. The I.C.I. process uses a bacterium, identified as *Methylophilus methylotropha*, as described by Hamer in Chapter 11. Other companies used methanol-utilizing yeasts, such as *Candida boudini* (Cardini *et al.*, 1976).

Interest in another alcohol, ethanol, as a raw material for S.C.P. manufacture has steadily increased over the years. Ethanol has many advantages as a starting material, including its complete miscibility with water, and its easy acceptance since it is traditionally associated with another branch of the food industry, namely the production of alcoholic beverages. It is, however, rather expensive, and this factor is undoubtedly a drawback to its more widespread use as an S.C.P. substrate. Probably the biggest plant for producing S.C.P. from ethanol is that owned by the Amoco Petrochemical Company, at Hutchinson in Minnesota, U.S.A. Their process uses a strain of *Candida utilis*, and the product is marketed under the name *torutein*. Also using ethanol as a raw material, the Mitsubishi Petrochemical Company in Japan has developed strains of *Candida* spp. which can grow at high temperatures; whilst in Spain, the Instituto Fermentaciones Industriales has pioneered a process using a strain of *Hansenula anomola*. Yet another project, jointly run by Exxon and Nestlé, uses an ethanol-utilizing bacterium, *Acinetobacter calcoaceticus*. This venture has pilot-plants in Linden, New Jersey, U.S.A. and in La Tour de la Pailz in Switzerland (Laskin, 1977).

Growing concern about the economics of the processes has been all too apparent in S.C.P. projects in recent years, as costs of raw materials, especially of alkanes, have escalated. It was natural, therefore, that consideration should be given to the cheapest of all potential substrates for S.C.P. production, namely industrial and domestic wastes. The range of waste materials which, taking into account microbiological and biochemical considerations, could be used as S.C.P. substrates is considerable. In a valuable review of the problem, Wimpenny (1975) proposed certain criteria which make a waste material acceptable as a raw material for S.C.P. production. Accepting these, he listed three classes of waste, namely agricultural wastes, including cattle and pig manure, poultry droppings, cereal and vegetable crop residues, and vegetable and potato-processing wastes; municipal wastes, such as sewage sludge and solid domestic refuse; and finally forest products such as forestry and sawmill wastes, and pulpmill effluents. One of the oldest S.C.P. processes uses a forest product, namely sulphite waste liquor from pulp mills. This waste continues to be so used, and probably the most recent advance on this sector was the setting up of the 'Pekilo' process, operated in Finland, and using the filamentous fungus *Paecilomyces variotti* (Romantschuk, 1975). The operating company has a plant at Jämsänkoski in central Finland, but this suffers from periodic shortages of the raw material as paper manufacturers increasingly turn to the Kraft or sulphate process.

The basic forest product, cellulose, is often hailed as the ideal substrate for S.C.P. manufacture, since it is continually renewed in the biosphere. Unfortunately, cellulose is rather resistant to microbial attack, there being few efficient producers of extracellular (or indeed cell-bound) cellulases. Certain filamentous fungi have been intensively studied, particularly at the United States Army Natick Laboratories. Other groups have turned to bacteria, such as species of *Cellulomonas*. These developments are described more fully by Callihan and Clemmer in Chapter 9.

Polysaccharide-containing wastes abound, and several processes have been developed to produce S.C.P. from some of these raw materials. One of the oldest is the Symba process, developed jointly by the Swedish Sugar Company and Chemap of Switzerland, in which the yeast *Endomycopsis fibuliger* hydrolyses starch in a liquid waste, while *C. utilis* grows on the hydrolysis products. The British firm of Rank, Hovis and McDougall, in collaboration with the Dupont Company in the United States of America, have also used starch-containing wastes, but to grow *Fusarium graminearum*. These, and other similar developments, are discussed by Forage and Righelato in Chapter 10.

Finally, there is whey, a regularly available waste material from the processing of milk to produce cheese. Production of S.C.P. from whey, considered by many to be the most promising waste raw material for S.C.P. production, is dealt with by Bayer and Meyrath in Chapter 8.

Reference has already been made to foods which contain appreciable quantities of microbial biomass that increase the nutritional value of the food. As information has been gathered regarding the nutrition of people in many underdeveloped countries of the World, it has come to be realized that the protein in micro-organisms contained in these foods often makes a substantial contribution to the total daily protein intake. As a result, several groups of researchers throughout the World have studied these traditional fermented foods with the object of discovering the nature of the microbes concerned, and of optimizing the contribution which consumption of the foods can make to the human diets.

The Indonesians, many centuries ago, developed a solid-substrate fermentation process in which soybeans are soaked, dehulled, partially cooked, and inoculated with strains of *Rhizopus* sp. The beans are knitted into a tight compact cake by the fibrous mould mycelium in the first three days after inoculation. The final cake, which contains over 40% protein, can be sliced or fried in deep fat or used as a meat substitute in soups. It is known as *tempe*, and in many respects is the archetypal solid-substrate fermented food. There are innumerable counterparts in the Orient and in countries with similar cultures and social scenes. Nigerian *ogi* is a breakfast food made from fermented maize. In the Sudan, moistened sorghum flour is allowed to ferment, and then baked into thin sheets which are known as *bisra*. These are just a few examples of the fermented foods which have traditionally nourished underdeveloped populations over the centuries.

Microbiological and biochemical research on these traditional fermented foods has, as yet, been the preoccupation of just a minority of workers in developed countries over the past decade or so, notably Clifford Hesseltine and Keith Steinkraus in the United States of America, and Robert Stanton, first in Great Britain and more recently in Malaysia. A brief cameo by Stanton and Wallbridge (1969) remains a most readable introduction to traditional fermented foods. They list several examples of cheeses made from starchy materials, cereals and beans. Readers are also recommended to peruse the article by Batra and Millner (1976). In recent years, interest in developing and upgrading these traditional fermented foods has come from several of the larger international food firms. Their contribution, together with that still being made from the fermented foods produced in underdeveloped countries, is chronicled by Djien and Hesseltine in Chapter 5 of this volume.

II. PRINCIPLES OF MICROBE CULTIVATION

A. Types of Culture

Following Robert Koch's development of techniques for growing micro-organisms in pure culture in the latter part of the last century, many different systems for growing micro-organisms in small (up to 100 litre) and large-scale (up to several 1000 litre) conditions have been devised. A recently published monograph by Pirt (1975) brought together, for the first time, information on the development of microbial growth systems, while the text edited by Dawson (1974) reproduces, with a useful commentary, the main relevant publications which contributed, over the first three quarters of this century, to our present understanding of the principles of microbial growth.

1. Batch Cultures

Following Koch's pioneer discoveries, there were many attempts to describe growth of microbial cultures in quantitative terms. To begin with, these inevitably were concerned with growth of microbes in portions or batches of medium, contained in single vessels and inoculated with a small population of micro-organisms. This is the so-called *batch culture*; a study of the growth of microbes in batch culture led, following research by Lane-Claypon (1909) and Slator (1917) among others, to the recognition of the classically accepted phases of microbial growth in these cultures. These phases are depicted in Figure 2.

Batch cultures of micro-organisms are now the commonly recognized method for culturing organisms in the laboratory under defined conditions, unacceptable though these may be to the student of microbial ecology. A major advance in our understanding of the mathematics of growth of micro-organisms in batch culture came from the monograph published by the late Jaques Monod in 1942. He introduced terms such as specific growth rate and yield of biomass, and pointed to the importance of the concentration of growth-limiting substrate, considerations which are now routinely used in any description of the growth of microbes in batch culture. Pirt (1975) has derived an equation describing the shape of the typical batch-culture curve shown in Figure 2.

In spite of the attractions offered by continuous-flow cultures, batch cultures are used in the majority of large-scale industrial microbial

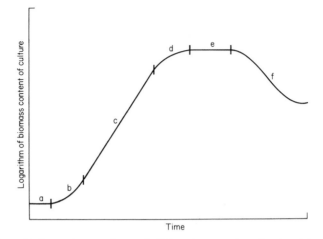

Fig. 2. Generalized growth curve for a unicellular micro-organism, showing six growth phases. Phase 'a' is the lag phase, 'b' the accelerating growth phase, 'c' the exponential growth phase, 'd' the decelerating growth phase, 'e' the stationary phase of growth, and 'f' the decline phase.

processes. This holds, even for production of such bulk products as beers. The principal reason for the continued preference for batch cultures is the versatility that these systems provide; a versatility which is demanded in processes ranging from brewing of beer to production of antibiotics.

There are basically two types of microbial culture, usually described as 'open' and 'closed'. A closed system is one in which some essential component of the system cannot both enter and leave it. The simple batch culture, consisting of a limited and unreplenished portion of nutrient medium, is an example of a closed system. A modification to the batch culture system, in which inoculum and medium are mixed on entry and the culture flows at a constant rate through the culture vessel without mixing, gives rise to a plug-flow culture; this is an open system. Plug-flow cultures have not been used as yet in industrial microbiological processes, although they have been examined on a pilot-plant scale for brewing beer. Pirt (1975) offers some suggestions for exploiting plug-flow cultures.

2. Continuous-Flow Cultures

During growth of micro-organisms in batch culture, the medium becomes depleted of nutrients and there is an accumulation of waste products of metabolism. Either or both of these factors lead to deceleration of growth rate, with the culture ultimately entering the

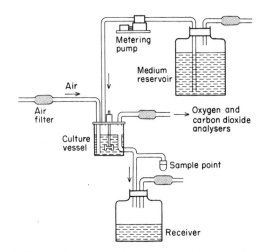

Fig. 3. Schematic representation of a simple laboratory chemostat.

stationary phase of growth (Fig. 2). The prospect of prolonging growth of microbes in culture has been discussed for over half a century (Řičica, 1968), and in the 1950s this interest led to the development of techniques for growing micro-organisms in continuous-flow culture. Nevertheless, the basic importance of continuous-flow microbial culture became clear only after the basic theory had been published by Monod (1950) and Novick and Szilard (1950). Pirt's (1975) text should be consulted for an account of this basic theory and its subsequent development.

The fundamental feature of a continuous-flow culture is that medium is added to the culture at a rate that is equal to that at which culture is removed from the vessel, thereby maintaining a constant volume of culture. Removal of culture is effected usually by an overflow or weir device, as shown in Figure 3. This type of continuous-flow culture is referred to as a *chemostat*, and is the most commonly used type for culturing micro-organisms continuously. A variant of the chemostat, in which the density of microbial biomass in the culture is maintained constant by addition of fresh medium actuated by a device which continuously monitors the turbidity of the culture, is known as a *turbidostat*. However, apart from certain experiments carried out in studies in microbial physiology, turbidostats are rarely used.

Large-scale production of microbial biomass is one of the few industrial microbiological processes which employ chemostats. Looking at industrial microbiological processes as a whole, it is tempting to assume that, because of the scale on which they are carried out and the demand for their products year-round, they would inevitably turn to chemostat cultures. However, other considerations, such as the need to use fully plant that is already installed (and that invariably means batch-culture plant), to have maximum versatility in use of that plant, and to have plant that can be guaranteed to run aseptically, have militated against the adoption of chemostat cultures. In production of microbial biomass— particularly that used for food and fodder—these considerations are less compelling. This is especially true of large-scale production of bacteria and yeasts from hydrocarbons, where contamination problems are not as great as with other types of culture, and where the pilot-scale developments have been carried out by organizations that traditionally employ continuously operating plant.

B. Microbial Nutrition

If it is to grow, a micro-organism must have available in its environment chemical compounds—nutrients—which can furnish all of the necessary elements that go to make up microbial cell material, as well as a utilizable form of energy. The bewildering array of compounds that can act as microbial nutrients, coupled with the diverse nutritional requirements which different micro-organisms have, present a complicated picture. Fortunately, it is possible to classify microbes into various nutritional groups, and a frequently used classification is that based on their energy-yielding metabolism. On this basis, there are two main classes of organism. First, there are *phototrophs* which use, directly, the energy in solar radiation, and secondly there are *chemotrophs*, which use chemical compounds as a source of energy. Both classes are further subdivided. Phototrophs can be divided into *photolithotrophs*, for which growth is dependent on exogenous inorganic electron donors, and *photo-organotrophs* which require, instead, exogenous organic electron donors. Chemotrophs, on the other hand, are subdivided into *chemolithotrophs*, which grow by oxidizing exogenous inorganic compounds, and *chemo-organotrophs*, growth of which depends on oxidation of, or fermentation of, exogenous organic compounds. Further information can be obtained from the texts by Pirt (1975), Dawes and Sutherland (1976) and Rose (1976).

From the point of view of the manufacturer of microbial biomass, there are two principal nutritional considerations.

1. *Ability to Grow on Cheap Raw Materials*

The cheapest of all sources of energy for micro-organisms is sunlight, and when the ability to harness solar radiation is coupled with the capacity to synthesize all cell constituents from carbon dioxide, one would seem to have the ideal organism to be mass produced for feed and fodder. The algae are just such organisms. Unfortunately, there are a limited number of regions in the world where the flux of solar radiation is sufficiently great to make mass cultivation of algae in the open an efficient process. Added to which, there is frequently a need to increase the partial pressure of carbon dioxide in the atmosphere above the algal culture. For these reasons, the economics of algal mass cultivation present problems, as indicated by Benemann and his colleagues in Chapter 7.

When developing processes for mass cultivation of micro-organisms for S.C.P., most workers have confined their attention to chemo-organotrophic microbes. A quarter of a century ago, beet and cane sugar molasses were considered suitable cheap raw materials for these processes. Not so today, for the costs of molasses have risen steeply. Moreover, in order that S.C.P. can compete on a cost basis with soybean and fishmeal, the manufacturer has been forced to consider much cheaper raw materials. Many micro-organisms, particularly well represented among filamentous fungi, can hydrolyse cellulose and starch and, since these polysaccharides are often found in waste materials, particularly agricultural wastes, these microbes have been researched as a S.C.P. Whey is another inexpensive material and, since some micro-organisms have the ability to utilize lactose, it too has been considered as a S.C.P. raw material.

Two other raw materials that have been used as substrates for S.C.P. production are petroleum and methane (see Section III.B, p. 26). Following earlier reports by Hass and Bushnell (1944) and Just and Schnabel (1948), it fell to Beerstecher in his text published in 1954, and Davis and Updegraf in their 1954 review, to focus attention on the ability of various micro-organisms to utilize alkanes as a carbon and energy source. The capacity to oxidize alkanes is found in bacteria, yeasts and filamentous fungi. However, the ability to utilize methane similarly is probably confined to bacteria. Again, there were early reports of this capacity among bacteria (Kaserer, 1905; Söhngen, 1905) which were not followed up until Foster and his colleagues extended research in this area (Foster, 1963).

2. Oxygen Supply

Dating from 1861 when Pasteur found that transferring yeast from an anaerobic to an aerobic environment led to an acceleration of growth rate, it has been appreciated that providing an adequate supply of oxygen is essential in the mass cultivation of most micro-organisms. However, technical difficulties delayed the full exploitation of this finding. Important contributions in the development of an understanding of oxygen supply to microbial cultures came from the development by Kluyver and Perquin of the shake-flask technique (Kluyver and Perquin, 1933), and the invention of efficient instruments for measuring the dissolved oxygen tension in microbial cultures.

C. Composition of Microbes

Micro-organisms harvested from exponential-phase cultures usually consist of 2–3% DNA, 10–15% RNA, 40–60% protein, 15–20% polysaccharide and 10–15% lipid, on a dry weight basis. There are about 3 μl of water for each mg dry weight in a micro-organism. In addition, there are pools of low molecular-weight compounds which are mainly precursors of the major cell polymers, as well as quantitatively minor components, such as vitamins and coenzymes. While overall analyses of the organic composition of micro-organisms were regularly reported in the earlier literature, such analyses now rarely appear as microbial biochemists have concentrated increasingly on the composition of subcellular organelles rather than whole cells. However, useful compilations of the contents of various components in micro-organisms have been assembled in the text edited by Laskin and Lechevalier (1973).

Changing virtually any chemical or physical factor in the environment alters, to a certain extent, the chemical composition of a micro-organism. When grown in bulk for commercial purposes, microbes are cultivated under conditions that are nutritionally favourable and which lead to rapid growth of the organisms. High rates of growth are accompanied by an increased protein content. At the same time, the RNA content of cells increases, and this can be a cause for concern in production of S.C.P. (see Section I.C.1, p. 11).

III. CURRENT STATUS OF INDUSTRIAL PRODUCTION OF MICROBIAL BIOMASS

A. Production of Enzymically Active Biomass

Processes, both operated and as yet only contemplated, for large-scale manufacture of microbial biomass can be placed in one of two groups. The first of these includes processes, the aim of which is to produce microbial biomass that is viable and metabolically active. The oldest of these is production of *Sacch. cerevisiae* for the leavening of dough in bread making. The development of this industry has already been described in this chapter in some detail (Section I.B, p. 3), and more recent developments are recounted by Burrows in Chapter 2.

Compared with other industries that manufacture enzymically active biomass, baker's yeast manufacture is positively enormous. As Burton (see Chapter 3) and Bulla and Yousten (see Chapter 4) indicate, production of bacteria for azofication, and as insecticides, are processes that are still in their infancy. These processes, however, lead to the production of bulk quantities of viable micro-organisms, which are placed in environments where their growth and metabolic activity are economically desirable.

B. Single-Cell Protein

The majority of contributions in this volume deal with the second group of processes: those that lead to production of microbial biomass which, either by its presence in a foodstuff or because it is added as a food or fodder supplement, is destined to increase the nutritional value of that food. All of the processes described are, at the present time, critically poised. Some are already making a measurable contribution to the quality of animal fodder. But whether the contribution made by these processes will increase, and whether they will be joined by S.C.P. produced in other ways, depends on a variety of technical and economic considerations.

Processes for producing S.C.P. fall into three classes. Firstly, there are those which involve encouraging microbial growth in foods. As already indicated, several groups of workers throughout the World are currently attempting to optimize production of fermented foods. However, progress is slow, largely because there is a tremendous in-built resistance in most countries of the World to proposals for changing, no matter how slightly, the composition of national dishes. Moreover, any plan to construct a production plant for manufacturing these foods must be taken only after extensive planning and research, in view of the costs involved, in what are usually poorly developed countries from an economic standpoint.

A second class of process includes those that use cheap or waste materials as substrates. Economically attractive though many of these processes may appear, such is the low and variable cost of alternative sources of protein that even these processes are hardly viable economically. They suffer, too, from the intermittent availability of many of the raw materials used, and from the slow rate at which some polymeric substrates, especially cellulose, can be converted into microbial nutrients.

Bearing in mind these reservations, it is amazing that a third class, which comprises processes based on relatively expensive raw materials, principally petroleum, methane and methanol, ever came to be considered, quite apart from being developed. However, because they became associated with large industrial combines, equipped with very considerable engineering expertise and a capacity to call on capital not available to smaller concerns, processes for S.C.P. production from petroleum and methane have been researched and developed to a considerable extent. Clearly, the raw materials used in these processes are expensive, and getting more so, and it is very doubtful if all of the processes that have been developed are, or are likely to become, economically viable. Indeed, many petroleum companies who entered the field have, in the past year, abandoned research on S.C.P. from hydrocarbons. These include B.P., who were pioneers in the field in the 1960s. Withdrawal of B.P. from the research and production of S.C.P., however, was largely a consequence of their inability to commission the plant that they had built, in collaboration with ANIC the Italian state-owned petroleum company, at Sarroch in Sardinia. The reasons for B.P.'s inability to commence S.C.P. production at Sarroch were mainly a result of Italian government officials finding traces of n-paraffins in the back fat of pigs fed *toprina*. The objection was vigorously contested by B.P., but to no avail. The saga, leading up to B.P.'s decision to withdraw from S.C.P., has been chronicled by Done (1978).

REFERENCES

Abercrombie, J. (1779). 'The Garden Mushroom: its Cultivation'. Lockyer Davis, London.
Allegro, J. M. (1970). 'The Sacred Mushroom and the Cross'. Hodder and Stoughton, London.
Batra, L. R. and Millner, P. D. (1976). *Developments in Industrial Microbiology* **17**, 117.
Beerstecher, E. (1954). 'Petroleum Microbiology'. Elsevier, Amsterdam.
Borgstrom, G. (1973). 'World Food Resources'. Intertext Books, Aylesbury, England.
Braude, R. (1942). *Journal of the Institute of Brewing* **48**, 206.
Burrows, S. (1970). *In* 'The Yeasts', (A. H. Rose and J. S. Harrison, eds.), Vol. 3, pp. 349–420. Academic Press, London.
Cardini, G., Di Fiore, L. and Zotti, A. (1976). *Abstracts of the Fifth Fermentation Symposium*, (H.-W.-Dwelleg, ed.), p. 202. Institut für Garungsgewerbe und Biotechnologie, Berlin.
Castro, A. C., Sinskey, A. J. and Tannenbaum, S. R. (1971). *Applied Microbiology* **22**, 422.
Champagnat, A., Vernet, C., Laine, B. and Filosa, J. (1963). *Science, New York* **197**, 13.
Currie, J. A., Dunnill, P. and Lilly, M. D. (1972). *Biotechnology and Bioengineering* **14**, 725.
Davis, J. B. and Updegraf, D. M. (1964). *Bacteriological Reviews* **18**, 215.
Davis, P. ed. (1974). 'Single Cell Protein'. Academic Press, London.
Dawes, I. W. and Sutherland, I. W. (1976). 'Microbial Physiology'. Blackwell Scientific Publications, Oxford.
Dawson, P. S. S. ed. (1974). 'Microbial Growth'. Dowden, Hutchinson and Ross, Inc., Stroudsberg, Pennsylvania, U.S.A.
Delbrück, M. (1910). *Wochschrift für Brauerei* **27**, 375.
Done, K. (1978). *Financial Times*, London, July 20th.
Dunnill, P. and Lilly, M. D. (1975). *In* 'Single-Cell Protein', (S. R. Tannenbaum and D. I. C. Wang, eds.), pp. 179–207. M.I.T. Press, Cambridge, Massachusetts, U.S.A.
Foster, J. W. (1963). *Antonie van Leeuwenhoek* **21**, 210.
Frey, C. N. (1930). *Industrial and Engineering Chemistry* **22**, 1154.
Gounelle, de Pontanel, H. ed. (1973). 'Proteins from Hydrocarbons'. Academic Press, London.
Goyco, J. A. and Asento, C. F. (1947). *Puerto Rico Journal of Public Health* **23**, 471.
Harris, E. E. (1949). 'Food Yeast Production from Wood Processing By-Products'. United States Department of Agriculture Forest Service, No. D 1754.
Hass, J. F. and Bushnell, L. D. (1944). *Journal of Bacteriology* **48**, 219.
Hayduck, F. (1913). *Zeitschrift für Spiriusindustrie* **36**, 233.
Hayes, W. A. and Nair, N. G. (1975). *In* 'Filamentous Fungi', (J. E. Smith and D. R. Berry, eds.), Vol. 1, pp. 212–248. Edward Arnold, London.
Hetherington, P. J., Follows, M., Dunnill, P. and Lilly, M. D. (1971). *Transactions of the Institution of Chemical Engineers* **49**, 142.
Just, F. and Schnabel, W. (1948). *Branntweinwirtschaft* **2**, 113.
Kaserer, H. (1905). *Zentralblatt für Bakteriologie, Parasitenkunde, Infektionskrankheiten und Hygiene, Abteilung II* **15**, 573.
Kihlberg, R. (1972). *Annual Review of Microbiology* **26**, 427.
Kluyver, A. J. and Perquin, L. H. C. (1933). *Biochemische Zeitschrift* **266**, 68.
Lane-Claypon, J. E. (1909). *Journal of Hygiene* **9**, 239.
Laskin, A. I. (1977). *In* 'Annual Reports on Fermentation Processes', (D. Perlman, ed.), Vol. 1, pp. 151–180. Academic Press, New York.

Laskin, A. I. and Lechevalier, H. eds. (1973). 'Handbook of Microbiology', Vol. 2. Chemical Rubber Company Press, Cleveland, Ohio.

Mateles, R. I. and Tannenbaum, S. R. eds. (1968). 'Single-Cell Protein'. M.I.T. Press, Cambridge, Massachusetts, U.S.A.

Mirocha, C. J., Pathre, S. V. and Christensen, C. M. (1969). *In* 'Economic Microbiology', (A. H. Rose, ed.), Vol. 3, pp. 468–522. Academic Press, London.

Monod, J. (1942). 'Recherche sur la Croissance des Cultures Bacteriénnes', 2nd edition. Hermann, Paris.

Monod, J. (1950). *Annales de l'Institut Pasteur, Paris* **79**, 390.

Novick, A. and Szilard, L. (1950). *Science, New York* **112**, 715.

Peterson, W. H., Snell, E. E. and Frazier, W. C. (1945). *Industrial and Engineering Chemistry* **37**, 30.

Pirie, N. W. (1976). 'Food Resources, Conventional and Novel'. 2nd edition. Penguin Books, London.

Pirt, S. J. (1975). 'Principles of Microbe and Cell Cultivation'. Blackwell Scientific Publications, Oxford.

Ramsbottom, J. (1953). 'Mushrooms and Toadstools'. Collins, London.

Rechcigl, M. ed. (1975). 'World Food Problem. A Selective Bibliography'. Chemical Rubber Company Press, Cleveland, Ohio.

Řičica, J. (1958). *In* 'Symposium on Continuous Cultivation of Micro-Organisms', (I. Malek, ed.), p. 75. Czechoslovak Academy of Sciences, Prague.

Romantschuk, H. (1975). *In* 'Single-Cell Protein II', (S. R. Tannenbaum and D. I. C. Wang, eds.), pp. 344–356. M.I.T. Press, Cambridge, Massachusetts, U.S.A.

Rose, A. H. (1976). 'Chemical Microbiology. An Introduction to Microbial Physiology', 3rd edition. Butterworths, London.

Sak, S. (1919). Danish Patent 28,507.

Scargill, I. (1974). 'The Dordogne Region of France'. David and Charles, Newton Abbot, Devon.

Schülein, J. (1937). 'The Brewer's Yeast as a Medicine and Feeding Stuff'. Verlag Steinkopf, Dresden.

Scrimshaw, N. S. (1975). *In* 'Single-Cell Protein II', (S. R. Tannenbaum and D. I. C. Wang, eds.), pp. 24–45. M.I.T. Press, Cambridge, Massachusetts, U.S.A.

Shacklady, C. A. (1975). *In* 'Single-Cell Protein II', (S. R. Tannenbaum and D. I. C. Wang, eds.), pp. 489–504. M.I.T. Press, Cambridge, Massachusetts, U.S.A.

Sinskey, A. J. and Tannenbaum, S. R. (1975). *In* 'Single-Cell Protein II', (S. R. Tannenbaum and D. I. C. Wang, eds.), pp. 158–178. M.I.T. Press, Cambridge, Massachusetts, U.S.A.

Slator, A. (1917). *Journal of Hygiene* **16**, 100.

Sobkowicz, G. (1976). *In* 'Food from Waste', (G. G. Birch, K. J. Parker and J. T. Worgen, eds.), pp. 42–57. Applied Science Publishers, London.

Söhngen, N. L. (1905). *Zentralblatt für Bakteriologie, Parasitenskunde, Infektionskrankheiten und Hygiene, Abteilung II* **15**, 513.

Stanton, W. R. and Wallbridge, A. (1969). *Process Biochemistry* **4**, 45.

Sucher, R. W., Robbins, E. A., Schuldt, E. H., Seeley, R. D. and Newel, J. A. (1973). *Proceedings of the 33rd. Annual Meeting of the Institute of Food Technologists, Paper No. 16*, Miami, Florida.

Tannenbaum, S. R. (1968). *In* 'Single-Cell Protein', (R. I. Mateles and S. R. Tannenbaum, eds.), pp. 343–356. M.I.T. Press, Cambridge, Massachusetts, U.S.A.

Tannenbaum, S. R. and Wang, D. I. C. eds. (1975). 'Single-Cell Protein II'. M.I.T. Press, Cambridge, Massachusetts, U.S.A.

Thaysen, A. C. (1957). *In* 'The Yeasts', (W. Roman, ed.), pp. 155–210. Academic Press, New York.
Tournefort, J. de (1707). *Memoirs de l'Academie de Science de l'Institut de France*, p. 58.
Verein der Spiritusfabrikanten in Deutschland (1919). German Patent, 300662.
Waslien, C. I., Calloway, D. H. and Margen, S. (1969). *Nature, London* **221**, 84.
Waslien, C. I., Calloway, C. H. and Margen, S. (1970). *American Journal of Clinical Nutrition* **21**, 892.
Weitzel, W. and Winchel, M. (1932). 'The Yeast, its Nutrieve and Therapeutic Value'. Verlag Rothgiese und Diesing, Berlin.
Wimpenny, J. W. T. (1975). *Reports in the Progress of Applied Chemistry* **59**, 383.
Wimpenny, J. W. T. (1967). *Process Biochemistry* **4**, 63.

2. Baker's Yeast

S. BURROWS

*Distillers Company Limited, Glenochil Technical Centre, Menstrie,
Clackmannanshire, Scotland*

I. INTRODUCTION

A. History

The earliest use of yeast for baking was in the form of leaven which is a dough of flour and water containing yeast and usually other organisms such as lactobacilli. The yeast multiplied in this medium and could be perpetuated and kept for later use. More recently, up to the beginning of the nineteenth century, top-fermenting brewer's yeast was collected from breweries and used for dough raising. The manufacture of yeast specifically for supply to bakers was started in the early nineteenth century, and about the middle of that century the use of pure yeast cultures commenced (Reed and Peppler, 1973). At that time the substrate was grain mash and an anaerobic process was used with the associated production of ethanol.

During the late nineteenth and early twentieth centuries, three major improvements to the process were introduced. These were, first, the introduction of aeration to improve the yield of yeast and decrease ethanol production, secondly the use of incremental feeding of nutrient which gave a further increase in yield, and thirdly, the change to molasses as substrate which was cheaper than grain and easier to store and prepare for use. Thus, the basic method of baker's yeast production today consists in seeding pure yeast cultures into large vessels and propagating using relatively high rates of aeration, together with the incremental feeding of molasses.

Especially since the Second World War, considerable developments have been made in the yeast strains used and in the equipment for aeration, sterilization, nutrient feeding, packing, storage and so on, so that today's yeast is more active, more stable and more consistent in properties than in the past. Another development during this period has been the introduction and gradual emergence of active dried yeast, of 90–96% dry matter content, as a more stable alternative to the moist compressed fresh yeast which has between 27% and 35% dry matter content. It was known over a century ago that compressed baker's yeast could be dried without completely losing its activity, but it was only established comparatively recently that, under suitable conditions, a dried material can be produced which contains little water and yet retains almost full viability when rehydrated properly. It was not until about 1920 that commercial drying of yeast was achieved, and since 1940 this type of yeast has become increasingly popular (Reed and Peppler, 1973).

B. General Considerations

The technology of the manufacture and use of baker's yeast has been discussed at length in recent years (Hoogerheide, 1969; Burrows, 1970; Harrison, 1971; Reed and Peppler, 1973; Holloway and Burrows, 1976; Rosen, 1977) and therefore, in this review, in addition to a brief description of commercial production, some special topics which have previously not received much attention will be considered in more detail.

Although during the past twenty years the study of single-cell protein production from waste materials or cheap substrates has advanced tremendously, the basic techniques of using highly aerobic, deep-tank homogeneous systems with incremental substrate feed have been in use in baker's yeast manufacture since the early part of the present century (Hayduck, 1919; Sak, 1919). Baker's yeast can in fact be regarded as a specialized form of single-cell protein but, as normally used, its contribution to protein nutrition is negligible, being of the order of 2 g protein per person per week in Western European countries. However, the total amount of baker's yeast produced is very large, being about 50 000 tonnes (moist cake) per annum in the United Kingdom and about 300 000 tonnes in the whole of Western Europe. In 1969 it was stated to be 700 000 tonnes per annum worldwide (Hoogerheide, 1969). The quantities of molasses (cane or beet) required to produce these amounts would be about 10 to 20 per cent higher than the weight of wet yeast produced.

Proposals have been made to increase the protein content of bread by adding inactive yeast, in which case modifications have to be made to the dough recipe to correct the adverse effect of such additions on loaf characteristics (Yanez et al., 1973; Volpe, 1976; Tanner et al., 1977). This procedure has the effect of improving the nutritional value of the lysine-deficient cereal protein owing to the high lysine content of yeast protein. However, problems of altered texture and flavour might be a serious disadvantage.

Recent nutritional studies have uncovered an unsuspected property of yeast in breadmaking. Phytic acid in flour has an adverse nutritional effect since it strongly binds essential trace metals making them unavailable. This is particularly the case with high-extraction (wholemeal) flours which contain a high content of phytic acid. Yeast fermentation results in a substantial decrease in phytic acid and consequent improvement in nutritional value of the bread (Lorenz and Lee, 1977).

In single-cell protein production, the main consideration is the yield of biomass although its protein and nucleic acid contents and digestibility must also be taken into account. In contrast to this, the paramount consideration in baker's yeast production is its quality as expressed by its fermentation characteristics in dough substrates, its storage stability and its ability to resist inhibitors such as sodium chloride and high concentrations of sugars. Although from an economic standpoint the yield of yeast per unit of growth substrate and per unit of manufacturing plant is of great importance, any batches which do not reach the desired standard of quality must be rejected. For these reasons, every effort must be made to control the process so that it conforms as closely as possible to the predetermined pattern, and strict cleanliness must be imposed from start to finish.

II. MANUFACTURE

A. Introduction

Baker's yeast production is carried out in most temperate countries of the world and in some tropical countries. The factories vary widely from modern sophisticated, highly automated, plants to more simple labour-intensive installations. The basis of the process used is, however, the same.

B. Nutrient Feeding

A mixture of beet- and blackstrap molasses, together with ammonia and phosphate, provides a balanced medium for yeast growth. However, where blackstrap molasses is difficult to obtain, beet molasses alone can be used with the addition of extra biotin. Similarly, where beet molasses is unobtainable at an economic price, blackstrap molasses can be used with the addition of some vitamins. Molasses contains colloidal and suspended matter and is also contaminated with unwanted yeasts and bacteria. In cold climates, the storage tanks have to be fitted with heating coils and the transfer lines lagged. When taken into process, the molasses mixture is diluted, centrifuged and pasteurized or sterilized. Pasteurization is usually carried out by boiling the molasses solution for between one and two hours; sterilization is performed in a continuous, short-time, high-temperature system with plate-type heat exchangers where careful process control is necessary to avoid overheating which results in blockages, caramelization and loss of nutrients. Molasses may also contain inhibitors which are difficult to detect by routine methods of analysis (Burrows, 1970). Once the presence of inhibitors has been confirmed, the only practical remedy is judiciously to blend away the batch concerned with satisfactory material.

Until fairly recently, the commonest method of providing the required incremental nutrient feeding was by the automatic filling and emptying of a small reservoir. The number of doses fed per hour could be varied and was controlled by a slotted metal tape which actuated a mercury switch. More versatile and reliable methods based on flow meters, electronic programmers and microprocessors are now coming into use.

As discussed in more detail later (Section III.B), the nutrient-feed programme used may be predetermined and varied according to the type and output of yeast required. Feed-back systems are possible, however, based on monitoring ethanol and minimizing its production, controlling the oxygen tension or following a set yeast-growth pattern. These methods help to maximize the yield, but control of yeast properties in this way is not at present possible since suitable methods of measurement and prediction of yeast quality at a sufficiently early stage in the propagation are not available.

Phosphate and assimilable nitrogen may be fed continuously or may be added batchwise at the start or at intervals during the process. It is important to maintain the correct ratios between molasses, phosphate and nitrogen feeds in order to arrive at the correct yeast composition and properties.

C. Aeration and Cooling

Large volumes of air are required for yeast propagation, and this has to be filtered to free it from dust, bacteria and yeasts. Fibrous filters or membrane filters of controlled pore size are available which perform this function very efficiently, but to keep the pressure drop at a low level large areas of filter are required together which efficient centrifugal air compressors.

Since the solubility of oxygen in water is low, a very efficient distribution and mixing system is required to effect a rapid transfer of oxygen to the yeast cell surface where it can be assimilated. A large number of different methods are now available and many patents have been taken out. A simple sparging system consisting of radiating perforated pipes covering the bottom of the vessel is still frequently preferred for simplicity and economy but its limited efficiency of oxygen transfer restricts the productivity for a vessel of given size. The shape of the vessel is important when this method is used since a tall narrow vessel provides a longer bubble path-length and hence greater absorption of oxygen. Stirrers and turbines of various designs are useful in small-scale work to aid air dispersion and are often used in the commercial-scale production of baker's yeast. Methods now favoured in yeast hydrocarbon fermentations which depend on increasing the bubble path-length by using tall vessels, baffles and recirculating systems may also be used. An additional advantage of the tall vessel is that a proportion of the liquid is subjected to a higher pressure, thus increasing oxygen solubility throughout a proportion of the fermenting liquor (Moo-Young, 1975).

Aeration rates may be controlled automatically by means of oxygen electrodes which measure the partial pressure of oxygen in the medium. Sterilizable electrodes which are sufficiently reliable for routine industrial use are now becoming available, but their full potential has probably not yet been reached. Measurement of ethanol concentration in the exit air may also be used in appropriate cases to control aeration (Patentauswertung Vogelbusch GmbH, 1961).

Cooling is achieved by the use of internal coils, external jackets or circulation of the fermenting liquid through external heat exchangers. The cooling liquid is usually mains water or pumped river water. Refrigeration systems have to be used in warm climates, and have the advantage that a sterile coolant is circulated. Even in temperate climates it is not always possible to attain the low cooling-water temperature required by recycling through cooling towers without a mechanical refrigeration stage.

The need for aeration and cooling and the requirement for a hygienic system to a large extent dictate the type of equipment used in such microbiological processes as yeast propagation. Thus, a stainless steel vessel of smooth construction and circular cross section is usually employed, the height being two or more times the diameter. Owing to the relatively high capital cost involved, pressure sterilization is not always employed, cleaning with hot water, detergents and chemical-sterilizing agents being used instead. This technique, combined with the use of pasteurized or preferably sterilized molasses, filtered air, and good housekeeping throughout the process, is adequate for production of baker's yeast, it not being necessary to use a completely aseptic process. Nevertheless, the avoidance of infection, particularly by fast growing wild yeasts, is of prime importance.

Another requirement resulting from the high aeration rates employed is the necessity to provide adequate antifoam equipment. Although molasses contains proteins and other foam-stabilizing materials, efficient and clean antifoam agents such as silicones are available and may be added automatically by antifoam dosage equipment to deal with this problem very effectively. Alternatively, cyclones may be installed in the air-exit pipelines from the fermenters to break the foam by centrifugal force (Solomons, 1969).

D. Production Stages

Having briefly discussed some of the requirements of baker's yeast manufacture, it is possible to consider the process as a whole. Typically, the initial stages are carried out in the microbiology laboratory where a few million cells of the desired strain of *Saccharomyces cerevisiae* in pure culture are swept from a nutrient agar plate into a suitable liquid medium. Over a period of several days, and by one or two transfers to fresh medium under strictly aseptic conditions, a few hundred grams of pure-culture yeast are prepared.

This pure culture is used to inoculate a special seed culture vessel in the factory, which has been sterilized with medium *in situ*. This provides enough yeast in 24 hours to seed a full-size commercial vessel of, say, about 100 m³ capacity. During this stage, full aeration and incremental feeding would be commenced and, for example, in about 20 hours enough yeast would be obtained to seed five more similar large vessels. This process may then be repeated to give 25 full-scale propagations producing in all between 100 and 500 tonnes of harvested moist yeast and requiring between 3500 and 7000 tonnes of air depending on the efficiency of the absorption system. It is common for the medium to be centrifuged after each commercial seed stage, and for the resultant yeast cream to be used as the inoculant. The whole process may be summarized by saying that the yeast cells multiply about 10^9 times in 10 to 12 days.

It is during the final propagation stage that most of the yeast is grown, and here the conditions used must be very carefully controlled in order to obtain optimum output and yeast quality. Much experience and development work are required to derive the conditions necessary for the consistent production of yeast of high quality.

At the end of the growth period, the medium may contain anything up to 80 g yeast dry matter per litre. This is concentrated to about 200 g per litre by centrifuging. Counter-current washing is also carried out by centrifugation and the final yeast cream is cooled and then filtered on specially designed rotary vacuum filters, or on more traditional filter presses, to give a yeast cake which is extruded, usually in the form of 1 lb, $\frac{1}{2}$-kg or 1-kg blocks, and wrapped in waxed paper by custom-built extruding, cutting and wrapping machines. The packets of compressed yeast are then placed in cartons, thoroughly chilled in the cold store, and distributed by refrigeration transport.

III. RECENT DEVELOPMENTS

A. Strain Stability

Production of baker's yeast starts in the microbiological laboratory and it is here that any departure from normal practice can seriously affect the quality and microbiological cleanliness of the ultimate product. The purity and genetic identity of the original inoculum must be beyond question. Thus, complete absence of bacteria, moulds or unwanted yeast strains is ensured by the stringent application of pure-culture techniques of proven reliability.

A more difficult assignment is the maintenance of the genetic identity of the strains in current use. Thus, it is well known that, although a yeast strain may appear to be reasonably stable over long periods when examined from the standpoint of its gross physiological properties, careful quantitative measurements under closely controlled conditions nearly always reveal slow but distinct changes after many generations of growth. This deterioration may be due to a number of different processes, but mitotic crossing over is possibly an important one (Emeis, 1965). In a culture which has only degenerated slightly, an examination of cultures derived from single-cell or single-colony isolates reveals the presence of a few cultures which give rise to yeast having properties which are widely divergent from those of the original strain (Fowell, 1951). This is therefore a sensitive method of detecting culture deterioration. It is possible that a single-colony isolate with an improved balance of properties might be found, but as a rule degeneration occurs (Emeis, 1965).

The problem of culture degeneration occurs most frequently with the older methods of culture maintenance which use some form of serial transfer (Reusser, 1963). Preservation on refrigerated, oil-covered slopes of nutrient agar has been recommended (Haynes *et al.*, 1955) since viability can be maintained for long periods with minimum growth and cell division. However, methods such as storage under liquid nitrogen or after freeze drying are becoming more popular since cell division is avoided completely after initial growth of the culture. Refrigeration in liquid nitrogen has been used successfully for obtaining standardized yeast inocula (Beezer *et al.*, 1976), but freeze drying is in some ways a more convenient technique and is now finding favour for brewing strains (Kirsop, 1974; Richards, 1975; Barney and Helbert, 1976).

Any method of culture maintenance must be backed up by testing in a pilot plant which copies as far as possible the factory conditions of growth and harvesting as well as the conditions of use in the bakery. Furthermore, if the commercial-scale propagation and the subsequent testing methods are sufficiently reproducible and sensitive, small changes can be detected and corrected by appropriate action before they are of significance from the point of view of customer acceptance.

B. Feeding of Assimilable Carbon

The propagation of baker's yeast by an aerobic, substrate-limited batch process is now the usual practice, although in some countries it is still produced in a semi-aerobic process in which ethanol is the main product (Kosikov *et al.*, 1975). In efforts to maximize cell yield, a number of different automatic feed-back systems for controlling substrate addition have been described, but these do not appear to have been widely adopted commercially, probably owing to the additional necessity of controlling the quality of the product. Those described depend upon measuring the residual sugar concentration (Chen and Gutmanis, 1976), respiratory quotient (Aiba *et al.*, 1976), dissolved oxygen concentration, ethanol, carbon dioxide or heat production, biomass concentration (Polivoda *et al.*, 1972) or a combination of these (Wang *et al.*, 1977). The feed-back control can then be applied to the nutrient (molasses) feed rate but, more usually, it is applied to the stirring rate, air-flow rate or the proportion of oxygen in the gas stream. There is no doubt that such methods are practicable (e.g. Patentauswertung Vogelbusch GmbH, 1961) but it is not clear whether they confer any advantage, either in terms of economics or the quality or consistency of the product.

When a predetermined programme is used, a large variety of patterns

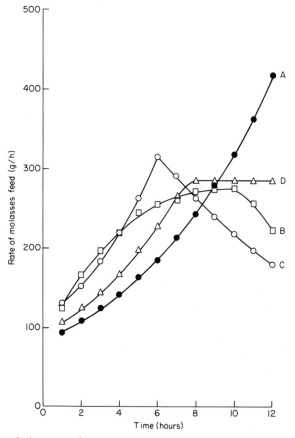

Fig. 1. Nutrient-feed patterns for compressed yeast manufacture. From Drews *et al.* (1962). See text for explanation.

for the molasses feed is possible and some of these have been described by Drews *et al.* (1962). Data from a selection of these were reproduced in graphical form by Reed and Peppler (1973) and a slightly different selection is given in Fig. 1. The most logical of these is the exponential pattern (curve A) where the specific growth rate of the yeast is theoretically maintained constant at a level below its maximum. In the case described by Drews *et al.* (1962), the yeast had a high fermentative activity accompanied with a high proportion of budding cells and poor keeping quality. When the feed rate decreased towards the end of the process (curves B and C), the yeast had no budding cells and good keeping quality, and was produced in higher yield than when using the previous feed pattern (curve A). Curve D shows an exponential increase followed by a constant rate of feed and this pattern gave yeast an intermediate yield with activity and stability intermediate between that in the other examples. These results show that certain principles govern the choice of feed pattern, of the limiting carbon source, and allow the operator, within limits, to modify the process to suit the requirements of the market or the demands of the yeast strain.

The mathematics of the growth of single-cell organisms have been dealt with on a number of occasions (e.g. Aiba *et al.*, 1973; Blanch and Dunn, 1974) and a few attempts have been made to describe nutrient-feed patterns in fed-batch processes in mathematical terms (Pirt, 1974; Yoshida *et al.*, 1973; Dunn and Mor, 1975). Approaches of this kind are valuable in designing propagation processes for baker's yeast. Some variables of interest are: (i) concentration of limiting nutrient in the feed; (ii) flow rates of nutrient feeds; (iii) fluid volume in the growth vessel; (iv) yeast concentration; (v) yeast growth rate; (vi) total yeast weight; and (vii) initial inoculum.

Jones and Anthony (1977) have derived mathematical equations relating these variables for various types of nutrient feed, assuming balanced growth. Thus, for a constant flow rate of growth-limiting nutrient:

$$\mu = \frac{FsY}{X_0 + FsYt}$$

when μ = specific growth rate, F = nutrient-feed flow rate, t = time, s = concentration of nutrient in the feed solution, X_0 = weight of cells at zero time, and Y = yield factor.

Also, for a constant rate of increase of nutrient-feed rate:

$$\mu = \frac{F_0 + Gt}{K + (F_0 + \frac{1}{2}Gt)t}$$

where $K = X_0/sY$, $F_0 =$ nutrient flow rate at zero time, and $G =$ rate of change of flow rate with time.

Equations such as these, together with knowledge of restraints imposed by the available aeration and heat-removal equipment and a knowledge of the effects of these variables on yeast yield and quality, may be used to design production processes with the minimum of experimentation. However, a mathematically programmed substrate feed may not be satisfactory during the first hour of a batch process owing to possible variations in the condition of the seed yeast and a variable lag before attaining balanced growth (Beran et al., 1961). Special provision for this period may therefore be necessary.

C. Dough Fermentation

Since the primary use of baker's yeast is in the fermentation of dough to make bread, it is essential that the manufacturer of baker's yeast should understand something of the mechanism of this process.

Although the fermentation of a simple sugar such as glucose by yeast results, after a few minutes, in the evolution of carbon dioxide at a constant rate, which depends upon the amount and activity of the yeast added, the fermentation of dough is far more complex. This is because flour contains a mixture of carbohydrates and growth-promoting substances. The complexity is well illustrated by the gassing-rate curves obtained using the zymotachygraphe (Flückiger, 1974) or by adapting the fermentometer test procedure (Burrows and Harrison, 1959) to provide rate curves instead of gas volumes. The curves shown in Fig. 2 were obtained using the fermentometer equipment with four different yeasting rates three of which corresponded with the conditions used in various bakery processes (Burrows, 1975). The flour used was a mixture of equal proportions of hard and soft flours (11.0 to 11.3% protein). No sugar was added, but sodium chloride and ammonium sulphate were used at the rate of 2.025 and 0.30g/100 g flour, respectively. The curves obtained show accelerating rates of gas production which, at low yeast levels, proceed for several hours due to the slower consumption of

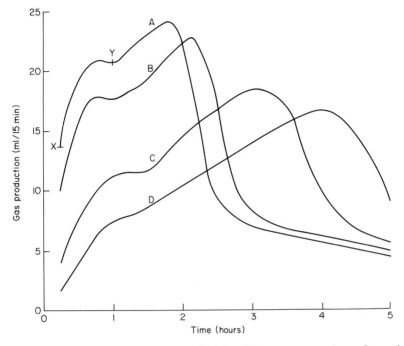

Fig. 2. Time-course of dough fermentation at 30°C using different processes and yeast-flour ratios. Curve A describes a fermentation in the fermentometer test (2.68% yeast : flour); B, the Chorleywood process (2.23% ratio); C, a two-hour bulk process (1.34% ratio); and D, a three-hour bulk process (0.89% ratio). Moist gas volumes were measured at 20°C and 760 mm pressure.

carbohydrates. This allows more time for the organization, activation and synthesis of the various systems involved in permeation of the cell by nutrients, and for their subsequent metabolism.

The area under the curves corresponds to the volume of gas evolved at any given time; this in turn corresponds to the amount of carbohydrate fermented. Thus, if the fermentation is taken almost to completion, a measure of the total fermentable carbohydrate in the dough can be obtained. Furthermore, if it is assumed that the different carbohydrates in dough are fermented sequentially and not simultaneously (Mackenzie, 1958), a simple calculation indicates roughly the time at which the various substrates are consumed. This calculation depends upon a knowledge of the fermentable carbohydrates in flour and here we have to rely upon some rather fragmentary data. The most recent and reliable results are those of Saunders *et al.* (1972) and are given in Table 1. Free glucose, fructose and sucrose together provide only sufficient

Table 1

Fermentable sugars in dough

| Sugar | Sugar content (g/100 g flour) | Equivalent gas volume (ml) for 18 g dry flour | | Approximate time (min) for fermentation (see curve A, Fig. 2) |
		Per individual sugar	Cumulative (less carbon dioxide dissolved)	
Glucose	0.02	1	0	0
Fructose	0.04	2	0	0
Sucrose	0.26	14	7	2
Glu_1-fru_2	0.40	22	29	20
Glu_1-fru_3	0.26	14	43	30
Higher glucofructans	0.72	40	83	60
Maltose (free)	0.12	7	90	65

Flour analyses are from Saunders *et al.* (1972). Gas volume was calculated as wet gas volume (saturated at 25°C) at 30°C and 760 mm pressure as measured in the fermentometer test (Burrows and Harrison, 1959); free water in dough was assumed to dissolve 10 ml carbon dioxide. The time was calculated from zero time in fermentometer test (13 min after mixing) but carbon dioxide liberated before zero time was estimated as 5 ml.

carbohydrate to support fermentation for a few minutes as shown by the calculations summarized in Table 1. The levosin or glucofructans (glu_1-fru_2, glu_1-fru_3, etc.) are believed to be hydrolysed by invertase (Tanaka and Sato, 1969; Medcalf and Cheung, 1971) although this was not confirmed by Biltcliffe (1972). In any case, the process is probably slower than sucrose hydrolysis and hence these carbohydrates are listed next in the table and their metabolism probably occurs mainly between points X and Y in curve A of Figure 1. Finally maltose, the main bulk of which is not present in the dry flour but is formed in the dough by the action of amylases on damaged starch, is fermented. It may be calculated from the curves that the total fermentable sugar is about 4.6% and the maltose formed from starch is about 2.8%, based on weight of dry flour.

It may now be seen that the inflection in curve A (high yeasting rate) occurs at the same time as the easily fermentable carbohydrates are running out. This appears to be due to the fact that, although maltose is present at this stage, the yeast cannot metabolize it as fast as it can ferment glucose. When the other sugars are consumed, however, the yeast begins to adapt its metabolism to ferment maltose. This probably involves permeation and other processes. When these are fully organized, the fermentation rate continues to accelerate again until ethanol accumulates and maltose is depleted.

This brief description of dough fermentation indicates that much more than glycolysis is involved and, although the glycolytic system must be the ultimate limiting factor, other limitations occur at different times. It may therefore be stated that the following yeast properties are required for a fast dough fermentation: (i) high potential glycolytic activity; (ii) ability to adapt rapidly to changing substrates; (iii) high invertase (or other appropriate enzyme) activity to hydrolyse the higher glucofructans as rapidly as possible; (iv) high potential maltose fermentation rate; and (v) ability to grow and synthesize enzymes and coenzymes under anaerobic conditions. Of these, the last is probably the most important and explains why maltose fermentation and dough-raising ability are not correlated, and also why the rate of dough fermentation, even at an early stage, is greater than the rate of fermentation of simple sugars in the absence of other nutrients (Lövgren and Hautera, 1977).

The properties just discussed contribute to the overall quality of the yeast and must be built into it by selecting the best strain available and growing it under optimum conditions. The complexity of the situation is also increased by other demands on yeast quality, such as stability, resistance to the effects of osmotic pressure, low pH values and other possible inhibitors (Flückiger, 1974). In addition other properties, such as good aerobic growth characteristics, are required by the yeast manufacturer to ensure good yields and trouble-free production.

IV. ACTIVE DRIED YEAST

A. Drying Methods

In spite of the fact that dried yeast is more costly to produce than moist compressed yeast, it continues to be much in demand and increasing efforts are being made to improve its quality. This interest is due to two main factors, namely its greater stability and its low water content which results in higher activity on a weight-for-weight basis compared with compressed yeast. Both of these factors give it a considerable advantage in terms of transport and storage costs. Its disadvantages are the expensive drying process, sensitivity to rehydration in cold water and low activity on a dry weight basis compared to compressed yeast.

Methods of drying have been extensively studied with a view to speeding up the process and producing a more active and more stable product. Thus, spray drying is a rapid and convenient process for the commercial scale but adverse cell-viability kinetics make it difficult to operate in this application (Elizondo and Labuza, 1974). Various materials may be added as protective agents to decrease damage under the rapid drying conditions imposed (Toyo Jozo Co. Ltd., 1967; Kyowa Hakko Kogyo Co. Ltd., 1969). Partial spray drying followed by fluid-bed drying has also been proposed (Rennie, 1976). However, apart from the difficulty of retaining a sufficiently high level of fermentative activity, these methods require higher inputs of energy to remove excess water in the yeast suspension than is required for drying crumbled or extruded yeast.

The fluidized-bed method of drying has recently found considerable favour since it is potentially much more rapid than band or drum drying. It is claimed that drying times as short as 20 minutes can be achieved with minimal loss of yeast activity (Langejan, 1971). A problem sometimes encountered using this method is that yeast particles tend to agglomerate during the early stages, making drying slow and uneven. A proposal to overcome this problem uses a continuous process in which the steady-state conditions within the fluidized bed are arranged so that the crumbled yeast being continuously fed in is quickly flashed past the sticky condition (Beran, 1975). Another method of overcoming this problem is to provide a previous stage in which the yeast is partially dried, either on a continuous belt (Goux, 1973) or by spraying the cream (Rennie, 1976).

An important feature of recent work has been the use of protective additives. These were first described by Mitchell and Enright (1957) and further developed by several other workers (e.g. Pomper and Akerman, 1968). The types of materials used include emulsifying agents (Chen and Cooper, 1960), edible oils (Kyowa Hakko Kogyo Co. Ltd., 1968) and oligosaccharides and other natural polymers (Toyo Jozo Co. Ltd., 1967). The additives are incorporated as homogeneously as possible by mixing with the yeast cream before filtration or with the moist yeast cake. The resulting cake is then crumbled or extruded and dried by the chosen method.

It is not certain whether these agents provide protection during drying or during the reconstitution which must take place before the product can be tested. However, it is known that some agents, such as fatty-acid esters and polyols, exert a certain degree of protection if merely added at the reconstitution stage (Lee, 1977a, b). When such additives are used, the fermentative activity of the dried yeast is higher than when they are omitted. This is associated with a lower leakage of nucleotides during reconstitution (Chen *et al.*, 1966). Since leakage has been shown to be closely associated with activity loss, it is possible that this loss is a direct result of increased membrane permeability resulting in decreased ability to retain intermediates and coenzymes within the cell at a sufficiently high concentration to provide full fermentative activity (Ebbutt, 1961). Certain additives may thus protect membranes from damage during drying or reconstruction and thereby minimize leakage of essential cell constituents.

In the absence of these additives, fermentative activity is reported to be seriously lowered when the moisture content of the yeast is decreased from 7 down to 4 per cent, but this procedure improves its stability. In the presence of additives, however, this effect on activity is not so serious, so that a better balance of properties can be achieved (Reed and Peppler, 1973).

B. Storage Stability

Apart from the inherent stability of the dried yeast itself, which depends upon the strain and method of manufacture, three factors determine its shelf life. These are: the prevailing temperature, the composition of the storage atmosphere, and the moisture content of the yeast. Loss of activity during storage appears to be connected, at least partially, with a loss of enzyme activity due to destruction of cocarboxylase (Chen and Peppler, 1956). Decrease in the activity of glycolytic and other enzymes has also been reported (Tokareva *et al.*, 1976).

The mechanism of these losses has not been studied but, at temperatures above 30°C, in the presence or absence of oxygen, conditions are suitable for the Maillard reaction to take place slowly. Thus free α-amino and ketonic groups in proteins, amino acids and coenzymes are probably present in close proximity, and these could well interact, with a consequent decrease in fermentative activity. An appreciation of this possibility is the basis of a patent in which amino acids or non-reducing sugars are added to minimize the damage done by carbonyl compounds during storage of biological materials (Commonwealth Science and Industry Research Organization, 1955).

C

At lower temperatures, the Maillard reaction is extremely slow and other reactions of an oxidative or hydrolytic nature may predominate. Even in the absence of air, adsorbed oxygen or oxygenated species may be present, resulting in limited damage during the early part of storage. Anti-oxidants may minimize this damage (Pomper and Akerman, 1969).

The part played by water appears to be similar to that in dehydrated foods in general. Thus, yeast exhibits typical water-sorption curves (Peri and De Cesari, 1974) indicating that, at normal temperatures, the monolayer condition is reached at a level of about 0.05 g water/g yeast dry matter. Above this value, free water is available for hydrolytic reactions and below it the underlying molecules may be more accessible to attack by molecular oxygen (Salwin, 1959). Improved stability in dried yeast by lowering the moisture content as far as four per cent has been reported (Chen et al., 1966) but data for drier material are not available.

V. STRAIN IMPROVEMENT

A. Historical

Over the past forty or more years, the science of yeast genetics has been developing rapidly and is being applied with increasing success to the needs of industry. Commencing with the pioneering work of Winge (1935) and the Lindegrens (1943), the techniques of sporulation and hybridization have been continuously improved (Fowell, 1969). On the theoretical side, advances have been made, for example, in the tedious work of chromosome mapping (Mortimer and Hawthorne, 1973; Sherman and Lawrence, 1974), while on the practical side several patents have been published describing new hybrids. Mutation has only been used to a limited extent in improving baker's yeast (e.g. Schultz and Swift, 1955) although its use in other industrial applications of yeast has been considerable (Borstel and Mehta, 1976; Kosikov and Medvedeva, 1976). Also, mutants have been used to facilitate the selection of rare hybrids obtained by standard mating procedures applied after the mutation step (Kosikov, 1977). Very recently, however, a patent has been issued giving details of methods of mutation and selection for producing strains having desirable combinations of properties as baker's yeasts (Société Industrielle Lesaffre, 1978).

Gilliland (1958) outlined the problems of industrial yeast genetics as being: first, the difficulty of obtaining hybrids from some yeast strains and, secondly, the problem of defining the requirements of industry, particularly in brewing. Problems common to applied genetics in general are the difficulties of genetic analysis where characters controlled by a whole series of genes are concerned, and also the fact that many properties of interest are of a quantitative nature and therefore not readily amenable to classical genetic analysis. This latter point is illustrated by the concern of the baker with rate of fermentation rather than whether a particular sugar is metabolized or not.

Fowell (1958) attempted to make a genetic analysis of dough fermentation in a paper in which he also outlined a practical breeding programme. Two parent strains were sporulated (Fowell, 1952) and the viable spores cultured and typed for mating reaction. These spore cultures were then propagated aerobically as baker's yeasts using six 20-litre vessels appropriately equipped to provide growth conditions similar to those used commercially (Burrows, 1970). The yield of yeast from a given amount of molasses and the fermentation rates at different stages of dough fermentation were measured. On the basis of these results, six spore cultures of each mating type from each parent were chosen and these were crossed to give two 6 × 6 squares of hybrids. The successful completion of the 72 matings required says much for the techniques and persistence of the geneticist. The hybrids were now grown and tested as before in blocks of six according to an appropriate statistical design. Analysis of the results indicated that the genetical control of gas production in dough by baker's yeast is very complex. A knowledge of the intricate biochemistry involved might have given a prior indication of this conclusion, but there was some hope that a few key controlling reactions might have resulted in a genetical situation which could be more easily dealt with.

A number of workers have produced hybrids of potential commercial value, and some have claimed improved performance of one kind or another. Thus, Bocharov (1958) crossed baker's and brewer's strains and obtained 41 hybrids, 40 per cent of which were faster fermenters than their parents. In a similar manner, Johnston (1965) obtained promising hybrids for brewing. Kosikov et al. (1975) and Kosikov (1977) also crossed baker's and brewer's yeasts and obtained polyploid hybrids with improved performance in molasses alcohol manufacture, making baker's yeast as a by-product. Hybrids for this purpose are also described by

Kirova and Shevchenko (1972). By crossing *Saccharomyces cerevisiae* with *Sacch. carlsbergensis*, a hybrid was obtained which behaved as well as a normal baker's yeast and also, unlike the latter, completely fermented raffinose (Koninklÿke Nederlandsche Gist-en Spiritusfabriek, 1969; Lodder *et al.*, 1969). This resulted in a slightly higher yield of yeast when using beet molasses as substrate for growth. This was because beet molasses contains between 0.5 and 4.2 per cent of raffinose although even normal baker's yeast will partly metabolize this trisaccharide, leaving melibiose. The potential advantage of this hybrid is, however, further diminished by the fact that, in many countries, between 20 and 100 per cent of the substrate used is cane molasses which contains no raffinose.

Several other claims for compressed yeast hybrids with improved properties have been made (Burrows and Fowell, 1961a, b; Koninklÿke Nederlandsche Gist-en Spiritusfabriek, 1965). The first two of these will be mentioned again in Section V (p. 58), and the last claims a hybrid which, when grown on molasses, will ferment maltose in the later stages of dough fermentation without the typical lag which normally occurs when switching from glucose to maltose fermentation (see Section IV, p. 44).

Special strains for preparing active-dried yeast to be used in baking have also been the subject of a few patent claims. Glutathione is excreted by dried yeast during reconstitution and has an adverse effect on dough structure (Jorgensen, 1945; Ponte *et al.*, 1960). This led to the selection of a mutant strain with a low glutathione content and hence improved bakery characteristics (Schultz and Swift, 1955). The flavour imparted to bread by dried yeast is greater than for compressed yeast, but may be decreased by the use of special strains (Burrows, 1968).

It is claimed that, by the combination of a special drying process and a new hybrid, compressed yeast was dried to produce a very active-dried yeast (Langejan and Khoudokormoff, 1973). This hybrid was first grown using a compressed-yeast type of propagation control to give harvested wet yeast of greater activity than normally used for drying. This material was then extruded to form small particles which were dried rapidly at below 50°C in the presence of swelling agents and surfactants. The resulting dried yeast could be safely reconstituted in cold water (28°C) in contrast to normal dried yeast which requires a temperature of 40°C for maximum activity. Two hybrids were described which behaved satisfactorily under these conditions.

Hoogerheide (1969) and Flückiger (1974), in discussing the optimum requirements   for   improved   strains,   stress   the   importance   of osmotolerance to resist the increased osmotic pressures resulting from the use of sugars and salts in various types of dough. Osmotolerance in yeasts has been extensively studied by Windisch and his associates (Koppensteiner and Windisch, 1972), and he claims to have succeeded in the difficult task of crossing osmotolerant yeast species with *Sacch. cerevisiae* (Windisch and Steckowski, 1970; Windisch *et al.*, 1976). However, it was possible to obtain only a very few hybrids and it is not certain if these would be suitable for use as baker's yeast. Osmotolerant strains for making active-dried yeast are also the subject of a patent claim (Société Industrielle Lesaffre, 1976).

B. Ploidy

Reports that industrial yeast strains are often aneuploid or polyploid are numerous (Emeis, 1961, 1967; Fowell, 1969; Lodder *et al.*, 1969; Windisch, 1972; Kosikov *et al.*, 1975; Johnston and Lewis, 1976). These deductions were sometimes based solely on observations of irregular segregation of genes for a single character, such as mating type, but other explanations for anomalies of this kind are possible (Fowell, 1956). The direct method of determining ploidy by counting chromosomes is difficult to apply to yeast using present techniques. However, Mundkur (1953) and Ogur (1957) have attempted to use methods of this nature, and Borodina and Meisel (1976) were able to count chromosomes in several yeast genera by staining with the fluorescent antibiotic olivomycin. Visualization of chromosomes was also achieved by Giemsa staining of air-dried protoplasts (Galeotti and Williams, 1978).

Four indirect methods were discussed by Winge and Roberts (1958), namely measurement of cell size, observation of segregation patterns of genes, construction of irradiation survival curves and measurement of DNA content. Ogur *et al.* (1952) compared cell size and DNA measurement, and concluded that the latter was the most reliable. Windisch (1961) relied on an analysis of the segregation of mating-type genes to determine ploidy, but this is inconclusive unless information on cell size or the segregation patterns of genes at a variety of sites on the chromosomes is also available. In a related series of strains of differing ploidy, cell size alone appeared to be a good diagnostic feature (Kosikov, 1977).

The question as to whether polyploidy in yeast confers any advantages has been considered by several authors (Windisch, 1961; Emeis, 1963; Scheda, 1963; Windisch, 1972; Kosikov *et al.*, 1975; Kosikov and Raevskaya, 1976). Scheda (1963) showed that, for glucose, the fermentation rate increased with ploidy and, for maltose fermentation, diploids and triploids were the best. Emeis (1963) constructed a series of completely homozygous strains, with ploidy increasing from one to eight by a combination of techniques including selection of spontaneous mutants for mating type, self-diploidization and inbreeding. Another series with a smaller range of ploidy was also formed from a different strain and, by crossing these two series, several heterozygous polyploids were obtained. In the two homozygous series, no differences in wort fermentation characteristics measured on a yeast weight basis could be found. In the heterozygous polyploids, however, triploids and tetraploids had greater fermentation efficiency than strains of higher or lower ploidy. These observations suggest that heterosis rather than ploidy was responsible for the improvement. Kosikov and Raevskaya (1976), using several different yeast strains ranging in ploidy from one to four, found that biomass production in simple Petri dish tests increased with ploidy, the largest difference being between haploid and diploid strains.

Any advantages found for polyploids cannot be explained on the simple idea of gene dosage, since cell size or cell weight also increases with ploidy and hence, calculated on a dry weight basis, any such effects would tend to be counterbalanced (Lamprecht *et al.*, 1976). However, in some types of aneuploid, this might not be the case. On the other hand, as the cell size increases, the ratio of surface-area to volume decreases and, therefore, the specific water content and the rates of gas and solute diffusion across the cell surface may change. What effects these changes would have on yeast properties is difficult to predict. The increased degree of homozygosity in polyploids would certainly be expected to confer greater genetic stability (Emeis, 1965).

C. Technical Problems

It is apparent from the foregoing that a certain degree of success has been obtained in hybridization of yeasts for industrial purposes, but until recently (Société Industrielle Lesaffre, 1978), little indication of how these results have been achieved has usually been given, although it is probable that empirical methods have been used. As a tribute to the valuable work of the late Mr. R. R. Fowell in industrial yeast genetics, stretching over a period of more than twenty years, a brief account of his methods and some of the results achieved will now be given.

The methods used by Fowell in breeding, and by the author in screening, were largely shaped by the difficulties encountered as the work progressed. These were: (i) difficulties in producing a reasonable number of viable spores for mating purposes; (ii) absence of mating ability in a high proportion of viable spores; (iii) impossibility of using simple laboratory tests as ultimate selection criteria; (iv) poor reproducibility of results when using lengthy growth and testing procedures simulating actual conditions of commercial practice; (v) low frequency of promising hybrids. Such problems are usually not met with in academic work, since strains with good sporulating and mating characteristics can be chosen and characters used which are amenable to simple laboratory tests. It may, however, be possible in industrial work to use laboratory growth tests in a limited fashion to eliminate very poor hybrids and defective mutants.

D. Genetic Techniques

Many workers have investigated the factors controlling spore formation in *Sacch. cerevisiae*, and it is now possible to get good results with many strains which were previously difficult if not impossible to use. Fowell (1966) recommends a preliminary anaerobic growth stage on a rich medium, followed by washing and sporulation on potassium chloride–sodium acetate agar of defined composition. Strains not responding to this treatment require individual study, if of sufficient value to warrant it. The ascus wall is now ruptured mechanically or by the use of a special enzyme preparation (Johnston and Mortimer, 1959). The spores may then be removed by micromanipulation or separated from vegetative cells by shaking with paraffin oil, when the spores go into the oil phase. When spore viability is low, selected heat treatment without prior rupturing of the ascus provides some degree of concentration since the spores are more resistant than vegetative cells. Subsequent growth on malt agar allows viable spore cultures to be isolated.

Although several methods are available for forming hybrids, a convenient way is first to grow the spores up as vegetative cultures so that stocks can be held and used repeatedly if proved to be of value. In a combination of the technique of Lindegren and Lindegren (1943) and Jakob (1962), proposed by Fowell (1966), the spore cultures of opposite mating type are mixed, shaken aerobically and then centrifuged to provide intimate contact between cells. Further aerobic treatment with malt wort results in a high proportion of hybrids being formed. These can then be isolated by subculturing. These methods ensure a high success rate, so that the most desirable parent strains can be used and hence any required design of hybridization scheme can usually be carried out. Thus, blocks or strings of hybrids can be formed and tested according to a previously determined statistical design, which facilitates assessment of the value of spore cultures and the screening of the resulting hybrids.

E. Screening Methods

Hybrids obtained as outlined above were grown in a plant which consisted of six 20-litre vessels (later increased to eight). The precise control of pH value, temperature, and medium-feed rate was found to be of great importance, and methods for doing this were gradually improved over the years in order to minimize the variability referred to earlier (Burrows, 1970).

Much of the hybridization work was concerned with the improvement of two main characteristics of baker's yeast. These were, first, the ability of the yeast to ferment dough at a rapid rate at the desired stages in the bakery process and, secondly, the ability to retain this property during storage for several days at ambient temperature. A third property, namely the amount of yeast produced from a given amount of molasses, was also important since, for economic reasons, it was necessary to maintain, and if possible improve, the yield at the same time as other characteristics were being improved. At a later stage in this work, a smaller and less versatile apparatus was constructed which allowed sixteen cultures to be grown side-by-side under identical conditions and which was much more economical in labour and materials (Burrows, 1970).

The fermentative activity of the harvested yeast was originally measured by the time taken for dough, prepared in a standard manner, to rise to a given height. During the course of the work, here described, this was replaced by a gas-volume test (Burrows and Harrison, 1959) which proved to be simple to operate and much more reproducible than the

earlier test. The stability was measured by incubating the yeast in a
standard manner before subjecting it to the dough fermentation test. In
some cases several other tests, such as rates of fermentation of maltose and
glucose in simple and complex media, were measured using the Warburg
technique. These tests did not, however, give any information of practical
use that was not obtained from the standard dough fermentation method,
and it was concluded that, in general, empirical tests based on practical
requirements were as good as, or better than, less empirical biochemical
or chemical tests.

F. Early Hybrids

When hybridization techniques were far from satisfactory, several hard-
won hybrids were obtained by crossing baker's with brewer's or wine
yeasts, and by inbreeding. Unfortunately, the most promising of these
turned out to possess the property of 'insolubility', that is, resistance to
resuspension after harvesting. This was usually not detected in small-
scale work and only became apparent in commercial-scale operation. It
appeared that some of the parents were heterozygous for this recessive
character, and it was not until more was understood as to the reasons for
its non-appearance in small-scale work and reproducible methods for
testing were developed (Burrows, 1970) that this problem was overcome.

G. Fermentative Activity

In the first large breeding scheme undertaken when hybridization
methods were sufficiently advanced, an existing commonly used strain of
commercial baker's yeast (A) was crossed with a brewing strain (B)
(Fowell, 1958). The latter sporulated reasonably well and, in
preliminary work, had given promising hybrids. Over 100 spore cultures
of the baker's strain and 26 of the brewer's strain were obtained which
were suitable for growing in their own right as baker's yeasts. Six spore
cultures from strain A and 6 of opposite mating type from strain B were
then chosen on the basis of dough fermentation tests after aerobic
propagation. These were crossed to form a 6 × 6 square of hybrids which
were grown and tested by standard methods. On the basis of the
properties of their corresponding hybrids, two spore cultures from strain

B were selected and crossed with a number of spore cultures of strain A. These hybrids, in general, had fermentative activities which were about 10 per cent higher than that of the original baker's yeast. The best of these was selected and, after commercial-scale trials, was used regularly in the early 1950s; its particular characteristic was its ability to change rapidly from hexose to maltose metabolism during dough fermentation without the lag associated with strain A (Mackenzie, 1958).

This selected hybrid (H1) was then crossed with another baker's strain (C) which had a faster initial fermentation rate in dough than strain A but showed an even more pronounced lag. This lag could be removed by preliminary growth or fermentation in a medium containing maltose, but for economic reasons it was not possible commercially. In the preliminary part of this breeding scheme, a hybrid (H2) was found which had excellent fermentation rate both in the early and later stages of dough fermentation. Subsequently, a more systematic examination of spore cultures and hybrids in this series failed to produce any strain with better properties and therefore, after suitable trials, hybrid H2 was selected and introduced commercially in the 1950s.

Statistical analyses of the results in the two schemes which produced hybrids H1 and H2 showed that the properties of the spore cultures themselves were only weakly correlated with those of their corresponding hybrids, so that propagation and testing of these were not very profitable. Secondly, real differences between spore cultures could generally be confirmed statistically, and their value as breeding stock could be reliably assessed by examining the properties of their corresponding hybrids. However, there were interactions between spore cultures of opposite mating reaction, with the result that the best hybrid was usually not the result of the crossing of the two apparently best spore cultures. The strength of the correlations and the magnitude of the differences between spore cultures and their interactions would depend upon the genetical characteristics of the breeding stocks, but these generalizations were valuable in designing future breeding experiments.

H. Keeping Quality

In the next stage, improvements in keeping quality and yield were sought by going back to the parents and crossing strain B with strain C. Spore cultures of known value from the latter strain were available, but those from strain B were of doubtful value having produced some hybrids of poor stability. New spore cultures from strain B were therefore obtained which were then selected on the basis of the properties of their corresponding hybrids produced when crossed with selected strain C spore cultures. In this way, a few selected spore cultures of each strain were chosen and new hybrids obtained from them if they were not already available. As the work progressed, hybrids were selected and reselected until only about six remained. These were then subjected to a variety of propagation conditions with variations in temperature and pH value programmes until the one which behaved best could be selected.

This hybrid (H3) was more stable on storage and gave a higher yield than previous selected hybrids, but had a lower activity and was therefore not used commercially. It was crossed with strain A with a view to improving this feature whilst retaining good keeping quality. For this breeding scheme, eight newly-equipped 20-litre propagation vessels were brought into use and the newly developed fermentometer test was introduced for routine use (Burrows and Harrison, 1959).

Two spore cultures of known value from strain A were crossed with 32 spore cultures from H3 to give two series of hybrids. Examination of these hybrids by aerobic propagation, followed by testing as baker's yeasts, enabled spore cultures to be selected for further breeding. In a similar manner, by crossing two of these selected strains with 32 spore cultures of strain A, eight of the latter were chosen on the basis of the properties of their corresponding hybrids.

Seven of the hybrids already produced and tested were compared with H3, and, since several had significantly higher fermentative activity, the breeding scheme was proceeded with. Thus, two 4 × 4 squares of hybrids were constructed using selected mater strains. Each square of hybrids was tested in a 4 × 4 balanced lattice design in blocks of four with five replicates. As well as giving estimates of spore culture effects and their interactions, this arrangement enabled all the individual hybrids to be compared with each other.

Very good discrimination between hybrids was obtained, and a few were selected from each square for more detailed examination. However, the best hybrids in this scheme all came from a H3 spore culture previously known to be valuable but not included in the 4 × 4 squares, since many hybrids had already been obtained from it for the preliminary screening of strain A spore cultures. This examination entailed further propagations in 8 × 2, 4 × 4 and 2^5 factorial experiments in which eight, four and two hybrids, respectively, were grown under a variety of conditions. For instance, in a 2^5 experiment, two selected hybrids were compared using two levels of aeration, two different temperature and pH value programmes, and two rates of feed of assimilable nitrogen.

In all, 160 new hybrids were grown involving about one thousand propagations including preliminary seed stages. Most hybrids were only grown once, some were grown several times and the hybrids finally selected were grown on more than 40 separate occasions. Three hybrids were eventually chosen, namely H4, H5 and H6, all of which had a better balance of properties than either parent and were better in one respect or another than existing commercial strains. Hybrid H6 had the best combination of activity and keeping quality, and was used commercially for several years (Burrows and Fowell, 1961b).

Hybrid H5 was of lower activity than the other two, but was very stable (Burrows and Fowell, 1961a). This advantage was maintained under conditions of continuous process working where temperature and pH value were maintained at lower levels than in batch-process practice (Burrows, 1970).

I. Summary

The four breeding schemes described are shown in diagrammatic form in Figure 3. This makes it clear that the hybrids H4, H5 and H6 were obtained by interbreeding the same three parent strains A, B and C which produced H2. The superior balance of properties of H4, H5 and H6, compared with H2, was not due to the route used in their formation, since the same genetical material was available, but to the improved methods of testing and screening employed. This work shows that empirical hybridization can be successful providing it is carried out on a large enough scale, and that the testing and selection procedures are carefully chosen.

VI. CONCLUSIONS

This selective review of progress in baker's yeast technology indicates that production methods and quality are continually being improved so that bakers today enjoy the advantages of having an excellent product of good

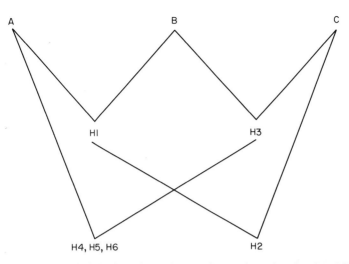

Fig. 3. Breeding schemes for baker's and brewer's yeasts. See text for explanation. A and C indicate baker's yeast strains, B a brewer's yeast strain, and H1–H6 selected hybrids.

stability, consistent quality and reasonable price. This achievement has been largely due to a combination of strain and process improvement resulting from a better understanding of the complex factors involved in producing baker's yeast with the desired balance of properties. There is every indication that improvements will continue to be made as sciences such as genetics, microbiology and biochemical engineering advance at an ever increasing rate.

We may therefore confidently expect that the fast growing science of genetical engineering will eventually lead to the construction of yeast strains with advantageous combinations of characters not previously available, such as high osmotolerance and fast dough fermentation. Also, if a normal baker's yeast strain could be modified by the inclusion of genetic structures which enabled it to produce high extracellular amylase activity, this would be of value in allowing it to be grown on starchy substrates. This property would also probably be of advantage in fermentations of substrates derived from starchy materials. The ability to grow on carbon sources which are less polluting than molasses, together with a diminished requirement for vitamins, would likewise be of commercial value. In this connection, the transference to baker's yeast of the ability to metabolize lactose would be a desirable aim.

The present burgeoning of microprocessor technology will have as profound an effect on yeast technology as on other manufacturing processes, making practical economic sense of integrated programming of propagation, harvesting, packing and cleaning operations. Off-line and on-line data processing with feed-back control, where it is practicable, will be widely adopted in the interests of improved yeast yield and quality.

REFERENCES

Aiba, S., Humphrey, A. E. and Millis, N. F. (1973). *'Biochemical Engineering'*. 2nd edition, p. 110, Academic Press, London and New York.
Aiba, S., Nagai, S. and Nishi-Zawa, Y. (1976). *Biotechnology and Bioengineering* **18**, 1001.
Barney, M. C. and Helbert, R. (1976). *Proceedings of the American Society of Brewing Chemists* **34**(2), 61.
Beezer, A. E., Newell, R. D. and Tyrrell, H. J. V. (1976). *Journal of Applied Bacteriology* **41**(2), 197.
Beran, K., Hospodka, J. and Hauba, L. (1961). *Folia Microbiologica* **6,** 86.
Beran, Z. (1975). British Patent 1 410 793.
Biltcliffe, J. O. (1972). *Journal of Food Technology* **7,** 63.
Blanch, H. W. and Dunn, I. J. (1974). *In* 'Advances in Biochemical Engineering', (T. K. Ghose, A. Fiechter and N. Blakeborough, eds.), Vol. 3, p. 127. Springer-Verlag, Berlin.
Bocharov, S. N. (1958). *Trudy Instituta Genetiki Acadamemiya Nauk S.S.S.R.* **24**, 251.
Borodina, V. M. and Meisel, M. N. (1976). *Microbiology (Eng. Trans. of Microbiologiya)* **45**(2), 323.
Borstel, R. C. von and Mehta, R. D. (1976). *Microbiology (Washington D.C.)*, p. 507.
Burrows, S. and Harrison, J. S. (1959). *Journal of the Institute of Brewing* **65**, 39.
Burrows, S. and Fowell, R. R. (1961a). British Patent 868 621.
Burrows, S. and Fowell, R. R. (1961b). British Patent 868 633.
Burrows, S. (1968). British Patent 1 135 418.
Burrows, S. (1970). *In* 'The Yeasts', (A. H. Rose and J. S. Harrison, eds.), Vol. 3, p. 349. Academic Press, London and New York.
Burrows, S. (1975). *Proceedings of the Sixth International Conference on Cereal Technology and Cereal Chemistry*, Potsdam, p. 295. Institut für Getreideverarbeitung, D.D.R.
Chen, S. L. and Peppler, H. J. (1956). *Archives of Biochemistry and Biophysics* **62**, 299.
Chen, S. L. and Cooper, E. J. (1960). United States Patent 3 041 249.
Chen, S. L., Cooper, E. J. and Gutmanis, F. (1966). *Food Technology* **20**, 1585.
Chen, S. L. and Gutmanis, F. (1976). *Biotechnology and Bioengineering* **18**, 1455.
Commonwealth Science and Industry Research Organization (1955). British Patent 799 644.
Drews, B., Specht, H. and Herbst, A. M. (1962). *Branntweinwirtshaft* **102**, 245.
Dunn, I. J. and Mor, J. R. (1975). *Biotechnology and Bioengineering* **17**, 1805.
Ebbutt, L. I. K. (1961). *Journal of General Microbiology* **25**, 87.
Elizondo, H. and Labuza, T. P. (1974). *Biotechnology and Bioengineering* **16**, 1245.
Emeis, C. C. (1961). *Proceedings of the Eighth Congress of the European Brewery Convention, Vienna*, p. 205. Elsevier, Amsterdam.
Emeis, C. C. (1963). *Proceedings of the Ninth Congress of the European Brewery Convention, Brussels*, p. 362. Elsevier, Amsterdam.
Emeis, C. C. (1965). *Proceedings of the Tenth Congress of the European Brewery Convention, Stockholm*, p. 156. Elsevier, Amsterdam.

Emeis, C. C. (1967). *Proceedings of the Eleventh Congress of the European Brewery Convention, Madrid*, p. 139, Elsevier, Amsterdam.
Flückiger, R. (1974). *Getreide, Mehl und Brot* **28**(9), 230.
Fowell, R. R. (1951). *Journal of the Institute of Brewing* **57**, 180.
Fowell, R. R. (1952). *Nature, London* **170**, 578.
Fowell, R. R. (1956). *Comptes Rendus des Travaux du Laboratoire Carlsburg, Serie Physiologique* **26**(8), 117.
Fowell, R. R. (1958). *Society of Chemical Industry Monograph*, no. 3, p. 116.
Fowell, R. R. (1966). *Process Biochemistry*, **1**(1), 25.
Fowell, R. R. (1969). *In* 'The Yeasts', (A. H. Rose and J. S. Harrison, eds.), Vol. 1, p. 303. Academic Press, London and New York.
Galeotti, C. L. and Williams, K. L. (1978). *Journal of General Microbiology* **104**, 337.
Gilliland, R. B. (1958). *Society of Chemical Industry Monograph*, no. 3, p. 103.
Goux, J. (1973). German Patent 2 147 715.
Harrison, J. S. (1971). *Journal of Applied Bacteriology* **34**, 173.
Hayduck, F. (1919). German Patent 300 662.
Haynes, W. C., Wickerham, L. J. and Hesseltine, C. W. (1955). *Applied Microbiology* **3**, 361.
Holloway, J. W. and Burrows, S. (1976). *Chemical Engineer*, June, No. 310, p. 435.
Hoogerheide, J. C. (1969). *Antonie van Leeuwenhoek* **35** (suppl.), 71.
Jakob, H. (1962). *Comptes Rendus de l'Academie des Sciences, Paris* **254**, 3909.
Johnston, J. R. (1965). *Journal of the Institute of Brewing* **71**, 130.
Johnston, J. R. and Lewis, C. W. (1976). *In* 'Proceedings of the Second International Symposium on the Genetics of Industrial Microorganisms', (K. D. Macdonald, ed.), p. 339. Academic Press, London.
Johnston, J. R. and Mortimer, R. K. (1959). *Journal of Bacteriology* **78**, 292.
Jones, R. C. and Anthony, R. M. (1977). *European Journal of Applied Microbiology*, **4**, 87.
Jorgensen, H. J. R. (1945). 'Studies in the Nature of the Bromate Effect', Oxford University Press, London.
Kirova, K. A. and Shevchenko, A. M. (1972). *Fermentnaya i Spirtovaya Promyshlennost, U.S.S.R.* **5**, 38.
Kirsop, B. (1974). *Journal of the Institute of Brewing* **80**, 565.
Koninklÿke Nederlandsche Gist-en Spiritusfabriek (1965). British Patent 989 247.
Koninklÿke Nederlandsche Gist-en Spiritusfabriek (1969). British Patent 1 152 286.
Koppensteiner, G. and Windisch, S. S. (1972). *Archives of Microbiology* **83**, 193.
Kosikov, K. V. (1977). *Microbiology (Eng. Trans. of Mikrobiologiya)* **46**(2), 311.
Kosikov, K. V. and Medvedeva, A. A. (1976). *Microbiology (Eng. Trans. of Mikrobiologiya)* **45**(2), 327.
Kosikov, K. V., Raevskaya, O. G., Khoroshutina, E. B. and Perevertailo, G. A. (1975). *Microbiology (Eng. Trans. of Mikrobiologiya)* **44**(4), 615.
Kosikov, K. V. and Raevskaya, O. G. (1976). *Microbiology (Eng. Trans. of Mikrobiologiya)* **45**(6), 890.
Kyowa Hakko Kogyo Co. Ltd. (1968). British Patent 1 132 793.
Kyowa Hakko Kogyo Co. Ltd. (1969). British Patent 1 146 367.
Lamprecht, I., Schaarschmidt, B. and Welge, G. (1976). *Radiation and Environmental Biophysics* **13**, 57.
Langejan, A. (1971). British Patent. 1 230 205.
Langejan, A. and Khoudokormoff, B. (1973). British Patent 1 321 714.
Lee, J. L. (1977a). British Patent 1 468 515.
Lee, J. L. (1977b). British Patent 1 470 378.

Lindegren, C. C. and Lindegren, G. (1943). *Proceedings of the National Academy of Sciences of the United States of America* **29**, 306.
Lodder, J., Khoudokormoff, B. and Langejan A. (1969). *Antonie van Leeuwenhoek* **35**, Suppl. F9.
Lorenz, K. and Lee, V. A. (1977). *Critical Reviews in Food Science and Technology* **8**(4), 383.
Lövgren, T. and Hautera, P. (1977). *European Journal of Applied Microbiology* **4**, 37.
Mackenzie, R. M. (1958). *Society of Chemical Industry Monograph no. 3*, 127.
Medcalf, D. G. and Cheung, P. W. (1971). *Cereal Chemistry* **48**, 1.
Mitchell, J. H. and Enright, J. J. (1957). *Food Technology* **11**, 359.
Moo-Young, M. (1975). *Canadian Journal of Chemical Engineering* **53**, 113.
Mortimer, R. K. and Hawthorne, D. C. (1973). *Genetics* **74**, 33.
Mundkur, B. D. (1953). *Experientia* **9**, 373.
Ogur, M. (1957). *Journal of Bacteriology* **73**, 360.
Ogur, M., Minckler, S., Lindegren, G. and Lindegren, C. C. (1952). *Archives of Biochemistry and Biophysics* **40**, 175.
Patentauswertung Vogelbusch GmbH (1961). British Patent 897 166.
Peri, C. and De Cesari, L. (1974). *Lebensmittel-Wissenschaft und Technologie* **7**(2), 76.
Pirt, S. J. (1974). *Journal of Applied Chemistry and Biotechnology* **24**, 415.
Polivoda, A. I., Baum, R. F., Mosin, V. F., Yakovenko, A. Z. and Fisher, P. N. (1972). British Patent 1 284 500.
Pomper, S. and Akerman, E. (1968). United States Patent 3 410 693.
Pomper, S. and Akerman, E. (1969). United States Patent 3 440 059.
Ponte, J. G., Glass, R. L. and Geddes, W. F. (1960). *Cereal Chemistry* **37**, 263.
Reed, G. R. and Peppler, H. J. (1973). 'Yeast Technology', p. 53, AVI Publishing Co. Inc., Westport, Connecticut.
Rennie, S. D. (1976). British Patent 1 459 211.
Reusser, F. (1963). *Advances in Applied Microbiology* **5**, 189.
Richards, M. (1975). *Proceedings of the American Society of Brewing Chemists* **33**, 1.
Rosen, K. (1977). *Process Biochemistry* **12**(3), 10.
Sak, S. (1919). Danish Patent 28 507.
Salwin, H. (1959). *Food Technology* **13**, 594.
Saunders, R. M., Ng, H. and Kline, K. (1972). *Cereal Chemistry* **49**, 86.
Scheda, R. (1963). *Archiv. für Mikrobiologie* **45**, 65.
Schultz, A. S. and Swift, F. R. (1955). United States Patent 2 717 837.
Sherman, F. S. and Lawrence, C. W. (1974). *In* 'Handbook of Genetics', (King, R. C., ed.), Vol. 1, p. 359. Plenum Press, New York.
Société Industrielle Lesaffre (1976). Belgian Patent 843 792.
Société Industrielle Lesaffre (1978). Belgian Patent 862 191.
Solomons, G. L. (1969). 'Materials and Methods in Fermentation', p. 133. Academic Press, London.
Tanaka, Y. and Sato, T. (1969). *Journal of Fermentation Technology* **47**(9), 587.
Tanner, R. D., Souki, N. T. and Russell, R. M. (1977). *Biotechnology and Bioengineering* **19**, 27.
Tokareva, R. R., Semikhatova, N. M., Gusera, L. I. and Pimenova, T. I. (1976). *Khlebopekarnaya i Konditerskaya Promyshlennost* **3**, 34.
Toyo, Jozo Co. Ltd. (1967). British Patent 1 064 212.
Volpe, T. (1976). *Baker's Digest* **50**(6), 48.
Wang, H. Y., Cooney, C. L. and Wang, D. I. C. (1977). *Biotechnology and Bioengineering* **19**, 69.
Windisch, S. (1961). *Wallerstein Laboratory Communications* **24**(85), 316.

Windisch, S. (1972). *Monatsschrift für Brauerei* **25**(8), 201.
Windisch, S. and Steckowski, U. (1970). *Gordian*, **70**, pp. 336, 381, 382, 434, 435.
Windisch, S., Kowalski, S. and Zander, I. (1976). *European Journal of Applied Microbiology* **3**, 213.
Winge, O. (1935). *Comptes Rendus des Travaux du Laboratoire Carlsberg, Serie Physiologique* **21**, 77.
Winge, O. and Roberts, C. (1958). *In* 'The Chemistry and Biology of Yeasts', (A. H. Cook, ed.), p. 93. Academic Press, New York.
Yanez, E., Wulf, H., Ballester, D., Fernandez, N., Gattas, V. and Monckeberg, F. (1973). *Journal of the Science of Food and Agriculture* **24**, 519.
Yoshida, F., Yamane, T. and Nakamato, K. (1973). *Biotechnology and Bioengineering* **15**, 257.

Note added in Proof
The Substrate Problem

The availability and quality of molasses, the traditional substrate for baker's yeast manufacture, is now an increasing cause for concern. This substrate has been the raw material of choice for over fifty years owing to its availability, microbiological stability and content of a range of nutrients which is nearly sufficient for balanced yeast growth. However, several factors now make it more difficult to use although a suitable substitute has yet to be found.

Manufacturers of sugar, particularly those using sugar beet, are gradually increasing the efficiency of sucrose extraction, thus resulting in smaller quantities of molasses of poorer quality being available. This shortage is aggravated by the increasing demand for molasses for animal feeding. A further problem is the progressively greater impact of government legislation on effluent control. Since only about 70 to 80 per cent of the total carbon in molasses is utilized by baker's yeast (Skogman, 1979) the residual fermentation liquor has a very high oxygen demand (B.O.D. value) so that it cannot be discharged into waterways. In some cases, the local sewage works may receive and treat the effluent on payment of a premium. If this is not possible the effluent must be treated on site.

Package systems for aerobic or anaerobic digestion are available (Skogman, 1979) but these must have a very high capacity to cope with large volumes of effluent of high B.O.D. value and therefore are expensive in terms of capital expenditure and space requirement. Another possibility is evaporation and subsequent sale of the concentrated residue as fertilizer or animal feed (Lewicki, 1977). With rapidly increasing energy costs, this course is becoming more and more uneconomic.

It is therefore not surprising that other possible substrates containing a higher proportion of utilizable carbon are being considered. One possibility is starch although this must be first hydrolyzed before baker's yeast strains at present available can utilize it (Reed and Peppler, 1973). In addition, trace elements, growth factors and other nutrients must be added. Also the dilute solutions obtained after hydrolysis are microbiologically unstable and must therefore be used immediately or dried for future use. Under ideal conditions this raw material should give a growth substrate of reasonably constant composisiton and an effluent of low B.O.D. value.

There is some possibility that, in the future, manufacturing units will be set up in favourable locations for processing cellulosic waste materials into useful chemicals (Brown and Fitzpatrick, 1976; Detroy and Hesseltine, 1978; Ladisch *et al.*, 1978; Ladisch, 1979). Fermentable carbohydrates and ethanol from such plants could be used for making baker's yeast. The latter has the advantage of microbiological stability and ease of transport but its method of use in this application has still to be worked out.

Much work has been done on the production of single-cell protein and ethanol from deproteinated whey (e.g. Coten, 1976; Reesen and Strube, 1978) but baker's yeast strains cannot metabolize this lactose-containing by-product of the cheese industry. Economic methods of making baker's yeast from this plentiful substrate are perhaps worthwhile aims for the future (Dion *et al.*, 1978).

References

Brown, D. E. and Fitzpatrick, S. W. (1976). *In* 'Food from Waste', (G. G. Birch, K. J. Parker and J. T. Worgan, eds.), p. 139 Applied Science Publishers, London.
Coton, S. G. (1976). *In* 'Food from Waste', (G. G. Birch, K. J. Parker and J. T. Worgan, eds.), p. 221 Applied Science Publishers, London.
Detroy, R. W. and Hesseltine, C. W. (1978). *Process Biochemistry* **13**(9), 2.
Dion, P., Goulet, J. and Lachance, R. A. (1978). *Canadian Institute of Food Science and Technology Journal* **11**(2), 78.
Ladisch, M. R., Ladisch, C. M. and Tsao, G. T. (1978). *Science, New York,* **201,** 743.
Ladisch, M. R. (1979). *Process Biochemistry* **14**(1), 21.
Lewicki, W. (1977). *Zeitschrift für die Zuckerindustrie* **27**(5), 302.
Reed, G. R. and Peppler, H. J. (1973). 'Yeast Technology', p. 55. A.V.I. Publishing Co. Inc., Westport, Connecticut.
Reesen, L. and Strube, R. (1978). *Process Biochemistry* **13**(11), 21.
Skogman, H. (1979). *Process Biochemistry* **14**(1), 5.

3. Bacteria for Azofication

JOE C. BURTON

Nitragin Co., Milwaukee, Wisconsin, U.S.A.

I. INTRODUCTION

The atmosphere above every hectare of the earth's surface contains around 77 000 metric tons of nitrogen; yet nitrogen is the element that most frequently limits food production. The vast majority of plants cannot utilize elemental nitrogen (N_2). Nitrogen must first be 'fixed' or combined with other elements before it can be assimilated.

Nitrogen can be fixed chemically under high pressure and temperature (Haber-Bosch process) or it can be fixed biologically. According to Hardy (1975) about two-thirds of the nitrogen available today for food production (about 175 million metric tonnes) is fixed annually by biological systems. With the expanding world population and a dwindling supply of fossil fuel for manufacturing nitrogenous fertilizers, an even greater emphasis will have to be placed on biological nitrogen fixation. Dinitrogen is fixed by algae and by bacteria. Certain of these bacteria are free-living while others work symbiotically with plants. This chapter is concerned only with bacteria, and major emphasis will be on organisms which Man has learned to harness for nitrogen fixation.

II. *RHIZOBIUM* SPECIES

A. Early History

Rhizobium species are identified by their ability to infect roots of leguminous plants, form nodules, and work symbiotically, in most instances with their host, in fixing molecular nitrogen. Culture of rhizobia in the laboratory for use in agriculture began shortly after Hellriegel and Wilfarth revealed the secret of leguminous nodules and their effect on growth of leguminous plants in 1886. The causal bacterium, *Rhizobium* sp., was isolated from leguminous nodules in 1888 by Beijerinck, a Dutch microbiologist. Culture of *Rhizobium* species in the laboratory began soon afterwards, but early attempts to use this new tool to increase crop production were disappointing. Fred *et al.* (1932) described the frustrations of this period and also some key developments in learning to use rhizobia to increase crop production.

B. *Rhizobium* Species

The nodule bacteria (*Rhizobium* sp.) are one of two genera in the family Rhizobiaceae, and are characterized by their ability to infect root hairs of leguminous plants and induce nodule formation. *Rhizobium* species are differentiated by the kind of leguminous plants they are capable of nodulating. Six species of *Rhizobium* are designated in Bergey's manual, eighth edition (1974). *Rhizobium meliloti* for *Medicago, Melilitos* and *Trigonella* plant species; *R. trifolii* for *Trifolium* spp; *R. leguminosarum* for *Pisum, Vicia, Lens* and *Lathyrus* spp; *R. phaseoli* for *Phaseolus vulgaris* and *P. coccineus*; *R. japonicum* for *Glycine max*; and *R. lupini* for *Lupinus* and *Orthinopus* spp. However, there are many kinds of rhizobia that have not been given a species name. After so many irregularities and incongruent reactions were noted in Rhizobium-leguminous plant associations, the validity of the concept of species designation was seriously questioned. No new species has been designated in the past 40 years. Rhizobia are simply designated by the name of the parent host from which they were isolated.

C. Strain Selection

The family Leguminosae consists of more than 12 000 species of leguminous plants but fewer than 100 of these are cultivated for food or forage. Nonetheless, *Rhizobium* strains must be selected for each leguminous host because of the high degree of specificity. In selecting rhizobia, several factors must be considered, namely: (1) nitrogen-fixing capacity with the specific host; (2) competitiveness with other rhizobia which may be present as well as with other micro-organisms in the soil; and (3) growth ability in culture media, in the carrier, and in the soil. *Rhizobium* strain selection is one of the inoculant manufacturer's prime responsibilities. Wide-spectrum strains are preferred over narrow-spectrum strains. Selection techniques are described in detail by Date (1976) and Burton (1976a, b, c).

D. Nutritional Requirements

Rhizobia are aerobic Gram-negative rods, measuring 0.5–0.9 μm × 1.2–3.0 μm in size. The cells occur singly or in pairs and are motile while young. Older cells stain unevenly in bands because of granules of poly-β-hydroxybutyrate, a reserve food in the cell which does not stain. Rhizobia are divided into so-called 'fast-growers' and 'slow-growers'. *Rhizobium meliloti, R. trifolii, R. leguminosarum, R. phaseoli*, birdsfoot lotus, *Astragalus, Sesbania* and other varieties of rhizobia are termed 'fast-growers' because they usually produce turbidity in broth or visible

Table 1.

Composition of media for growing dinitrogen-fixing bacteria

1. *Rhizobium* spp. (Burton, 1967)

Distilled water	1000 ml
Sucrose	10.00 g
Mannitol	2.00
K_2HPO_4	0.50
$MgSO_4.7H_2O$	0.20
NaCl	0.10
$CaCO_3$	0.25
Yeast autolysate	1.00

2. *Azotobacter* sp. (Ashby, 1907)

Distilled water	1000 ml
Mannitol	15.0 g
K_2HPO_4	0.2
$MgSO_4.7H_2O$	0.2
NaCl	0.2
$CaSO_4.2H_2O$	0.1
$CaCO_3$	5.0

3. *Azotobacter* sp. (Burk, 1934)

Distilled water	1000 ml
Sucrose	20.0 g
K_2HPO_4	0.64
KH_2PO_4	0.16
$MgSO_4.7H_2O$	0.20
NaCl	0.20
$CaSO_4.2H_2O$	0.05
Na_2MoO_4	0.001
$FeSO_4$	0.003

4. *Azotobacter chroococcum*
   (Dobereiner *et al.*, 1972)

Distilled water	1000 ml
Sucrose	5.00 g
$MgSO_4.7H_2O$	0.125
$CaCl_2$	0.200
$Fe_2(SO_4)_3$	0.025
$MnSO_4.H_2O$	0.025
$Na_2MoO_4.2H_2O$	0.005

colonies on solid media in 3–5 days at 28°C. 'Slow-growers', *Rhizobium japonicum, R. lupini*, cowpea, and big trefoil lotus cultures require 8–10 days to attain an equal amount of growth.

Rhizobium species are not particularly fastidious in their nutritional requirements. They utilize monosaccharides and disaccharides readily, and to a lesser extent trisaccharides, alcohols and organic acids (Allen and Allen, 1950; Vincent, 1975). Pentoses, such as arabinose and xylose, are the preferred sources of carbon for the 'slow-growers' (Wilson and Umbreit, 1940; Graham and Parker, 1964). Strains within a species of *Rhizobium* vary in their ability to utilize carbohydrates. This should not be too surprising in the light of the known preferences of strains of rhizobia for specific host plants which undoubtedly vary widely in their chemical composition and metabolic activity. The ability of rhizobia to utilize a carbon source will depend to some extent on the basal medium, nitrogen source, oxidation–reduction potential, size of inoculum and possibly other factors. In inoculant manufacture, sucrose and mannitol are used most commonly. Yeast extract or autolysate is a source of nitrogen and

Table 1—*continued*

5. *Azotobacter paspali*
   (Dobereiner and Campelo, 1971)

Distilled water	1000 ml
Brown cane sugar	20.00 g
K_2HPO_4	0.05
KH_2PO_4	0.15
$MgSO_4.7H_2O$	0.20
$CaCl_2$	0.02
$FeCl_3$	0.01
$NaMoO_4.2H_2O$	0.002
$CaCO_3$	1.00
Bromothymol blue 0.5%	5 ml

6. *Spirillum lipoferum*
   (Dobereiner and Day, 1974)

Distilled water	1000 ml
Sodium malate	5.00 g
KH_2PO_4	0.40
K_2HPO_4	0.10
$MgSO_4.7H_2O$	0.20
NaCl	0.10
$FeCl_3.6H_2O$	0.01
$Na_2MoO_4.2H_2O$	0.002

7. *Spirillum lipoferum* (Okon *et al.*, 1977)

Distilled water	900 ml
DL-Malic acid	5.00 g
$MgSO_4.7H_2O$	0.20
NaCl	0.10
$CaCl_2$	0.02
NH_4Cl	1.00
NaOH	3.00
K_2HPO_4	6.00
KH_2PO_4	4.00
(mixed in 100 ml water and autoclaved separately)	
Yeast extract (Difco)	0.10
Minerals: Fe, Mo, Mn, B, Co, Zn	Traces

8. *Derxia* sp. (Dobereiner and Campelo, 1971)

Distilled water	1000 ml
Starch	20.0 g
K_2HPO_4	0.05
KH_2PO_4	0.15
$MgSO_4.7H_2O$	0.20
$CaCl_2$	0.02
$FeCl_3$	0.01
$Na_2MoO_4.2H_2O$	0.002
$NaHCO_3$	1.00
Bromothymol blue 0.5%	5 ml

growth factors. Most strains of rhizobia can use ammonium ions and some can use nitrate ions as a source of nitrogen, but a small amount of yeast extract is beneficial.

Strains of *R. leguminosarum*, *R. trifolii* and *R. phaseoli* may require one, all, or any combination of biotin, thiamin and pantothenic acid (Wilson, 1940; Allen and Allen, 1950; Graham, 1963). It is good practice to have all of them in the medium. Strains of *R. meliloti*, *R. lupini*, *R. japonicum* and the cowpea rhizobia require only biotin, and not all of them require it. The nutritional requirements of rhizobia for many of the uncommon or exotic legumes deserve more study.

The role of inorganic salts in nutrition of the rhizobia is not well understood. Plant extracts have generally been used in culturing rhizobia. The mineral content of these has probably been adequate for most rhizobia. It is generally conceded that rhizobia need small concentrations of iron, magnesium, calcium, potassium, manganese, zinc and cobalt (Vincent, 1977). Yeast extract, a common ingredient of the culture medium for growth of rhizobia, contains nutritionally significant concentrations of iron, calcium, magnesium, sodium, potassium and manganese (Steinberg, 1938). Medium no. 1 (Table 1) is commonly used in culture of rhizobia. Mannitol is often substituted for sucrose when growing *R. japonicum* and other 'slow-growers'.

E. Inoculant Manufacture

1. *Growing Rhizobia*

Methods of culturing rhizobia vary with the manufacturer, but the submerged-culture technique is generally employed. Selected strains of *Rhizobium* are first cultured on agar slants. These serve as inocula either for Roux culture vessels or shake-flasks containing sterile broth. The Roux or shake-flask cultures constitute the seed inoculum for the small fermenters. Three to four days incubation at 24°C is required for each stage of production.

Fermenters vary, from large bottles which can be autoclaved to stainless-steel vessels equipped with jackets or coils for rapid heating and cooling, mechanical agitators and various control devices. Rhizobia are aerobic, and fermenters must be equipped to furnish a continuous supply of compressed sterile air. Rhizobia do not require intense aeration; excellent growth is obtained at an oxygen partial pressure as low as 0.01 atmosphere (Wilson, 1940). A partial pressure of 0.15 atmosphere is optimum for respiration, but a lower concentration of oxygen is adequate for growth.

Rhizobia will grow in the range 15° to 35°C. Most species grow best in the range of 30° to 32°C, but *R. meliloti* prefers 35°C. With a good medium and adequate aeration, a cell population of 4 or 5 $\times$ 10^9 per ml can be attained in 96 hours with a one per cent inoculum. The incubation time can be decreased substantially by using a larger inoculum. In culturing rhizobia for use in making inoculants, it should be remembered that biomass is not a prime objective. A good concentration of active cells in the broth is desirable, but growth ability and retention of nitrogen-fixing capacity of rhizobia on the seeds and in the soil, until seedlings susceptible to nodulation develop, is the primary objective.

2. Types of Inoculants

a. *Seed inoculants* are for direct application to leguminous seeds. They are designed to stick to the seed coat. Six types of inoculant are currently being produced for application to seeds: (1) bottle or agar slope culture, (2) peat-base, (3) liquid or broth, (4) lyophilized or freeze-dried, (5) frozen concentrates and (6) oil-dried cultures in a vermuculite base. The peat-base inoculum is the most common and is generally considered the most dependable. Peat gives better protection to the bacteria both in the package and on the seed (Burton, 1967; Vincent, 1975; Brockwell, 1977). The peat must be of good quality, however, and it should be sterilized before use. The desired qualities in a carrier are: (a) freedom from toxins which might affect rhizobia, (b) good absorption qualities, (c) high content of organic matter, (d) ease in drying and pulverizing, (e) good adhesion to seeds, and (f) availability in quantity at reasonable prices. The qualities of a peat should be tested first with live rhizobia because the true value cannot be determined by chemical analysis or physical measurements. Other carriers have proved satisfactory where quality peat is not available. Dube and Mandeo (1973) report good success with pulverized lignite. Strijdom and Deschodt (1976) used a mixture of coal dust, bentonite and lucerne powder. Corby (1976) was very successful with a compost of corn cobs, limestone, single super-phosphate and ammonium nitrate, but six months are required for composting, drying and processing the mixture. Khatri *et al.* (1973) found rice husks satisfactory.

Two approaches are used in the large-scale manufacture of solid-carrier inoculants. In Australia and Europe, rhizobia are first grown in broth, and the liquid culture is injected into a pliable film package containing the sterilized carrier. The amount of broth culture needed is predetermined to assure adequate moisture for optimum growth. The rhizobia proliferate in the package which is used to convey the inoculant to the farmer. This method has the advantage of eliminating or greatly decreasing the number of contaminants, but it is not readily adaptable to large-scale operations because of the high labour cost. Also, it imposes difficulties in carrying out an effective quality-control programme.

The method favoured in the United States is to grow the rhizobia in large fermenters to a minimum of 1×10^9 bacteria per millilitre. A predetermined amount of extremely fine calcium carbonate (sufficient to bring the pH value of the peat to 6.8) is mixed with the broth culture to form a suspension which is sprayed onto the peat while it is being agitated in a paddle, or ribbon-type, mixer. The inoculant should have a moisture content of about 38% (wet basis) after mixing. The inoculant is then placed in thin layers in roller tubs and incubated at 26 to 28°C for 24 to 48 hours. During this time, the heat of wetting is dissipated and the moisture is absorbed uniformly. The inoculant is then milled to break up lumps and effect a more homogeneous product. The pliable film package is used almost exclusively. The rhizobia increase about tenfold during the first two to three weeks after packaging when stored at 28 to 30°C.

   b. *Soil inoculants*. Inoculation of leguminous seeds is not practical under the following conditions: (1) when the seeds are treated with fungicides or insecticides which are toxic to the rhizobia; (2) seeds such as peanuts have a very thin fragile seed coat; adding even a small amount of water could loosen the coat and greatly decrease the extent of germination; and (3) the new type air planter works only with seeds which have a smooth even surface. Any particulate matter on the seed coat, such as the *Rhizobium* inoculant, could cause malfunctioning of the drill. Soil inoculation is compulsory when the success of the crop depends upon effective nodulation. Seed and soil inoculants are shown in Figure 1.

   Soil inoculants are of three types. The granular peat-base type is designed to facilitate growth of rhizobia in the peat particles, and for easy

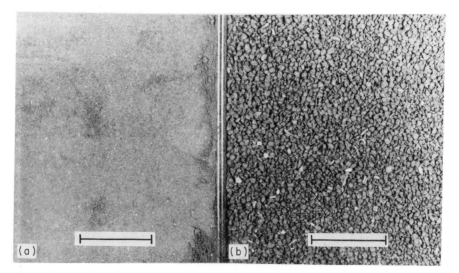

Fig. 1. Types of peat-base *Rhizobium* inoculants. Left: powder for seed; Right: granular for soil application. The bar is 1 cm in length.

application through a drill attachment built to metre out accurately as little as 4 to 5 kilograms per hectare of the inoculant in the row with the seed (Burton, 1976a). The inoculant must consist of fine granules (14 to 40 mesh) which will flow freely. A second type, prepared by coating granular solid non-absorbant cores with an adhesive solution and powdered inoculant, is described by Taylor (1972).

The third type of soil inoculant is a frozen concentrate of *Rhizobium* cells prepared by centrifuging the broth culture and quickly freezing the concentrated cream of paste until ready for use. The frozen inoculant is thawed when ready for use, and diluted with sufficient water to facilitate distribution of a trickle of the rhizobial suspension in the seed furrow with the seed at time of planting (Leffler, 1976). Usually from 15 to 30 litres of bacterial suspension are needed for one hectare.

Soil inoculants have the following advantages: (1) permit application of large numbers of rhizobia that can be applied to the seed; (2) compatibility with chemical fungicides and insecticides applied to seeds; (3) ease of application; and (4) they facilitate inoculation of live stands of leguminous plants which for some reason are not effectively nodulated.

3. Longevity

In peat-base inocula, maximum *Rhizobium* populations are attained in about 14 days at 24°C with the 'fast-growers' and 28 days with the 'slow-growers'. In order to maintain high populations of viable cells, the inoculants should be held at 4°C, or lower, after peak populations have been attained. With higher temperatures, the number of viable rhizobia declines. A decrease in viable cell content is directly related to temperature (Fig. 2). This applies to all forms of inoculant with the possible exception of lyophilized cultures. According to Vincent (1975), lyophilized inoculants have a slightly longer shelf life than peat cultures at 28–30°C, but death rate after application to seed is much higher.

The date beyond which an inoculant is considered unsatisfactory for use is usually printed or stamped on the package. In Australia and India, inoculants are dated to expire in six months. In the United States, the indicated effective period is approximately one year, but the use period does not overlap two planting seasons. The net effect is the same, but the six-month dating period is more realistic. Ideally, expiration dates should relate to storage temperature as well as time.

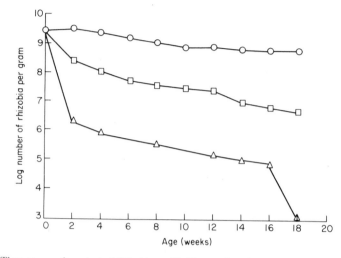

Fig. 2. Time-course of survival of *Rhizobium meliloti* in peat-base inoculum contained in a sealed polythene package and stored at 21°C, (○), 32°C (□) or 43°C (△).

4. Quality Standards

Quality standards should specify the minimum level of viable rhizobia considered necessary to assure effective nodulation of specific legumes planted under normal soil and weather conditions. The number of rhizobia required to bring about effective nodulation will depend upon age and form of inoculant, strain of *Rhizobium*, method of application, type of seed, weather, temperature, moisture, type of soil, seed bed preparation, and possibly other factors. Quality standards should be sufficiently high to compensate for a reasonable range of variation in these factors.

In Australia, a central organization (A.I.R.C.S.; Australian Inoculant Research and Control Service) is responsible for quality control. University or Department of Agriculture personnel test and select rhizobial strains, maintain and issue stock cultures of the selected strains to manufacturers, check quality of broth culture before it is mixed with the peat, and then collect samples of the finished product at various distribution points for analysis. A minimum of 10^8 viable rhizobia per gram up to the expiration date is considered satisfactory. This has recently been increased from 10^6. A similar control system is now in operation in India (Sahni, 1977). Canada requires a minimum of 10^3 viable rhizobia per seed up to the expiration date of the inoculant. In the

United States, manufacturers are responsible for isolating, testing and selecting their own strains of rhizobia as well as setting their own standards in manufacture. Four or five U.S. State Departments of Agriculture collect and test inoculant samples for quality. Testing methods and standards vary from state to state.

F. Inoculant Use

It has often been pointed out (Burton, 1967, 1975) that leguminous seeds are inoculated because of convenience; it is an easy method of placing rhizobia in the soil where the seedling roots will grow. Leguminous seeds differ widely in size, shape, colour, nature of seed coat, and other qualities. Generally, leguminous seeds are not compatible with rhizobia. Inoculation methods must be adjusted to the seed and to the system of planting, and should be carried out as near to planting as possible. One of the most dependable methods of inoculating seeds is to make a slurry with the peat-base inoculant and then mix thoroughly with the seed (Fig. 3).

About 5 g of inoculant are applied to one kilogram of seed; the amount of water which can safely be used will vary with seed size. Caution must be

Fig. 3. Soybean seed, left inoculated with a slurry of a peat-base inoculant, right: uninoculated.

taken to avoid planter damage to wet seeds. The inoculum should provide a minimum of 10^3 viable rhizobia per seed, and many more are needed for certain legumes (Burton and Curley, 1965).

Farmers generally dislike moistening or wetting seeds, and prefer to mix the dry inoculant with the seed in the planter box. Not wanting to lose customers, inoculant manufacturers have not taken a firm stand against this practice even though it is not dependable. The inadequacy of the method can best be demonstrated in fields low in nitrogen which have not previously grown the particular crop.

Seeds in general are poor carriers for rhizobia. The death rate is rapid even when large inocula are applied by the best known methods. Adhesives are used to stick the inoculant to the seed, and certain sugars are beneficial in increasing the longevity of rhizobia on the seed, both before and after planting. Burton (1972, 1975) and Date and Roughley (1977) discuss this subject in detail. Certainly, the potential benefits which can be attained from inoculating leguminous crops with effective rhizobia justify a maximum effort.

III. *AZOTOBACTER* SPECIES

A. Early Development

Beijerinck, the Dutch microbiologist who isolated *Rhizobium* sp. from leguminous nodules in 1888, was the first to demonstrate, in 1901, that free-living *Azotobacter chroococcum* and *A. agilis* are also capable of fixing atmospheric nitrogen. This discovery created a surge of interest in the azotobacters and their potential for supplying nitrogen to higher plants. Studies were initiated in many countries, but early attempts to use *Azotobacter* spp. to increase crop production were disappointing. Field results were erratic and not reproducible.

Certain European workers, notably Stoklasa in Czechoslovakia, Bottomley in England and Markrinov in Russia, had better success. Sizable increases in yields of several non-leguminous crops were obtained (Rubenchik, 1960). An *Azotobacter* sp. preparation, under the name of 'Azotogen' and later called 'Azotobacterin', was introduced in U.S.S.R. by the All-Union Research Institute of Agricultural Microbiology in 1937. While this product has undergone many changes since it was introduced, it is still being manufactured today. Sufficient

'Azotobacterin' is produced annually in Russia to inoculate 10 to 15 million hectares. The product is produced at various locations in Russia and by various methods, but the same strains of *Azotobacter* are used throughout the country.

B. Genera of Azotobacteriaceae

Unlike the rhizobia which comprise a single genus, the Azotobacteraceae are grouped into four genera (Bergey's Manual, 1974):

1. *Azotobacter* spp.—large ovoid cells, cysts formed.
   Azotobacter chroococcum—uses starch, mannitol; contains a black, water-insoluble pigment.
   Azotobacter vinelandi—uses mannitol; has a green fluorescence.
   Azotobacter beijerincki—non-motile; does not use starch or mannitol; has a cinnamon colour.
   Azotobacter paspali—motile; does not use starch or mannitol.
2. *Azomonos* spp.—large ovoid cells; cysts not formed.
   Azomonos agilis—white fluorescence; habitat fresh water; slime formed.
   Azomonos insignis—no colour; habitat fresh water; no slime formed.
   Azomonos macrocytogenes—white fluorescence; uses mannitol; soil habitat.
3. *Beijerinckia* spp.—small rods; forms a tenacious gum; catalase-positive.
   Beijerinckia indica—old colonies are pink; motile; uses lactose.
   Beijerinckia mobilis—old colonies have an amber colour; motile; does not use lactose.
   Beijerinckia fluminensis—old colonies are pink; motile.
   Beijerinckia derxii—shows a buff fluorescence; non-motile; prefers lactose and casein.
4. *Derxia* sp.—small rods; forms tenacious gum; catalase-negative.
   Derxia gummosa—old colonies are brown; may form long filaments.

C. Strain Selection

Selection of *Azotobacter* strains poses two main questions: (a) Should strains be selected for the soil where they will be used or for the plant to be inoculated? (b) Which criteria are meaningful—growth potential, nitrogen-fixing capacity, ability to synthesize auxins and vitamins, or carbohydrate utilization.

In Russia, the All-Union Research Institute for Soil Microbiology selects as follows (Rubenchik, 1960). Strains with the largest cells are first tested for growth rate. Presumably, the *Azotobacter* sp. is predominantly *A. chroococcum*. Strains with the highest growth rates are next tested for nitrogen-fixing ability. The nitrogen-fixing strains are then tested for capacity to produce auxins and vitamins. Strains which survive the laboratory tests are screened for their ability to increase crop yields in pot and field experiments. The influence of the soil and of the plant on strain effectiveness is limited to the final tests.

A preference of *Azotobacter* spp. for certain plant hosts in a relationship which might be considered a type of symbiosis has been noted by some workers. Hardy (1975) uses the term 'associative symbiosis' to describe this relationship. According to M. B. Petrenko, *Azotobacter* strains associated with certain plants exhibit a high nitrogen-fixing activity, a resistance to acid and an ability to utilize root excretions which does not occur with other plants (Rubenchik, 1960). Dobereiner *et al.* (1972) report a similar situation. Certain *Azotobacter* spp. (*A. paspali* in association with *Paspalum notatum* 'batatis', a teraploid cultivar) fixed a significant quantity of nitrogen (90 kg nitrogen/ha/yr). This same *Azotobacter* strain, in association with 'pensacola', a diploid cultivar, fixed very little nitrogen. *Azotobacter paspali* grows very poorly in soil and only in association with a few varieties of *Paspalum notatum*.

D. Nutritional Requirements

Azotobacter species are easy to culture. A wide variety of carbon sources, including monosaccharides, disaccharides, trisaccharides, organic acids, alcohols and some aromatic compounds, can be utilized. However, species and even strains within a species may differ in preference for carbon compounds. *Azotobacter chroococcum* is the only species which is known to use starch. Some of the media used in growing *Azotobacter* spp. are given in Table 1 (p. 68).

A distinguishing characteristic of *Azotobacter* species is their ability to use dinitrogen. However, these bacteria can also assimilate ammonia nitrogen. In fact, a small concentration of ammonium salts is often incorporated into the medium to attain more rapid growth. Aspartic and glutamic acids are readily utilized. *Azotobacter* species must be supplied with a small amount of phosphorus (inorganic or organic), sulphur as sulphate, calcium and magnesium. Molybdenum is required both for nitrogen fixation and ammonia assimilation. Traces of boron, manganese and iron are also required. Organic growth factors are not considered necessary, but more rapid growth often occurs when they are added to the medium.

E. Inoculant Manufacture

1. Growing the Bacteria

Russia has been the major producer of *Azotobacter* inoculants for many years. Currently *Azotobacter* inoculants are being produced by two manufacturers in India, namely 'Bafelab' and 'Bactogin' (S. V. Sahni personal communication), and one in Germany.

The *Azotobacter* cells are grown in three ways, namely shallow layers in Roux bottles, shake-cultures and in fermenters. Sucrose and molasses are the best carbon sources. With vigorous aeration and a 10% starter, cell populations of 10^8 to 10^9 per millilitre can be attained in 48 hours at 28°C to 30°C. Aeration is of prime importance. Ten litres of air per 100 litres of medium per minute are recommended. Lee and Burris (1943) grew *A. vinelandi* in submerged culture using Burk's medium re-inforced with 2% sucrose. The steps in production were: (1) growth for 24 hours in flasks containing 15 ml medium; (2) transfer to Roux flasks containing 100 ml medium and incubation for 24 hours at 30°C; (3) contents of two Roux flasks (200 ml) were then transferred to one 10-litre bottle containing six litres of medium; (4) after 24 hours incubation and aeration in the large bottles, the contents of six bottles (36 litres) were used to inoculate a 200- to 300-litre fermenter. High concentrations of *Azotobacter* cells can be obtained in submerged culture, but loss of viability is rapid, whether kept in broth, lyophilized or dried at higher temperatures under vacuum.

2. Carrier Media

In preparing *Azotobacter* inocula in Russia, the solid materials used are referred to as drying materials rather than carriers, but the function is the

D

same S. P. Norkina attempted to prepare a dry inoculum by adding
cells to sterile organic soil containing carbonates (quoted by Rubenchik,
1960). A vacuum and temperature of 35°C were used to speed up drying.
Only 13% of the cells were alive immediately after drying, and only 4%
survived for as long as 20 days.

In order to solve the problem, Mishustin (1970) recommended
preparation of 'Azotobacterin' locally using a marsh peat as the base. A
tonne of peat is mixed with 10 kg of sugar, 1 to 2 kg superphosphate, 20 kg
calcium carbonate and a starter of *Azotobacter* cells harvested from the
agar surface of 25 Roux flasks. Initially, the starter is mixed thoroughly
with one-half of a tonne of the mixture. After mixing, the moist, freshly
prepared inoculant is spread in thin layers, 20 to 40 cm deep, and
incubated at 20–25°C for three days. The balance of the substrate is next
mixed thoroughly with the inoculant containing *Azotobacter* cells. After
another three days of incubation, the tonne of product is ready for use.

3. Inoculant Use

Azotobacter inoculants are generally applied to seeds as an aqueous
solution. Potatoes and seedlings are soaked in a suspension of the
inoculant for a few hours just before planting. With liquid preparations,
200 ml and 500 ml, respectively, are applied to the seed or potatoes for one
hectare. With peat inoculants, 3 kg and 6 kg, respectively, are used on
seeds or potatoes (Rubenchik, 1960). Granular azotobacter preparations
are considered superior to the liquid suspensions because the cells remain
together as colonies rather than being dispersed as single cells, and are
better able to survive an antagonistic soil microflora. In preparing the
granular type, an organic soil, lime, superphosphate and humus are
blended thoroughly and mixed with 5–30% sawdust and rye straw. The
mixture is then inoculated with a culture of *Azotobacter* sp. or
'Azotobacerin' to form a thick paste. This is extruded to form cylindrical
granules (0.5 × 4 cm) in size which are then dried by coating with
powdered limestone (Rubenchik, 1960). It is claimed that *Azotobacter* sp.
in these granules can survive for a long time even in podzol soils.

F. Outlook

With the success the Russians claim in increasing crop production from
use of *Azotobacter* sp. fertilizers (Azotobacterin), one wonders why other

countries have been unable to show benefit from using *Azotobacter* inoculants in crop production. After exhaustive laboratory and greenhouse studies, Allison (1947) concluded as follows: 'On the basis of present knowledge, the explanation that *Azotobacter* spp. provide nitrogen to their companion can be discarded for the following reasons: (a) no one has demonstrated that *Azotobacter* thrives in the rhizosphere or that it is present in sufficient numbers to fix appreciable nitrogen; (b) root excretions could provide energy for only a trace of nitrogen; (c) where crop residues are present, most of the energy sources are not available to *Azotobacter* spp. Even with sugar, *Azotobacter* spp. fixed only one pound of nitrogen per 100 lb of sugar.'

Clark (1948) inoculated two soils used to grow tomato seedlings with *A. chroococcum* and *A. vinelandi*. One of the soils contained a flora of *Azotobacter* sp. naturally, whereas the other was free from these bacteria. In both soils, there was a rapid decline in numbers of *Azotobacter* sp. and there was no effect on growth of the tomato plants. The Russian preparation 'Azotobacterin' included in the experiment also caused no change in the microflora of the tomato roots and had no beneficial effect on growth. The American literature on the effect of *Azotobacter* sp. on crop production is extensive, and is predominatly negative (Gainey, 1925; Katznelson, 1940; Vandecaveye and Moodie, 1943; Allison, 1947, 1973).

In a recent article from Rothamsted Experimental Station in England, Dart and Day (1975) concluded that *Azotobacter* spp. are usually present in small numbers or not at all on root surfaces even after inoculation. The plant growth-stimulating effects, when present, are believed to be due to hormone production (gibberellic acid) rather than nitrogen fixation. Exceptions to the diminished nitrogen-fixing role for *Azotobacter* spp. appear in the Near East and possibly in tropical climates. Abd-el Malek (1971) found that both *Azotobacter* spp. and nitrogen-fixing *Clostridium* spp. are ubiquitous soil inhabitants in Egypt and Iraq, and that they occur in high numbers except where barrenness, accumulation of sodium chloride or other depressing factors exist. Their response to energy materials is marked. Sizable gains in soil nitrogen are linked to the growth of *Azotobacter* spp.

Dobereiner *et al.* (1972) reported fixation of 90 kg nitrogen per hectare by *A. paspali* in association with Bahia grass *Paspalum notatum* cv. 'batatais'. Seemingly, this particular cultivar was able to exude energy materials which the *Azotobacter* sp. was able to use. Another cultivar,

'Pensacola', did not benefit from association with *A. paspali*. Hardy (1975) summarizes the situation as follows: 'The contribution of fixed N by free-living bacteria to crop production is small. Limitations include poor coupling to crop photosynthesis and low rate of N_2 fixed to carbohydrate utilized. Use of such organisms as a biological N-fertilizer factory does not appear attractive without improvement in efficiency.'

IV. *SPIRILLUM* SPECIES

A. Discovery

Beijerinck, the Dutch microbiologist who isolated Rhizobium sp. from legume nodules in 1888 and *Azotobacter* sp. from soil 13 years later, was the first to describe another type of nitrogen fixer, namely *Spirillum lipoferum*, in 1925. The genus *Spirillum* is one of seven in the family Pseudomonadaceae (Bergey's Manual, 1948). Beijerinck considered *Spirillum lipoferum* a transitional form between species of *Spirillum* and *Azotobacter*.

The demonstration of the nitrogen-fixing ability of *S. lipoferum* in association with a tropical grass, *Digitaria decumbens*, cv. '*Transvala*', was one of the highlights of the 1st International Symposium on Dinitrogen fixation held at Pullman, Washington, U.S.A. (Dobereiner and Day, 1974). However, Becking described a nitrogen-fixing organism which could have been *Spirillum lipoferum* back in 1963. Using $^{15}N_2$, he observed that the organism incorporated labelled dinitrogen when supplied with a small amount of yeast extract.

Dobereiner postulated that the micro-organism enters into a symbiotic establishment within the cortical tissues of the root, and that photosynthetic products are furnished through the phloem, which has direct interchange with the cells. According to Dobereiner (1975), establishment of the organism in the cortical root tissue of the corn plant was confirmed by Burris at the University of Wisconsin. Minerals absorbed by the roots are transferred through the cortex into the xylem. Malate and aspartate, the primary C_4 products of photosynthesis in grasses, are the preferred carbon sources for *Spirillum lipoeferum* (Table 1, medium 6, p. 69). This same mechanism of associative symbiosis could also function in other grass-micro-organism nitrogen-fixing associations.

B. Occurrence

Spirillum lipoferum was found in almost all of 76 soil samples collected in Germany and Austria (Dobereiner and Day, 1974). More than half of the grass roots and soil samples collected in tropical countries (Africa and Brazil) contained abundant populations of *S. lipoferum*. The organism was more common on roots of *Digitaria* sp. than on other grasses, and the organisms were generally uniform in size. Less than 10% of the samples collected in temperate countries (Kenya, U.S.A. and South Brazil) contained the organism (Dobereiner *et al.*, 1976). The organism was more common in neutral than in acid soils. Roots of corn, sorghum, wheat and rye harboured the organism more frequently than leguminous plants.

Bacteria similar to *Spirillum* spp. were found in Florida soils (Schank *et al.*, 1976). Barber and Evans (1976) reported a Gram-negative motile aerobic bacterium associated with the roots of *Digitaria sanguinolis* which was capable of fixing nitrogen. Biochemical and physiological studies indicated that the organism was related to the Azotobacteriaceae, but it differed from any described members of this family.

C. Strains or Specific Associations

Knowledge of the preferences of *Spirillum lipoferum* strains for particular host plants or *vice versa* is meagre. Dobereiner (1976) showed one cultivar of digitgrass, 'Transvala', benefited from association with *S. lipoferum*, whereas another, 'Pensacola', was not benefited. Smith *et al.* (1976) studied a large number of temperate and tropical grasses. Corn, sorghum and millet showed a high nitrogen-fixing potential as compared with tropical grasses. Yields of Pearl millet and guinea grass inoculated with strain 13t of *S. lipoferum* were significantly higher than those from uninoculated plots, but fertilizer nitrogen was needed to stimulate the response to inoculation. Dobereiner (1976) and Dart and Day (1975) agree that breeding of plants for an improved associative symbiosis offers better prospects than *S. lipoferum* strain selection.

D. Growing the Bacteria

So far as the author knows, *Spirillum lipoferum* is not being cultured for use in agriculture, although it is being cultured in fermenters on a small scale for field research plots. Schank *et al.* (1976) cultured *S. lipoferum* strain 13t in 10-litre fermenter batches using a nitrogen-free medium containing 0.5% malate as the energy source. Cultures were sparged with a gas mixture consisting of 5% oxygen and 95% dinitrogen and were grown at 30°C.

Okon *et al.* (1976) describe a method for growing *Spirillum lipoferum* in a 180-litre fermenter for field research (Table 1, medium 7, p. 69). Dobereiner and Day's medium (Table 1) was modified to provide increased buffering capacity, micronutrients and a limited concentration of ammonium chloride (1.0 g/l) and yeast extract (0.05 g/l Difco) to initiate aerobic growth. The phosphate buffer was autoclaved separately and added to the cooled medium with a pH value of 6.8.

The initial inoculum was a 24- to 48-h slant culture suspended in 5 ml 0.05 M sterile phosphate buffer. The inoculum was added to 200 ml sterilized medium in Erlenmeyer shake flasks. The flask cultures served as starters for sterilized medium in 10-litre bottles with spargers and vigorous mixing. For subsequent culture in greater volume, the inoculum used was 10% (v/v) of a culture of *S. lipoferum* in its late exponential stage of growth containing yeast extract and 10^9 cells per ml. The micro-organisms can be grown with 0.2 atmosphere oxygen.

Okon *et al.*, (1976) show that *S. lipoferum* grows vigorously on malate, succinate, lactate and pyruvate. The organism is highly aerobic, and has doubling times as low as two hours when grown on ammonia. Dinitrogen is reduced only under micro-aerophilic conditions. Cells grown with fixed nitrogen converted rapidly to use of dinitrogen upon exhaustion of fixed nitrogen and incubation at a partial pressure of 0.005 to 0.007 atm oxygen. Thus, it appears that the organism could be grown easily with equipment used to produce inocula of *Rhizobium* and *Azotobacter* species.

E. Outlook

Nitrogen-fixing bacteria are found in most soils and especially tropical ones. Growth of these organisms is stimulated by certain cultivars of tropical grasses, but not by others (Dobereiner and Day, 1974; Dart and Day, 1975). The widespread occurrence of *S. lipoferum* in tropical soils led Dobereiner (1976) to suggest that more relevant information would be obtained from plant genetics rather than the traditional bacterial inoculation approach. Similarly, Dart and Day (1975) concluded that the prospect of benefit from inoculation of seeds with asymbiotic nitrogen-fixing bacteria is poor. Availability of carbohydrate in the plant root seems to be the limiting factor to dinitrogen fixation.

Barber *et al.* (1976) inoculated, in greenhouse studies, several lines of corn and sorghum with cultures of *S. lipoferum* from Brazil. No benefit was obtained from inoculating any of 16 inbred lines of corn or 6 hybrids with the Brazilian cultures of *S. lipoferum*. Plants grown without added nitrogen, either with or without inoculum, exhibited severe symptoms of nitrogen deficiency. Since the temperate climate soils of Oregon and Florida in U.S.A. harbour fewer *S. lipoferum* than tropical soils, one might expect some benefit from inoculation, but none was apparent in these studies. At this point, breeding of plants with an improved dinitrogen-fixing ability appears to offer more promise, but the bacteria should not be forgotten. Strains of *S. lipoferum* may well vary in their adaptability to specific hosts and in their ability to establish themselves in the root tissues. This variation may be manifested only on the improved plant lines.

V. OTHER DINITROGEN-FIXING BACTERIA

A. Types and Significance

The list of soil bacteria capable of fixing dinitrogen is long and continues to increase (Stewart, 1966; Dart and Day, 1975; LaRue, 1977). More than 47 species of bacteria from 12 families have been shown to fix dinitrogen (Dart and Day, 1975). In addition to those already mentioned, bacteria of the following genera also fix dinitrogen: *Clostridium, Pseudomonas, Bacillus, Aerobacter, Achromobacter, Klebsiella, Methanobacterium, Desulfovibrio, Chromatium, Rhodospirillum, Rhodopseudomonas, Rhodomicrobium, Chlorobium, Thiocoposa* and *Corynebacterium* (LaRue, 1975). It is difficult to assess the amount of nitrogen fixed by these bacteria and its influence on food production, but certainly the clostridia fix significant amounts of nitrogen under anaerobic conditions (Abd-el-Malek, 1971; Rinaudo *et al.*, 1971; Paul *et al.*, 1971). *Klebsiella* spp. are considered significant in pastures, grasslands and in rice fields (Koch and Oya, 1974; Paul *et al.*, 1971; Day *et al.*, 1975); with proper conditions, the organisms proliferate to significant populations and demonstrate strong nitrogenase activity.

Soil and environmental conditions determine the bacterial species which can grow and gather dinitrogen. There appears to be a widespread distribution of nitrogen-fixing organisms in soil. Dart and Day (1975) postulate that the prospect of benefit from inoculating seeds is remote; the main limiting factor appears to be the availability of carbohydrate in the plant root. Inoculation of seed or soil with asymbiotic nitrogen-fixers other than those described already would have little chance of success in improving plant growth or adding to the soil nitrogen supply. The writer knows of no attempt to produce inocula with these species of bacteria.

VI. OVERALL PROSPECTS

A great amount of study has been applied to free-living nitrogen-fixing bacteria in soils with the development of an enormous literature. Encouraging results have been obtained in many instances, but increasing crop yields by inoculating seed or soil with asymbiotic nitrogen-fixing bacteria remains an uncertainty, even with *Azotobacter* sp., the organism which has received the greatest amount of study.

After studies over a period of years, Ridge and Rovira (1968) concluded as follows: 'Bacterization experiments rarely showed yield increases surpassing the 10% required for real significance. The greatest increases were obtained on fertile soils. There appears to be no immediate likelihood that any regular sound scientific inoculation practice will be developed.' The conclusions of Mishustin (1970) are particularly interesting because Russia has produced inoculants of *Azotobacter* sp. in large quantities for a half century. Mishustin (1970) states that beneficial effects on field crops are obtained in only about one-third of the tests, and that the increases in yield are too small to be considered significant. Beneficial effects from *Azotobacter* inoculation have been noted with vegetables grown in heavily manured soils. Inoculation of seeds brings about a 20 to 30% increase in yields and earlier maturity, and he attributes this effect to synthesis of biologically active compounds by *Azotobacter* sp. Hardy (1975) presents a dark picture for all of the free-living nitrogen-fixers. He states: 'The contribution of fixed N by free-living bacteria to crop production is small. Limitations include poor coupling to crop photosynthesis and low di-nitrogen fixed in proportion to carbohydrate consumed. Soil addition of *Azotobacter* or *Clostridium* that are constitutive in nitrogenase probably will be of little agronomic significance'.

In sharp contrast to the free-living nitrogen-fixing bacteria, the *Rhizobium*-leguminous plant association will certainly assume an ever increasing role in supplying food protein. The challenge is to breed plants with improved photosynthetic efficiency and match them properly with *Rhizobium* strains which have the ability to fix large quantities of dinitrogen. Certain leguminous hosts, such as the bean *Phaseolus vulgaris*, cowpea *Vigna unguiculata*, chickpea *Cicer arietinum*, mung bean *Vigna radiata* and possibly the peanut *Arachis hypogeae*, need to be developed for improved symbiotic efficiency. They need to live more amiably with rhizobia in their nodules and fix larger quantities of elemental nitrogen. This can best be accomplished by a good symbiotic relationship between microbial and plant geneticists as well as between bacteria and plants. Geneticists need to devote more attention to fixation of nitrogen by each of the symbionts and they need to work together.

REFERENCES

Abd-el-Malek, Y. (1971). *Plant and Soil, Special Volume*, p. 423.
Allen, E. K. and Allen, O. N. (1950). *Bacteriological Reviews* **14,** 273.
Allison, F. E. (1947). *Soil Science* **64,** 413.
Allison, F. E. (1973). 'Soil Organic Matter and its Role in Crop Production', p. 184. Elsevier Scientific Publishing Co., New York.
Ashby, S. F. (1907). *Journal of Agricultural Science* **2,** 38.
Barber, L. E. and Evans, H. J. (1976). *Canadian Journal of Microbiology* **22,** 254.
Barber, L. E., Tjepkema, J. D., Russell, S. A. and Evans, H. J. (1976). *Applied and Environmental Microbiology* **32,** 108.
Becking, J. H. (1963). *Antonie van Leeuwenhoek* **29,** 326.
Beijerinck, M. W. (1888). *Botanik Zeitung* **46,** 726.
Beijerinck, , M. W. (1901). *Zentralblatt für Bakteriologie, Parasitenkunde, Infektionskrankheiten und Hygiene* **7,** 561.
Beijerinck, M. W. (1925). *Zentralblatt für Bakteriologie, Parasitenkunde, Infektionskrankheiten und Hygiene* **63,** 353.
Bergey's Manual of Determinative Bacteriology (1948). (R. E. Buchanan and N. E. Gibbons, eds.), 6th edition. The Williams and Wilkins Co., Baltimore, Maryland, U.S.A.
Bergey's Manual of Determinative Bacteriology (1974). (R. E. Buchanan and N. E. Gibbons eds.), 8th edition. The Williams and Wilkins Co., Baltimore, Maryland, U.S.A.
Brockwell, J. (1977). *In* 'A Treatise on Dinitrogen Fixation, Section IV, Agronomy and Ecology', (R. W. F. Hardy and A. H. Gibson, eds.), p. 277. John Wiley and Sons, New York.

Burk, D. (1934). *Ergebnisse der Enzymforschung* **3,** 25.

Burton, J. C. (1967). *In* 'Microbial Technology', (H. J. Peppler, ed.), p. 1. Reinhold Publishing Co., New York.

Burton, J. C. (1972). *In* 'Alfalfa Science and Technology', (C. H. Hanson, ed.), American Society of Agronomy. Monograph no. 15, p. 29. American Society of Agronomy.

Burton, J. C. (1975). *In* 'Symbiotic Nitrogen Fixation', (P. S. Nutman, ed.), International Biological Programme Synthesis, Volume 2, p. 175. Cambridge University Press, London.

Burton, J. C. (1976a). *In* 'World Soybean Research 170', (L. W. Hill, ed.). Interstate Printers and Publishers Inc., Danville, Illinois, U.S.A.

Burton, J. C. (1976b). Proceedings First International Symposium on Nitrogen Fixation, (W. E. Newton and C. J. Nyman, eds.), Volume 2, p. 429. Washington State University Press, Pullman.

Burton, J. C. (1976c). Proceedings of the 33rd Southern Pasture and Forage Crop Improvement Conference, (H. D. Wells, ed.), p. 43. Agricultural Research Service, Tifton, Georgia.

Burton, J. C. and Curley, R. L. (1965). *Agronomy Journal* **57,** 379.

Clark, F. E. (1948). *Soil Science* **65,** 193.

Corby, H. D. L. (1976). *In* 'Symbiotic Nitrogen Fixation in Plants', (P. S. Nutman, ed.), p. 169. Cambridge University Press, London.

Dart, P. J. and Day, J. M. (1975). *In* 'Soil Microbiology a Critical Review', (N. Walker, ed.), p. 255. John Wiley and Sons, New York.

Date, R. A. (1976). *In* 'Symbiotic Nitrogen Fixation in Plants', (P. S. Nutman, ed.), p. 137. Cambridge University Press, London.

Date, R. A. and Roughley, R. J. (1977). *In* 'A Treatise on Dinitrogen Fixation, Section IV, Agronomy and Ecology', (R. W. F. Hardy and A. H. Gibson, eds.), p. 243. John Wiley and Sons, New York.

Day, J., Harris D., Dart, P. and Van Berkom, P. (1975). *In* 'Nitrogen Fixation by Free-living Microorganisms', (W. D. P. Stewart, ed.), International Biological Programme, Volume 6. Cambridge University Press, London.

Dobereiner, J. (1976). *Science, New York* **189,** 368.

Dobereiner, J. and Campelo, A. B. (1971). *Plant and Soil, Special Volume,* p. 457.

Dobereiner, J. and Day, J. M. (1974). *In* 'Proceedings of the First International Symposium on Nitrogen Fixation', (W. E. Newton and C. J. Nyman, eds.), Volume 2, p. 518. Washington State University Press, Pullman.

Dobereiner, J., Day, J. M. and Dart, P. J. (1972). *Journal of General Microbiology* **71,** 103.

Dobereiner, J., Marriel, I. E. and Nery, M. (1976). *Canadian Journal of Microbiology* **22,** 1464.

Dube, J. N. and Mandeo, S. L. (1973). *Research Industry* **18,** 94.

Fred, E. B., Baldwin, I. L. and McCoy, E. (1932). 'Root Nodule Bacteria and Leguminous Plants'. University of Wisconsin Press, Madison.

Gainey, P. L. (1925). *Soil Science* **20,** 73.

Graham, P. H. (1963). *Journal of General Microbiology* **30,** 245.

Graham, P. H. and Parker, C. A. (1964). *Plant and Soil* **20,** 383.

Hardy, R. W. F. (1975). *In* 'Proceedings of the First International Symposium on Nitrogen Fixation, (W. E. Newton and C. J. Nyman, eds.), Volume 2, p. 693. Washington State University Press, Pullman.

Hellriegel, N. and Wilfarth, H. (1886). 'Tagelblatt Versemmlung Deutcher Naturforscher und Aerzte', Berlin.

Khatri, A. A., Choksey, M. and D'Silva, E. (1973). *Science Culture* **39,** 194.

Katznelson, H. (1940). *Soil Science* **49,** 21.

Koch, B. L. and Oya, J. (1974). *Soil Biology and Biochemistry* **6,** 363.

LaRue, T. A. (1977) *In* 'A Treatise on Dinitrogen Fixation, Section III; Biology, (R. W. F. Hardy and W. S. Silver, eds.), p. 19. John Wiley and Sons, New York.

Lee, S. B. and Burris, R. H. (1943). *Industrial and Engineering Chemistry* **35,** 354.

Leffler, W. A. (1976). U.S. Patent 3 976 017.

Mishustin, E. N. (1970). *Plant and Soil* **32,** 545.

Okon, Y., Albrecht, S. L. and Burris, R. H. (1976). *Journal of Bacteriology* **127,** 1248.

Okon, Y., Albrecht, S. L. and Burris, R. H. (1977). *Applied and Environmental Microbiology* **33,** 85.

Paul, E. A., Myers, R. J. K. and Rice, W. A. (1971). *Plant and Soil, Special Volume*, p. 495.

Ridge, E. H. and Rovira, A. D. (1968). *Transactions of the 9th International Congress of Soil Science* **3,** 473.

Rinaudo, G., Balandreau, J. and Dommergues, Y. (1971). *Plant and Soil, Special Volume*, p. 471.

Rubenchik, L. I. (1960). *In* 'Azotobacter and its Use in Agriculture', Translated from Russian by A. Artman. Published for the National Science Foundation, Washington, D.C.

Sahni, V. P. (1977). *In* 'Proceedings Workshop Exploiting the Legume—Rhizobium Symbiosis in Tropical Agriculture', (J. M. Vincent, A. S. Whitney and J. Bose, eds.), Miscellaneous Publication 145, Department of Agronomy and Soil Science. University of Hawaii.

Schank, S. C., Smith, R., Queensberry, K. H. and Bouton, J. H. (1976). *In* 'Proceedings of the 33rd Southern Pasture and Forage Crop Improvement Conference', (H. D. Wells, ed.), p. 57. Agricultural Research Service, Tifton, Georgia.

Smith, R. L., Bouton, J. H., Schank, S. C., Queensberry, K. H., Tyler, M. E., Milam, J. R., Gaskins, M. H. and Littell, R. C. (1976). *Science, New York* **192,** 1003.

Steinberg, R. A. (1938). *Journal of Agricultural Research* **57,** 461.

Stewart, W. D. P. (1966). 'Nitrogen Fixation in Plants'. The Athlone Press, Western Printing Services, Ltd., Bristol.

Strijdom, B. W. and Deschodt, C. C. (1976). *In* 'Symbiotic Nitrogen Fixation in Plants', (P. S. Nutman, ed.), p. 151, Cambridge University Press, London.

Taylor, G. G. (1972). U.S. Patent 3 672 945.

Vandecaveye, S. C. and Moodie, C. D. (1943). *Soil Science Society of America Proceedings* **7,** 229.

Vincent, J. M. (1975). *In* 'The Biology of Nitrogen Fixation', (A. Quispel, ed.), p. 265. North Holland Publishing Company, Amsterdam.

Vincent, J. M. (1977). *In* 'A Treatise on Dinitrogen Fixation, Section III; Biology', (R. W. F. Hardy and W. S. Silver, eds.), p. 27. John Wiley and Sons, New York.

Wilson, J. B. and Umbreit, W. W. (1940). *Proceedings of the Soil Science Society of America* **5,** 262.

Wilson, P. W. (1940). 'The Biochemistry of Symbiotic Nitrogen Fixation'. The University of Wisconsin Press, Madison.

4. Bacterial Insecticides

LEE A. BULLA, Jr. AND ALLAN A. YOUSTEN

*U.S. Grain Marketing Research Laboratory, Science and
Education Administration, U.S. Department of Agriculture,
Manhattan, Kansas, U.S.A.*
and *Department of Biology, Virginia Polytechnic Institute and
State University, Blacksburg, Virginia, U.S.A*

I. INTRODUCTION

There are a variety of bacteria associated with insects. Most of these
bacteria belong to the families Pseudomonadaceae, Enterobacteriaceae,
Lactobacillaceae, Micrococcaceae and Bacilliaceae (Table 1). Members

of these families may be obligate or non-obligate pathogens depending upon their host association in nature (Bucher, 1973). Obligate pathogens, generally, are fastidious and are restricted to a specific host or group of hosts, whereas non-obligate pathogens may be transient insect associates living free of any host most of the time. What characterizes a bacterial insect pathogen is difficult to define. Certainly, any bacterium

Table 1

Selected examples of bacteria that have been determined insect pathogens

Pseudomonadaceae	
Pseudomonas aeruginosa	Grasshoppers
Pseudomonas septica	Scarabaeid beetles, striped ambrosia beetle
Vibrio leonardii	Wax moth,European cornborer
Enterobacteriaceae	
Serratia marcescens,	Variety of butterflies,
Escherichia coli	moths, skippers
Aerobacter aerogenes	Grasshoppers, various butterflies, moths, and skippers
Proteus vulgaris, Proteus mirabilis, Proteus rectgeri	Grasshoppers
Salmonella schottmuelleri var. *alvei*	Honeybees, wax moth
Salmonella enteritidis, Salmonella typhosa, Shigella dysenteriae	Wax moth
Lactobacilliaceae	
Diplococcus and *Streptococcus* spp.	Cockchafer, silkworm, gypsy moth, processionary moth
Streptococcus faecalis	Wax moth
Micrococcaceae	
Micrococcus spp.	June beetle, sawflies, houseflies, various lepidopteran insects including: nun moth, European corn borer, cutworms
Bacilliaceae	
Bacillus thuringiensis, Bacillus cereus	Variety of butterflies and moths
Bacillus popilliae, Bacillus lentimorbus	Scarabaeid beetles
Bacillus sphaericus	Mosquitoes
Bacillus larvae	Honeybees
Clostridium novyi, Clostridium perfringens	Wax moth

4. BACTERIAL INSECTICIDES

that causes demonstrable harm may be considered pathogenic, but identification as a disease agent must be based on strict adherence to Koch's canons; (i) the organism is always found in infected hosts but is not present in healthy ones; (ii) the organism can be grown in pure culture outside the host; (iii) pure cultures, when inoculated into susceptible insects, initiate the characteristic disease symptom(s); and (iv) the organism can be re-isolated from the experimental insect and recultured artificially, retaining the characteristics of the originally isolated organism.

For practical purposes, a *bacterial insecticide* is defined here as a bacterium, bacterial product, or activity of a bacterium that results in death of an insect, and *bacterial control* is defined as the adverse influence of a bacterium on an insect, the bacterium being introduced or applied by

Table 2

Subspecies of *Bacillus thuringiensis*

Variety name	H Antigen serotype[a]	Esterase type
berliner	1	berliner
finitimus	2	finitimus
alesti	3a	alesti
kurstaki	3a3b	ND[b]
sotto	4a4b	sotto
dendrolimus	4a4b	dendrolimus
kenyae	4a4c	kenyae
galleriae	5a5b	galleriae
canadensis	5a5c	ND
entomocidus	6	entomocidus
entomocidus-limassol	6	entomocidus
subtoxicus	6	entomocidus
aizawai	7	galleriae
morrisoni	8	morrisoni
tolworthi	9	tolworthi
darmstadiensis	10	ND
toumanoffi	11	toumanoffi
thompsoni	12	thompsoni
pakistani	13	ND
ostvinia	ND	ND
wuhanensis	ND	ND
israelensis	ND	ND

[a] Based on the classification according to de Barjac and Bonnefoi (1973).
[b] ND indicates that the type was not determined.

man. Although a number of bacteria have been shown to be pathogenic to insects, only a relatively few have been successfully used as bacterial insecticides. The foremost candidates are *Bacillus thuringiensis*, *B. popilliae*, *B. lentimorbus* and *B. sphaericus*. All four of these bacteria are members of the family Bacilliaceae and are spore-formers; the former two are 'crystalliferous', i.e. they produce a discrete, characteristic parasporal crystalline inclusion within the sporulating cell in addition to the endospore; the latter two are acrystalliferous. *Bacillus popilliae* and *B. lentimorbus* are obligate pathogens, whereas *B. thuringiensis* and *B. sphaericus* are not true pathogens but facilitate various disease symptoms due to elaboration of certain toxic metabolites and end products.

A. *Bacillus thuringiensis*

Bacillus thuringiensis (Fig. 1) is one of the best-known crystalliferous bacteria. It is frequently considered a pathogenic variety of *B. cereus* although most insect pathologists and microbiologists consider it as a separate authentic species (Bulla *et al.*, 1975). It exhibits pathogenicity to lepidopteran insects primarily. The species is differentiated into 12 subspecies (Table 2) characterized mainly by serological analysis of

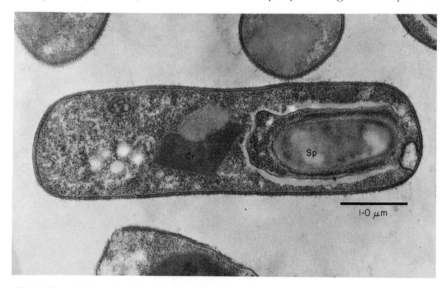

Fig. 1. Sporulating cell of *Bacillus thuringiensis*. Cr, parasporal crystal; Sp, spore. Marker bar indicates 1.0 μm.

flagellar antigens and, secondarily, by electrophoretic analysis of esterases from vegetative cells (de Barjac and Bonnefoi, 1973; Norris and Burges, 1965). The intracellular parasporal crystal is formed outside the exosporium during stages 3 to 5 of sporulation (Bechtel and Bulla, 1976). It contains a single glycoprotein subunit that has a molecular weight of approximately 1.2×10^5 (Bulla *et al.*, 1976, 1977). The carbohydrate consists of glucose (3.8%) and mannose (1.8%). Presumably, subunits are synthesized and progressively assembled to produce a crystalline protoxin that is activated after ingestion by an insect susceptible to the toxic component(s). At alkaline pH values, characteristic of most lepidopteran insect guts, the protoxin is apparently solubilized and activated by an autolytic mechanism involving an inherent sulphydryl protease that renders the protoxin insecticidal. Activation generates protons, degraded polypeptides, sulphydryl-group reactivity, proteolytic activity and insect toxicity. *In vivo* activation of the protoxin and proteolytic activity of the toxin could cause any one of the pathological disorders that have been ascribed to the crystal (Cooksey, 1971): (i) separation of gut cells and detachment from the basement membrane; (ii) enhancement of secretory activity of gut epithelial cells; (iii) increase in permeability of the gut wall to sodium ions with a slower rate of glucose uptake into the haemolymph; (iv) elevated potassium-ion concentration in the haemolymph; and (v) gut paralysis and sometimes general paralysis of the body.

B. *Bacillus popilliae* and *Bacillus lentimorbus*

Bacillus popilliae is another crystalliferous bacterium (Fig. 2) that is a pathogen of various scarabaeid beetles (Dutky, 1940; Rhodes, 1965; St. Julian and Bulla, 1973). The bacterium, when ingested by beetle larvae, invades the haemocoel wherein it undergoes vegetative proliferation and subsequent sporulation causing death of the larvae. The mass of spores that accumulates upon encroaching death of the insect is ultimately released to the surrounding soil and consequently the pathogen can survive for an extended period. These spores are eaten by newly hatched beetle larvae and, upon germination and outgrowth in the alimentary tract, begin the infectious process again. The name given to this infection is 'milky disease', because of the milky appearance of the haemolymph containing spores of *B. popilliae* or *B. lentimorbus*, a closely related

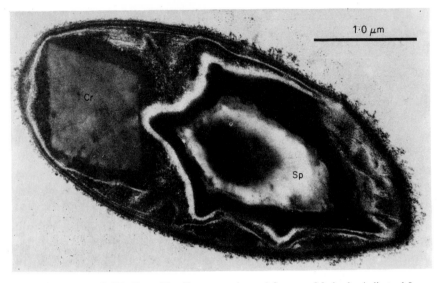

Fig. 2. Sporulated cell of *Bacillus popilliae*. Cr, parasporal crystal; Sp, spore. Marker bar indicates 1.0 μm.

acrystalliferous organism (Fig. 3). Theoretically, *B. popilliae* and *B. lentimorbus* represent persistent and perpetual microbial insecticides, although physical and chemical properties of soil as well as climatic conditions, agricultural and horticultural practices, and density of the larval population, influence their effectiveness in nature.

C. *Bacillus sphaericus*

Bacillus sphaericus is highly toxic to larvae of several mosquito species. The bacterium is a Gram-variable facultative anaerobe that is ubiquitous in terrestrial and aquatic habitats. It exhibits specificity to a number of medically important mosquitoes, and is inert to nontarget vertebrates and invertebrates that have been examined. Not all strains of *B. sphaericus* are highly insecticidal, and those isolates that have insecticidal activity have not been fully characterized. The potency of a *B. sphaericus* strain varies with the mosquito species, the *Aedes* species being the least susceptible and the *Culex* species being the most susceptible.

The toxicity of *B. sphaericus* SS II-1 is apparently not associated with sporulation (Fig. 4). Vegetative cells kill mosquitoes quite effectively. If a cell suspension of a specific age is adjusted to constant optical density, the amount of suspension required to kill 50% of mosquito larvae changes

Fig. 3. Third-instar Japanese beetle larvae (*Popillia japonica*). Healthy (uninfected) larva with transparent haemolymph (left) and milky (infected) larva laden with *Bacillus popilliae* spores (right).

very little during the cell cycle. Also, non-sporulating cells and oligosporogenic mutants are as toxic as the parental strain despite little or no sporulation. *Bacillus sphaericus* is not being produced commercially at this time.

II. COMMERCIAL PRODUCTION OF *BACILLUS THURINGIENSIS*

A. Currently Available Products

Insecticides containing *B. thuringiensis* as the active ingredient have been produced on a large scale in many countries including the United States,

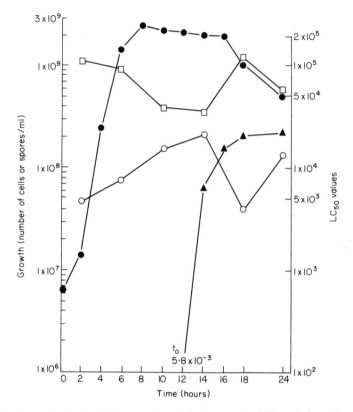

Fig. 4. Relationship of toxic activity and cell age of *Bacillus sphaericus* SSII-1 cells grown in a complex medium. Symbols: LC$_{50}$ values expressed in terms of dilution of cells (O); LC$_{50}$ values expressed as the number of viable cells present after dilution (□); growth curve to indicate the approximate stage of growth at which the cells were tested for toxic activity (●); spore counts to indicate the degree of sporulation of the culture when assayed for toxic activity (▲).

France, Germany, Czechoslovakia, Yugoslavia and the Soviet Union. These products have appeared under a number of different trade names (Table 3) including Dipel, Thuricide, Biotrol, Agritrol, Bakthane, Bactur, Biospor, Bactospeine, Baktukal, Bakthurin, Enterobacterin and Dendrobacillin. At the present time in the United States, two companies are producing these insecticides: Abbott Laboratories (Dipel™) with research and manufacturing in North Chicago, Illinois; and Sandoz, Inc. (Thuricide ®) with administrative, research and production facilities in San Diego, California, Homestead, Florida, and Wasco, California, respectively. Nutrilite Products (Biotrol ®) of Lakeview, California, has temporarily withdrawn its product from the market pending changes in

Table 3

Bacillus thuringiensis insecticide formulations available in the United States of America

Name[a]	Potency[c]	Form
Thuricide-HP	16 000 IU/mg	Wettable powder
Thuricide-16B	3430 IU/mg	Aqueous concentrate
Thuricide-HPC	4000 IU/mg	Aqueous concentrate
Thuricide-HP		Dust
144M dust	320 IU/mg	
90 M dust	200 IU/mg	
Thuricide		Granule bait for European corn borer
400 M granules	880 IU/mg	
Thuricide-HP	175 IU/mg	Cornmeal bait for tobacco budworm
Dipel-WP	16 000 IU/mg	Wettable powder
Dipel-SC	9000 IU/mg	Aqueous concentrate
Dipel-LC	4000 IU/mg	Aqueous concentrate
Dipel-HG	4320 IU/mg	Wettable powder
Bactur[b]		

[a]The formulations listed here each are recommended for specific uses by manufacturers.
[b]Distributed by the Thompson-Hayward Chemical Company; refer to the characteristics of Thuricide for comparison.
[c]$IU/mg = \dfrac{LC_{50} \text{ standard} \times \text{standard (IU/mg)}}{LC_{50}}$

Table 4

Some insect pests currently controlled with *Bacillus thuringiensis* preparations

Insect pest	Plants affected
Cabbage looper	Broccoli, cabbage, cauliflower, celery, lettuce, potato, melon
Imported cabbageworm	Broccoli, cabbage, cauliflower
Tobacco hornworm	Tobacco
Tobacco budworm	Tobacco
Tomato hornworm	Tomato
Alfalfa caterpillar	Alfalfa
Gypsy moth	Forest trees
European corn borer	Corn
Grape leaf-folder	Grape
Codling moth	Apples, pears
Green cloverworm	Soybeans
Orangedog	Citrus
Range caterpillar	Range grass
Sugarcane borer	Sugarcane
Cotton bollworm	Cotton
Spruce budworm	Forest trees
Indian meal moth	Stored grains

production methods to improve the potency of its product. Production and sale of *B. thuringiensis*-containing insecticides have increased greatly in the last ten years as the reliability of the product has been increased. Reliability has been improved by the selection of more toxic strains, by development of improved formulations, and by collection of information relating to better methods of application and field stability. The greatest use of the products continues to be on vegetable and field crops such as broccoli, cabbage, cauliflower, lettuce, melons, tomatoes, grapes, soybeans, tobacco and alfalfa (Table 4). An area of considerable interest is the use of these insecticides in forest areas against such insects as the gypsy moth, eastern spruce budworm and douglas fir tussock moth. For economic reasons, the use in forests has not yet been developed as fully as it may be in the future.

B. Product Protection and Development

As Hannay (1953) and Angus (1954) probed the nature of the pathogenicity of *B. thuringiensis* in the 1950s, Steinhaus (1951) and others were conducting limited trials to test the effectiveness of the bacterium under field conditions. The trials by these workers and by others were sufficiently encouraging to bring about initiation of commercial production of *B. thuringiensis*-containing insecticides in the United States about 20 years ago. At that time, producers attempted to protect their financial interests by obtaining so-called 'process patents' covering the techniques used in production. The patents of Megna (1963) and of Drake and Smythe (1963) describe submerged fermentations for production of *B. thuringiensis* and other insect pathogens for use as insecticides. The patent of Mechalas (1963) describes a semi-solid fermentation process using primarily wheat bran with smaller additions of soybean meal, fishmeal, dextrose and mineral salts as substrate. The semi-solid process was carried out in bins with humidified air circulated through them. Although these early patents provided the public and other industries with a general description of the processes, they have had little if any effect on the course of development of the *B. thuringiensis* insecticide industry. The processes described were readily changed and improved, and the bacterial strains used were easily isolated from the product itself which contained spores of the bacteria. Rather than depend upon patents for protection, the present manufacturers depend upon limited access to manufacturing methods to protect their share of the market. Key points in these methods are fermentation conditions to achieve large population of cells containing large amounts of the toxin, recovery to obtain most of the cells with the toxin undamaged, and formulation of the toxin-containing cells into forms which are stable and effective in the field.

C. Cultural Strains

The strains of *B. thuringiensis* used in commercial insecticide formulations have differed among the various companies, and change as new strains with improved insecticidal activity are isolated. Many of the early field trials were carried out with strains of subspecies *thuringiensis* (serotype I; serotypes are based on serological analysis of flagellar H antigens as developed by de Barjac and Bonnefoi, 1968, 1973). This serotype produces a heat-stable exotoxin (also called 'fly factor') in addition to the proteinaceous parasporal crystalline protoxin, otherwise known as δ-endotoxin. See Bulla *et al.* (1975) for a discussion of serotypes. Thus, the insecticidal activity in the field reflected a combination of these activities. Exotoxin-producing strains grown by submerged fermentation could be freed from the soluble exotoxin if the bacteria were recovered by centrifugation. However, exotoxin-producing strains produced by semi-solid fermentation, in which the bran and cells were simply dried and finely ground to a dust, contained exotoxin in the final product. During most of the years that the product known as Thuricide was produced by International Minerals and Chemical Corporation, several strains of the subspecies *galleriae* (serotype V; H antigens, 5a, 5b) were used. This product is now produced by Sandoz, Inc. and it, as well as Dipel, contains a highly toxic strain (HD-1) of subspecies *kurstaki* (serotype III; H antigens 3a, 3b). The HD-1 strain does not produce exotoxin and, therefore, these products do not contain the toxin. Isolation of the HD-1 strain by Dulmage (1970a) allowed a major advance in the effectiveness of *B. thuringiensis*-containing insecticides. These products have a higher level of toxicity than previous formulations.

D. Standardization

In the past, products containing *B. thuringiensis* were standardized by determining the number of viable endospores per unit volume or weight of the product. However, it became apparent that two strains with equal numbers of spores present could have very different levels of toxicity for any one species of insect in a bioassay (Dulmage, 1970b). As a result, toxicity is now reported in International Units, determined in a bioassay employing the cabbage looper, *Trichoplusia ni* (Dulmage *et al.*, 1971). The activity is compared to an international standard preparation, E-61, prepared at the Institut Pasteur, Paris, France, and assigned a potency of 1000 International Units/mg. Individual manufacturers prepare their own secondary standards based upon E-61.

E. Production

Production of *B. thuringiensis* on a large scale by submerged fermentation does not involve any unusual or exotic techniques. Stock cultures of the particular strain in use may be maintained by lyophilization, by freezing in liquid nitrogen, or by drying of spores on an inert support such as paper strips. The development of inoculum involves transfer through a tube of broth to shake-flasks, to one or more stirred fermentation vessels, and finally to the main fermenter vessel (Fig. 5). These final vessels may range in size from about 50 000 to 125 000 litres. The growth medium may be composed of any of a number of complex organic carbon and nitrogen sources. The Megna patent (1963) suggested the use of 1.86% beet or cane molasses, 1.4% cottonseed meal, 1.7% corn-steep solids and 0.1% $CaCO_3$. Bacterial populations were reported to reach $2-5 \times 10^9$/ml. The fermentation was for 16–20 hours, an optimum time to achieve good spore and parasporal crystal (protoxin) formation. The patent stressed the importance of exhausting both carbon and nitrogen sources from the medium at about the same time. This particular medium has not been used in commercial production for many years. Media more characteristic of present formulations might contain the following ingredients in various combinations: 3–5% fishmeal, 2–3% cottonseed meal (Proflo), 2–3% soybean meal, 2–4% molasses, 1% diammonium phosphate and 0.5–1.0% starch. The amounts and combinations used depend upon the level of toxicity achieved with a particular strain, and to some extent upon price fluctuations of the materials. Although the producers have carried out extensive studies of the effects of medium composition on toxicity of the resulting cell mass, rather little has been

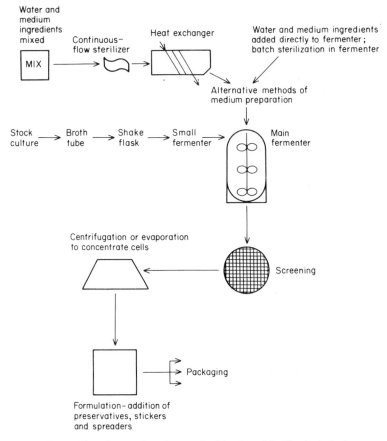

Fig. 5. Flow diagram for submerged cultivation of *Bacillus thuringiensis*.

published in this area. The studies by Dulmage (1970b, 1971) pointed out that: (i) some strains render a more toxic preparation than other strains when compared on an equal cell-weight basis; (ii) there may be variations in the level of toxicity among strains of a single serotype; and (iii) any one strain may yield a more toxic preparation when grown on one medium than on another medium when the cell masses are compared on an equal weight basis. Comparison of cell masses by weight rather than by equal volumes of fermentation broth is important because one medium may simply allow growth of a larger number of cells than another medium. The reason why one strain of *B. thuringiensis* is markedly more toxic when grown in one medium compared to another medium is unclear. Formation of a larger parasporal crystal or an extra crystal in each cell is an unlikely explanation for the 100-fold differences reported by Dulmage (1970b, 1971). Differences in toxicity among strains grown on the same medium have often been ascribed to different solubility properties of the parasporal crystal in the gut of the target insect. Insect pathologists have generally failed to provide a better explanation. If there is differential solubility, this phenomenon is probably the result of varying types of amount of proteins composing the parasporal crystal. However, comparative, quantitative biochemical or immunological studies of the protoxin are lacking.

Although there is not a direct correlation between spore count and toxicity when comparing two strains, there is a relationship between the degree of sporulation and crystalline protoxin formation when one strain is tested. Both spore and crystal are synthesized at about the same time, and fermentation conditions that affect synthesis of one frequently affect the other. For example, good aeration is required for sporulation and parasporal crystal formation by *B. thuringiensis*. Successful commercial cultivation requires aeration and agitation in the large vessels, and this results in a high frequency of spore and crystal formation. It has been shown by feeding experiments which included antibiotics in the test diets that the presence of spores, which can germinate and grow in the insect, contributes to the lethality of the preparation (Ignoffo *et al.*, 1977; Somerville *et al.*, 1970). Thus, cultivation conditions must be optimized to produce good sporulation as well as crystal formation.

Cultivation temperatures are usually maintained at 28°C–35°C although many strains of *B. thuringiensis* will grow at temperatures as high as 41°C. The pH value of the culture is not specifically controlled by acid or base addition. A typical cultivation runs for 20–30 hours, after which the medium is passed through one or more sizing screens to remove coarser material. The cell mass is recovered by centrifugation or concentrated by evaporation. Formulation involves adjustment of the product to a selected value of International Units per milligram, and addition of preservatives, sunlight-screening agents, spreaders (to allow uniform wetting of leaf surfaces) and stickers (to produce a more tenacious film on the leaves). The use of these additives with *B. thuringiensis*-containing insecticides has been presented in detail by Angus and Luthy (1971).

Bacteriophage infection of *B. thuringiensis* cultures has been a continuing problem for some manufacturers. *Bacillus thuringiensis* is attacked by a number of distinct lytic bacteriophages, some of which have been described (Norris, 1961; Chapman and Norris, 1966; Colasito and Rogoff, 1969; Ackerman et al., 1974; Van Tassell and Yousten, 1976). Prevention of phage infection has included extreme sanitation in plant facilities, and isolation of phage-resistant mutants as new phages appear. In at least one plant, a series of bacterial strains, each resistant to one or more phage types, has been rotated through successive culture runs in an attempt to avoid build-up of phages lytic for a particular bacterial strain. This technique has been only partially successful. The location of some fermentation facilities in the open air in very windy and dusty locations may contribute to phage problems. The ability of B. thuringiensis to 'package' lytic bacteriophages in a heat- and disinfectant-resistant form inside the spore (Van Tassell and Yousten, 1976) may also contribute to a high level of phage contamination of manufacturing sites, especially if the contents of infected fermenters are not disposed of carefully.

Production of B. thuringiensis by a semi-solid wheat-bran fermentation method had been carried out until recently by Nutrilite Products, Inc. The product, Biotrol XK, contained B. thuringiensis subsp. kurstaki HD-1 at a potency of 7500 International Units/mg of a wettable powder. This product is no longer produced, pending development of improvements in the process that will allow development of increased potency. It is at present questionable whether a semi-solid fermentation process can yield a product that is competitive in terms of potency with that produced by the submerged cultivation process. The semi-solid process was described in detail by Dulmage and Rhodes (1971).

III. COMMERCIAL PRODUCTION OF *BACILLUS POPILLIAE*

A. Currently Available Products

The primary commercial producer of a B. popilliae-containing insecticide is Fairfax Biological Laboratory, Clinton Corners, New York, U.S.A. The product is called Doom ® and, according to the package label, it contains active ingredients comprised of a 'mixed culture of not less than 100 million viable spores of resistant stages of either or both B. popilliae and B. lentimorbus per gram of inert powder'. Doom ® is manufactured under a licence granted by the United States Department of Agriculture. Bacillus popilliae and B. lentimorbus are highly effective against the

Japanese beetle which feeds on more than 275 different plants and annually destroys field crops, fruits and ornamentals worth millions of dollars. Dutky (1940) described and named *B. popilliae* as the causative agent for type A milky disease, and *B. lentimorbus* for the type B disease. Because *B. popilliae* is a much better control agent and is less fastidious in artificial culture than *B. lentimorbus*, research on this bacterium has been emphasized and spans more than 40 years since the discovery by Hawley and White (1935) that the 'white group' of beetle larvae contained a bacterial infection.

Two patents of Dutky (1942b, c) describe the method for control of Japanese beetles using *B. popilliae* and *B. lentimorbus* formulations and the process for propagating the bacteria. Although the causative agents of milky disease were identified 38 years ago, little progress has been made since that time toward sporulating the organisms in liquid artificial culture. The inability to sporulate efficiently the milky-disease organisms *in vitro* has restricted their use in controlling Japanese beetles because it is the spores, not the vegetative cells, that are disseminated artificially.

B. Cultural Strains

The most efficient sporulation of *B. popilliae in vitro* has been accomplished on a solid medium containing yeast extract and the ingredients of Müeller-Hinton (Difco), trehalose and phosphate (Sharpe and Rhodes, 1973; Sharpe *et al.*, 1970). Haynes and Rhodes (1966) did report, however, sporulation of *B. popilliae* in liquid medium containing activated carbon. In an attempt to isolate a strain of *B. popilliae* with the capacity to sporulate at high frequency, Sharpe *et al.* (1970) carefully examined laboratory strains that had been: (i) exposed to a number of media containing a variety of nutrients; (ii) serially passed through insect larvae and artificial media for an extended period of time; and (iii) grown in continuous culture and exhibited different colonial types. By careful selection of medium and cultural conditions, these workers were able to select a strain (N.R.R.L. B-2309M) that formed 10–20% spores in the total colony population cultured on solid medium (M.Y.P.T) containing beef infusion, casamino acids, soluble starch (with the Müeller–Hinton ingredients at 1%), 1% yeast extract, 0.3% K_2HPO_4 and 0.05% trehalose. As stated, this sporulation frequency is the highest reported for any natural isolate or laboratory strain of *B. popilliae* which includes N.R.R.L.

B–2309, N.R.R.L. B–2309S and N.R.R.L. B–2309N, in addition to N.R.R.L. B-2309M (Sharpe and Bulla, 1978).

In a similar liquid M.Y.P.T. medium, cells of strain B–2309M form 1% spores (about 7×10^6 spores/ml) during 12 days of static culture following a 24-hour shaken-culture period. The spores are similar in appearance to those formed in larvae and they are infective by intrahaemocoelic injection; the spores are not infective perorally (Schwartz and Sharpe, 1970). Sporogenic substrains of strain N.R.R.L. B–2309 (B–2309T and B–2309G) have been isolated by Sharpe and St. Julian (1969) that form 3 to 5% spores in colonies. These *in vitro* spores of substrains B–2309T and B–2309G resemble spores from Japanese beetle larvae more than spores of the other substrains and, following intrahaemocoelic injection, they produce a normal milky disease. Their peroral infectivity has not been tested. All of the above strains and substrains are contained in the culture collection of the Northern Regional Research Centre, Peoria, Illinois, U.S.A.

C. Production

In vivo production of the obligate pathogens *B. popilliae* and *B. lentimorbus* involves artificial infection of third-instar Japanese beetle larvae, incubation of the larvae at 28°C in soil seeded with rye and other appropriate grasses, harvesting of the larvae after 16–21 days, pulverization of the larvae, addition of stabilizing agents and packaging. The first process for obtaining *B. popilliae* spores in this manner was described by White and Dutky (1940) and Dutky (1942a) which entails injection of about 100 000 spores of *B. popilliae* into haemolymph of healthy larvae and the larvae being incubated until visible symptoms of milky disease develops. The diseased larvae are then ground to a powder, the spore content of the powder is determined, and the ground larvae are mixed with appropriate talc. The talc–spore mixture is standardized to contain 100 million spores per gram of the dust; the spore dust at this concentration is then ready for field distribution.

The infectious process with *B. popilliae*, described by St. Julian *et al.* (1970), develops in four phases (Fig. 6). There is an initial incubation phase (phase 1) of about two days in which relatively few bacterial cells are found. From day three to day five (phase II), vegetative cells rapidly proliferate, and by day five prespore forms occur as well as an occasional

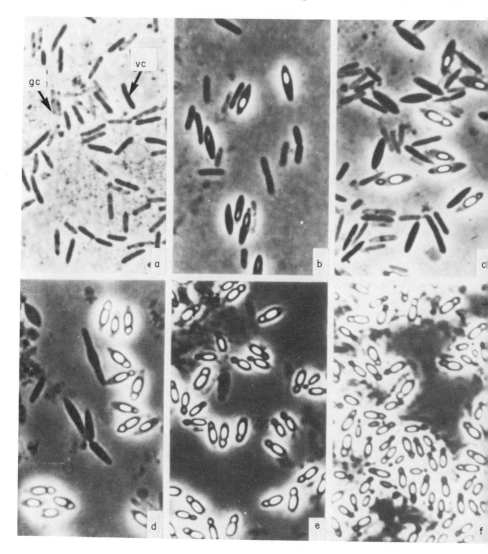

Fig. 6. Sporulation of *Bacillus popilliae* (milky disease development) in Japanese beetle larvae. (a) Viable vegetative cells (vc) and granular, non-reproductive cells (gc) of *Bacillus popilliae* in haemolymph of Japanese beetle larvae during day 3 of infection. (b, c, d) Haemolymph containing *Bacillus popilliae* vegetative cells, prespores, and spores, respectively. The larvae from which the haemolymph came were infected for 5, 7 and 10 days, respectively. (e, f) *Bacillus popilliae* spores and a few vegetative cells in larvae infected for 16 and 21 days, respectively.

spore. Phase III is the intermediate phase (days 5–10) in which there is a change from predominantly vegetative growth to prespore and spore development. Phase IV is the sporulation phase that terminates by day

14 to day 21 and involves massive sporulation and ensuing larval death. During the intermediate phase, prespore forms constitute 20–35% of the total population at days five and six, and they increase to 40–80% by the tenth day. Further development of prespores during phase IV is characterized by a decreasing number of vegetative cells and prespores, while large numbers of spores accumulate. In this terminal phase, an individual larva will contain approximately 5×10^9 spores (5×10^{10} per millilitre of haemolymph) and appears milky. Obviously, sporulation of *B. popilliae* within the Japanese beetle larva is not a synchronous process. There is no exponential phase of exclusive vegetative proliferation followed by a stationary phase and sporulation cycle, but vegetative growth and sporulation proceed concomitantly.

Bacillus popilliae and *B. lentimorbus* represent a unique category among bacteria because they are Gram-variable, facultatively anaerobic, and catalase-less in addition to being insecticidal. Whether the parasporal crystal of *B. popilliae* is toxic to larvae is not yet determined, although there appears to be some toxicity associated with the crystal. Also, characterization of the inclusion body is incomplete, although it is known that the crystal is predominantly protein and can be separated into three cathodic components when the solubilized crystal is subjected to high-voltage electrophoresis. It contains no lipid or nucleic acids but does contain carbohydrate (Weiner *et al.*, 1974).

Bulla *et al.* (1978) have reviewed the current status of the biology of *B. popilliae* and, therefore, we simply re-iterate here that 'what prompts efficient sporulation of the bacterium exclusive of, or within, its insect host is not known. Attempts to determine the factor(s) that regulates spore formation *in vivo* and to simulate the sporulation process in liquid artificial culture have been unsuccessful. Currently, the most efficient means of producing spores of *B. popilliae* for use as a commercial microbial insecticide is by artificially infecting individual larvae. Obviously, such an operation is costly and, therefore, large-scale fermentation production of the spores would considerably reduce the costs. A much clearer understanding of the metabolism of both the insect host and the bacterial pathogen, as well as some insight into the regulatory mechanisms that control bacterial sporogenesis during the infectious process are needed to facilitate the development of a successful industrial fermentation production of *B. popilliae* spores.'

IV. SAFETY, SPECIFICITY AND EFFICACY

The basic premise on which bacterial insecticide industry operates is that a bacterial agent must be active against the target species without adversely affecting man, other animals and plants. However, there are other parameters to be considered, namely: (i) convenience of application (applicable as a dust, liquid spray, or bait); (ii) integration with conventional chemical insecticides; (iii) storability; (iv) economics; (v) ease of production; (vi) virulence; and (vii) safety and aesthetics. The success of a bacterial insecticide depends upon a variety of factors such as environmental and climatic conditions, commodity to be protected (field crop or stored product), mode and timing of application, behaviour and habits of the target host(s), and defense mechanisms(s) of the insect.

Extensive field testing of the efficacy of *B. thuringiensis*, *B. popilliae* and *B. lentimorbus* has been done. None of these bacteria affects biocenosis nor do they threaten man and animals and the environment. What impact these bacteria have on the environment when used as insecticides is difficult to predict, because few data are available on persistence and concentration of bacterial agents in the food chain. Certainly, these particular bacteria are efficacious for their respective insect hosts, and all evidence indicates that they are harmless to vertebrates and other non-target life forms.

Although the risk of infecting man with an invertebrate bacterial pathogen is probably very low, there are tests designed to determine their safety for vertebrates. These are: (i) acute and chronic toxicity and pathogenicity for animals; (ii) primary irritation to eye or skin; (iii) teratogenicity and carcinogenicity; (iv) selection of mutants for possible pathogenicity; (v) specificity of pathogenicity for vertebrates; (vi) potential pathogenicity for plants; and (vii) *in vitro* assessments of cytopathic effects, haemadsorption, and interference (Ignoffo, 1973). *Bacillus thuringiensis*, *B. popilliae* and *B. lentimorbus* have been subjected to most of these tests, and they were found to be virtually harmless to vertebrates and other non-target organisms.

In 1960, the United States Food and Drug Administration granted full exemption from tolerance for *B. thuringiensis* on food and forage crops. This means that there is no specific residue limit on those crops for which the product is registered for use. It can be used up to the day of harvest if the user desires to do so. We will not review the many individual reports describing the experiments which determined the degrees of safety. The reader is referred to the discussion of Heimpel (1971) and the report prepared by Forsberg (1976). The latter monograph is directed primarily at the complex ecological implications of the large-scale application of *B. thuringiensis* against the spruce budworm in Canada. There seems to be no data to suggest that *B. thuringiensis*, used at the levels recommended by the producers, has any acute or chronic adverse effects upon man, or higher or lower animals. However, the Forsberg (1976) monograph suggests many areas where indirect effects on the ecosystem have not been fully investigated. In geographical areas of the United States where these products have been heavily used for almost 20 years, there is no evidence of adverse environmental effect or of the development of insect resistance.

From time to time, it has been suggested that there may be a danger of mutation or genetic exchange through transduction or transformation that would allow development of human pathogenicity in *B. thuringiensis*. This fear is usually associated with the fact that *B. thuringiensis* has some similarities to the human and animal pathogen, *B. anthracis* (Somerville and Jones, 1972). As a means of preventing the introduction of animal pathogens into insecticide products, each lot of spore preparation, prior to the addition of other materials, is tested by subcutaneous injection of at least 1×10^6 spores into each of five mice weighing 17–23 grams. Each test animal must show no evidence of infection or injury when observed for seven days following infection (Anonymous, 1976). This test is a control against the possibility, unlikely (perhaps impossible) as it may be, of a mutation to pathogenicity in the inoculum build-up or in the fermenter itself. Genetic conversion by transduction or transformation in the fermenter would seem to be unlikely since a pathogenic donor of DNA would have to be present. The presence of *B. anthracis* in a modern fermentation plant being run with a pure culture of another bacterium would not seem to be a serious concern. Certainly, the health of the plant personnel would be affected before a genetic conversion could occur in the fermenter. Once *B. thuringiensis* spores have been disseminated in the field, they might be exposed to *B. anthracis* or to its phages and therefore could conceivably be converted to a form with pathogenicity similar to *B. anthracis*. However, this scenario presupposes the presence of *B. anthracis* in the environment, and the conversion of a small number of *B. thuringiensis* cells would seem to constitute a negligible additional hazard. It must be added that there are no published reports of successful transformation or transduction within the species *B. thuringiensis* even under laboratory conditions. Thorne (1978), however, has reported transduction in *B. thuringiensis*. This area of research deserves much more attention than it has received.

V. SUGGESTED READINGS

Afrikian, E. G. (1973). Entomopathogenic bacteria and their significance. Academy of Sciences of the Armenian Soviet Socialist Republic, Erevan, U.S.S.R.

Bulla, L. A., Jr., Costilow, R. N. and Sharpe, E. S. (1978). Biology of *Bacillus popilliae*. *Advances in Applied Microbiology* **23**, 1–18.

Bulla, L. A., Jr., Rhodes, R. A. and St. Julian, G. (1975). Bacteria as insect pathogens. *Annual Review of Microbiology* **29**, 163–190.

Falcon, L. A. (1971). Use of bacteria for microbial control of insects. *In* 'Microbial Control of Insects and Mites', (H. D. Burgess and N. W. Hussey, eds.), pp, 67–95. Academic Press, New York.

Lecadet, M. M. (1970). *Bacillus thuringiensis* toxins—the proteinaceous crystal. *In* 'Microbial Toxins', (T. C. Montie, S. Kadis and S. J. Ajl, eds.), pp. 437–471. Academic Press, New York.

Prasad, S. S. V. and Shethna, Y. I. (1976). Biochemistry and biological activities of the proteinaceous crystal of *Bacillus thuringiensis*. *Biochemical Reviews* **47**, 70–77.

Rogoff, M. and Yousten, H. (1969). *Bacillus thuringiensis*: microbiological considerations. *Annual Review of Microbiology* **23**, 357–386.

St. Julian, G., Bulla, L. A., Jr., Sharpe, E. S. and Adams, G. L. (1973). Bacteria, spirochetes, and rickettsia as insecticides. *Annals of the New York Academy of Sciences* **217**, 65–75.

Somerville, H. J. (1973) Microbial toxins. *Annals of the New York Academy of Sciences* **217**, 93–108.

REFERENCES

Ackerman, H., Smirnoff, W. and Bilsky, A. (1974). *Canadian Journal of Microbiology* **20**, 29.

Anonymous (1976). *United States Federal Register*. Item 180.1011.

Angus, T. and Luthy, P. (1971). *In* 'Microbial Control of Insects and Mites', (H. Burgess and N. W. Hussey, eds.), p. 623. Academic Press, New York.

Angus, T. (1954). *Nature, London* **173**, 545.

Bucher, G. E. (1973). *Annals of the New York Academy of Sciences* **217**, 8.

Bechtel, D. B. and Bulla, L. A., Jr. (1976). *Journal of Bacteriology* **127**, 1472.

Bulla, L. A., Jr., Costilow, R. N. and Sharpe, E. S. (1978). *Advances in Applied Microbiology* **23**, 1.

Bulla, L. A., Jr., Kramer, K. J., Bechtel, D. B. and Davidson, L. I. (1976). *In* 'Microbiology–1976', (D. Schlessinger, ed.), p. 534. American Society for Microbiology, Washington D.C.

Bulla, L. A., Jr., Kramer, K. J. and Davidson, L. I. (1977). *Journal of Bacteriology* **130**, 375.

Bulla, L. A., Jr., Rhodes, R. A. and St. Julian, G. (1975). *Annual Review of Microbiology* **29**, 163.

Chapman, H. and Norris, J. (1966). *Journal of Applied Bacteriology* **29**, 529.

Colasito, D. and Rogoff, M. (1969). *Journal of General Virology* **5**, 367.

Cooksey, K. E. (1971). *In* 'Microbial Control of Insects and Mites', (H. D. Burgess and N. W. Hussey, eds.), p. 247. Academic Press, New York.

de Barjac, H. and Bonnefoi, A. (1973). *Entomophaga* **18**, 5.

de Barjac, H. and Bonnefoi, A. (1968). *Journal of Invertebrate Pathology* **11**, 335.

Drake, B. B. and Smythe, C. V. (1963). United States Patent 3 087 865.

Dulmage, H. T. and Rhodes, R. A. (1971). *In* 'Microbial Control of Insects and Mites', (H. D. Burgess and N. W. Hussey, eds.), p. 507. Academic Press. London and New York.

Dulmage, H. T., Boening, O., Rehnborg, C. and Hansen, G. (1971). *Journal of Invertebrate Pathology* **18**, 240.

Dulmage, H. T. (1970a). *Journal of Invertebrate Pathology* **15**, 232.

Dulmage, H. T. (1970b). *Journal of Invertebrate Pathology* **16**, 385.

Dulmage, H. T. (1971). *Journal of Invertebrate Pathology* **18**, 353.

Dutky, S. R. (1940). *Journal of Agricultural Research* **61**, 57.

Dutky, S. R. (1942a). United States Department of Agriculture, Bureau of Entomology Plant Quarantine, Publication no. E-192.

Dutky, S. R. (1942b). United States Patent 2 258 319.

Dutky, S. R. (1942c). United States Patent 2 293 890.

Forsberg, C. W. (1976). National Research Council of Canada, Publication no. NRCC 15385, Ottawa, Canada.

Hannay, C. (1953). *Nature, London* **172**, 1004.

Hawley, I. M. and White, G. F. (1935). *Journal of the New York Entomological Society* **43**, 407.

Haynes, W. C. and Rhodes, L. J. (1966). *Journal of Bacteriology* **91**, 2270.

Heimpel, A. M. (1971). *In* 'Microbial Control of Insects and Mites', (H. D. Burgess and N. W. Hussey, eds.), p. 469. Academic Press, London and New York.

Ignoffo, C. M. (1973). *Annals of the New York Academy of Sciences* **217**, 141.

Ignoffo, C., Garcia, C. and Couch, T. (1977). *Journal of Invertebrate Pathology* **30**, 277.

Mechalas, B. J. (1963). United States Patent 3 086 922.

Megna, J. C. (1963). United States Patent 3 073 749.

Norris, J. R. and Burgess, H. D. (1965). *Entomophaga* **19**, 41.

Norris, J. (1961). *Journal of General Microbiology* **26**, 167.

Rhodes, R. A. (1965). *Bacteriological Reviews* **29**, 373.

St. Julian, G. and Bulla, L. A., Jr. (1973). *In* 'Current Topics in Comparative Pathobiology', (T. C. Cheng, ed.), p. 57. Academic Press, New York and London.

St. Julian, G., Sharpe, E. and Rhodes, R. A. (1970). *Journal of Invertebrate Pathology* **15**, 240.

Schwartz, P. A., Jr. and Sharpe, E. S. (1970). *Journal of Invertebrate Pathology* **15**, 126.

Sharpe, E. S. and Bulla, L. A., Jr. (1978). *Applied and Environmental Microbiology* **35**, 601.

Sharpe, E. S. and Rhodes, R. A. (1973). *Journal of Invertebrate Pathology* **21**, 9.

Sharpe, E. S. and St. Julian, G. (1969). *Bacteriological Proceedings, American Society for Microbiology* **A111**, 17.

Sharpe, E. S., St. Julian, G. and Crowell, C. (1970). *Applied Microbiology* **19**, 681.

Somerville, H. J. and Jones, M. L. (1972). *Journal of General Microbiology* **73**, 257.

Sommerville, H., Tanada, Y. and Omi, E. (1970) *Journal of Invertebrate Pathology* **16**, 241.

Steinhaus, E. (1951). *Hilgardia* **20**, 359.

Thorne, C. B. (1978). *Applied and Environmental Microbiology* **35,** 1109.
Van Tassell, R. and Yousten, A. (1976). *Canadian Journal of Microbiology* **22,** 583.
Weiner, B. A., Bulla, L. A., Jr., Rhodes, R. A. and Yousten, A. A. (1974). *Proceedings of the Society for Invertebrate Pathology*, p. 27.
White, R. T. and Dutky, S. R. (1940). *Journal of Economic Entomology* **33,** 306.

5. Tempe and Related Foods

KO SWAN DJIEN AND C. W. HESSELTINE

Department of Food Science, Agricultural University, Wageningen,
The Netherlands and
Northern Regional Research Center,
U.S. Department of Agriculture, Peoria, Illinois, U.S.A.

I. INTRODUCTION

In Indonesia, the word *tempe*, pronounced *témpé*, is a collective name for a food product which is fermented by a *Rhizopus* species. It may be made of soybeans, groundnuts (peanuts), bengook (*Mucuna pruriens* Wall, a type of bean), lamtoro (*Leucaena leucocephala* Benth.), or from other substrates A well-made tempe is by definition a compact cake completely covered and penetrated by the white mould mycelium of *Rhizopus* sp. To indicate from which raw material a certain type of tempe is made, an additional word is placed after the word tempe. For example, *tempe kedele* (pronounced as *témpé kedele*) is made of kedele (soybeans) and *tempe lamtoro* is made of young lamtoro seeds. Many other types of tempe are known (Table 1).

Table 1.

Types of tempe

Name of product	Raw material	Reference
Tempe bengook	Seeds of *Mucuna pruriens* var. *utilis*	Oey and Lie (1964), Saono *et al.* (1974)
Tempe bongkrek (kelapa)	Partly defatted coconut	Saono *et al.* (1974), Vorderman (1902), Mertens and van Veen (1933), Darwis and Grevenstuk (1935), van Veen (1966)
Tempe bongkrek katjang	Groundnut press cake	Vorderman (1902)
Tempe enthoe	Partly defatted coconut	Vorderman (1902)
Tempe gembus	Soybean residue	Saono *et al.* (1974)
Tempe kedele	Soybeans	Boorsma (1900), Vorderman (1902), Ko and Hesseltine (1961), Saono *et al.* (1974)
Tempe koro	Seeds of *Phaseolus lunatis*	Sudarmadji (1975)
Tempe lamtoro	Seeds of *Leucaena leucocephala* (*L. glauca*)	Oey and Lie (1964), Saono *et al.* (1974)
Tempe morrie	Mixture of soybeans and partly defatted coconut	Vorderman (1902)
Tempe tjenggereng	Partly defatted coconut	Vorderman (1902)

Tempe kedele, made of the yellow species of soybeans, is the most important and popular type in Indonesia. The single word *tempe* usually refers to this type of fermented food and, accordingly, will be used in this review to indicate the fermented product made of soybeans.

For reasons of pronunciation, publications in English spell the name *tempeh* or *témpé*. However, in the Indonesian language, it is spelled *tempe*, and was so used by the first authors to write about this product (Prinsen Geerligs, 1895; Boorsma, 1900; Vorderman, 1902, and others).

A. Appearance and Preparation

Tempe of good quality is a white compact cake with a clean, fresh odour, but is not consumed raw. Often it is deep-fried in coconut oil or prepared in various other ways, including baking. After it is sliced or cut into pieces, it can be added as an essential part to many Indonesian dishes such as soups and various mixed dishes containing vegetables, meat, fish, tofu, coconut or other ingredients. With its approximately 40% protein content, on a dry basis, tempe is a valuable protein addition to rice, the principal food in Indonesia.

B. Production

Although tempe is a food product characteristic of Indonesia it is, surprisingly, unknown in surrounding countries like China and Japan, where the soybean is also traditionally an important part of the diet. In small quantities, tempe is made in Malaysia, Surinam, and in the Netherlands by immigrants from Indonesia.

The worldwide interest in tempe, particularly in the last decade, has been directed to its high protein content. It could be introduced as a source of low-cost protein food in countries with a protein shortage. In 1971, the soybean (and tempe) was introduced in the Republic of Zambia, Africa, for human nutrition (Thio, 1972). The time has been too short, as yet, to know whether this food is acceptable to the population.

Although most of the 130 million (approx.) Indonesians are consumers, tempe is still manufactured in a traditional way, at home or as a small-scale domestic industry. Efforts to produce tempe on an industrial scale in Indonesia, or any other country that has a shortage of protein foods, have not yet been successful. On the other hand, closely associated with the increasing interest of vegetarians in all foods of vegetable origin, there is a possibility of production on an industrial scale in the U.S.A.

C. Literature

Tempe has undoubtedly been produced in Indonesia for centuries, but as with many other traditional processes, there are no written records of its origin. At the turn of the century, tempe was mentioned in the reports of some Dutch medical doctors who worked in Indonesia. They published articles on Indonesian foodstuffs, or foods from soybeans in general, in which tempe was just mentioned. Prinsen Geerligs (1895) described some Chinese foods made from soybeans, and reported a method used in Java, Indonesia, to make tempe. Boorsma (1900) extensively described production methods and chemical analysis of tempe in his publication on the chemical investigation of native foods in the former Netherlands Indies.

Jansen (1923) and Jansen and Donath (1924) incorporated tempe in their rat-feeding experiments on the protein digestibility of some foods and established the content of vitamin A and the 'anti-beri-beri' vitamin. The B-vitamin content was investigated by van Veen (1932, 1935). During World War II, tempe was a very helpful food in many prisoner of war camps in Indonesia, where many Europeans were starving. Roelofsen (1946) reported the great shortage of protein in the camps and the important role played by tempe in reducing deaths. Patients in the hospitals suffering from gastro-intestinal diseases consumed tempe milk, prepared by grinding the cake with water. It is believed patients would not have survived on unfermented soybeans. Van Veen (1946) confirmed that even patients with dysentry and nutritional oedema were able to assimilate tempe.

After these reports on the important role that tempe played during World War II in prisoner of war camps, only two other publications appeared, until 1960. Tammes (1950) published an article on the manufacture of tempe, and van Veen and Schaefer (1950) studied chemical changes caused by the tempe fungus on soybeans during fermentation.

Approximately 15 publications on tempe appeared before 1960. Since then, about 60 additional scientific studies on the tempe fermentation have been published. This increased interest on tempe fermentation was initiated by two groups of American scientists at the New York Agricultural

Experiment Station, Geneva, New York, and at the Northern Regional Research Center, Peoria, Illinois. As a result, pure-culture fermentation methods were developed on both laboratory and pilot-plant scales. Chemical changes on soybeans during tempe fermentation and the nutritional value of tempe were studied in detail. The physiology and biochemistry of the tempe mould were also investigated (Hesseltine and Wang, 1972). Moreover, possibilities were studied for making other types of tempe from cereals like wheat or from combinations of soybeans and cereals (Wang and Hesseltine, 1966).

Meanwhile, in Indonesia, the attitude towards tempe has gradually changed over the last 15 years. Although most people like tempe, it was formerly considered as an inferior food, mainly because it is less expensive than other protein foods like meat, fish and eggs; another reason was that products of low quality were sometimes sold at the market. But, during the last decade through studies by universities as well as by government agencies, more attention has been paid to this product.

II. INOCULUM

A. Tempe Mould

Different mould species have been reported in the literature as the mould responsible for the fermentation of soybeans into tempe. Prinsen Geerligs (1895) mentioned the name *Chlamydomucor oryzae* Went et Geerligs (now known as *Amylomyces rouxii Calmette*); later on, this name was cited by Boorsma (1900). Burkill (1935) reported that *Aspergillus oryzae* (Ahlburg) Cohn was the organism needed for tempe fermentation. Obviously, these identifications of the mould species were not correct. But, in other reports, *Rhizopus oryzae* Went et Geerligs was mostly mentioned (Vorderman, 1902; Stahel, 1946; van Veen and Schaefer, 1950; Dupont, 1954; Steinkraus *et al.*, 1960). Now we know at least four species of *Rhizopus* including *R. oligosporus* Saito, *R. oryzae*, *R. stolonifer* (Ehrenb ex Fries) Lind, and *R. arrhizus* Fischer, which can be used to make tempe (Ko and Hesseltine, 1961). Later, Hesseltine (1965) successfully tested two additional species from Japan, namely *R. formosaensis* Nakazawa and *R. achlamydosporus* Takeda, where they were originally reported to have been isolated from tempe. He tentatively considered these cultures to be strains of *R. oryzae*. Two Philippine species of phycomycetes were reported as being used for making tempe in the laboratory (Diokno-Palo and Palo, 1968). One was a *Rhizopus* strain closely related to *R. niveus* Yamazaki, but it showed better sporangial formation. It could make a profusely moulded, compact and appetizingly flavoured tempe. The second species was *Cunninghamella elegans* Lendner, which often required 48 h for its mycelium to overgrow the soybean substrate.

In an extensive survey (Ko, 1965) to determine which mould species are generally used by traditional Indonesian manufacturers to make good tempe, 118 cultures were isolated from 81 tempe samples collected from markets in various parts of Indonesia. Tempe collections were made on the islands of Java and Sumatra, and ranged from locations at sea level to places in the mountains up to 1000 m above sea level, with mean temperatures ranging from 15°C to above 30°C.

Collection sites were selected intentionally to answer the question of whether or not mould cultures isolated from cooler regions would be different from those from warm places (Ko, 1965). Most of the cultures which could produce tempe in pure culture, however, turned out to be *R. oligosporus*. Obviously, *R. oligosporus* is the principal species used in Indonesia in the preparation of tempe (Hesseltine *et al.*, 1963; Hesseltine, 1965) and is not dependent on the place where tempe is produced. This conclusion is in accordance with Boedijn (1958), who reported that *R. oligosporus* can always be isolated from tempe.

Hesseltine (1965) characterized this strain of *R. oligosporus* by the sporangiospores which showed no striations and which were very irregular in shape under any condition of growth. The sporangiophores are short, unbranched, and arise opposite rhizoids that are very short in length and show less branching. All isolates show large numbers of chlamydospores, spores which function as sporangiospores.

B. Traditional Inoculum

Traditionally, the inoculum for tempe fermentation is obtained in several ways. It may be the mould residue left over in the wrappings of the previous tempe cake (Boorsma, 1900), or tempe itself may be broken into pieces and mixed with the prepared soybeans (Stahel, 1946). Pulverized dried tempe can also be used (Roelofsen, 1946). The surface of a tempe cake, where most of the mould mycelium is found, may be sliced and sun-dried to be used as inoculum.

Another traditional method is to place leaves of *Erythrina* sp., *Hibiscus similis* Bl, *Tectona grandis* Linn. or a crushed leaf of *Musa* sp. inside the package of inoculated soybeans. During fermentation the mould will grow on the inserted piece of leaf and cover it with a white layer of mycelium. The leaves are removed, sun-dried, and stored until they are needed for inoculation of a new batch of moist cooked soybeans.

These traditional methods of inoculum preparation had several disadvantages. Since viability decreases within a couple of weeks of storage, and since these inocula are not homogeneous in viable propagules, it is difficult to estimate how much is required for a given amount of soybeans. Furthermore, contamination with other micro-organisms led to fermentation failures. To improve traditional starters, Hermana and Roedjito (1971) inoculated steamed rice and a steamed mixture of cassava and soybean flour with 'laru tempe' (tempe inoculum); the origin and type were not mentioned. After two days incubation between two tampahs (plates made of plaited bamboo ribbons), which were placed with the hollow sides opposite each other and covered by a moistened cotton cloth, the moulded mass was sun-dried, pulverized, and preserved in a stoppered bottle.

A third type of inoculum was made from commercial tempe, which was sliced, sun-dried and pulverized. Each of these three types of inocula, made from different substrates, was mixed with various proportions of dried rice-, cassava- or wheat flour. These mixtures were used to inoculate prepared soybeans. The effects of type, quantity and age of each of the inocula on fermentation time and keeping qualities of the resulting tempe were studied. The inoculum made from rice appeared to be the best one. It was easy to prepare, less expensive than traditional inoculum, easy to measure by weighing or by means of a measuring spoon, and, provided it was kept dry, its activity did not decrease after it was kept six months at ambient temperature (25°C). No efforts were made to determine the types of moulds involved in each of the inocula.

C. Pure-culture Inoculum

In a number of laboratory studies, spore suspensions from pure cultures of *R. oligosporus* on agar slants were used for inoculation (Hesseltine *et al.*, 1963, 1967; Martinelli and Hesseltine, 1964; Wang and Hesseltine, 1966). However, for production of tempe on an industrial scale, none of the methods mentioned is suitable. The traditional inocula are not pure and sometimes causes failures due to contaminating bacteria, especially *Bacillus* sp.; preparing pure cultures on agar slants for mass production of spores is expensive and time-consuming.

In a pilot-plant study on production of dehydrated tempe, Steinkraus *et al.* (1965) used an inoculum of freeze-dried and pulverized *R. oligosporus* culture, which was grown on sterilized hydrated soybeans in batches of 500 g at 37°C for four days. However, with this method, Wang *et al.* (1975) experienced failure and uncertainties. Therefore, they studied the possibilities of mass production of *R. oligosporus* spores on several solid substrates including pearled wheat, cracked soybeans, polished rice, and wheat bran. The grains were mixed with various amounts of water in 300-ml Erlenmeyer flasks, sterilized and inoculated with a spore suspension. Based on the results of this study, they propose to make *R. oligosporus* spore preparations by fermenting either rice, a mixture of rice and wheat bran (4:1), or a mixture of wheat and wheat bran (4:1), at a substrate-to-water ratio of 10:6 for 4–5 days at 32°C. The fermentation mass should then be immediately freeze-dried and ground into fine powder. Since the beginning of 1976, the Northern Regional Research Center has distributed over 2500 lots of this inoculum in plastic bags to persons wishing to make tempe.

Although freeze-dried mould cultures should be very suitable for application in a modern factory, the requirements for equipment and technical personnel are not easily met in unsophisticated factories such as are commonly found in less industrialized countries. To avoid these obstacles, Rusmin and Ko (1974) developed a simple method for preparation of a semi-pure culture inoculum for tempe fermentation. Cooked rice was inoculated with a spore suspension and spread to a loose layer approximately 1 cm thick in a covered and perforated aluminium tray. It was then incubated at 37°C.

Concomitant with fungal growth and formation of sporangia, the initial substrate moisture content of 67% gradually decreased to 5% or less during the incubation period. The moulded, dried rice was then pulverized. Before incubation, the inoculated rice contained 10^5 to 10^6 mould spores per gram; this number became 10^8 to 10^9 in the final product. One gram of this inoculum was used for every kilogram of the original air-dried soybeans. It was found that the best storage conditions for preservation of the inoculum were at a low temperature (4°C) and low relative humidity (near 0%). However, when refrigeration is not available, the inoculum can be stored in sealed dry containers at room temperature. The methods of inoculum preparation and application were successfully tested by one of us in a pilot plant producing 75 kg of tempe daily for more than one year (1967–1968). This method of inoculum preparation does not require sophisticated and expensive equipment like a freeze drier. It is adaptable to small-scale factory production, and is basically applicable for large-scale tempe production. A review of the literature in the preparation of mould spores for food fermentation was recently published by Hesseltine et al. (1976).

III. PRODUCTION METHODS

A. Basic Procedure

Like traditional methods for preparing any other indigenous foods, the procedures for manufacturing tempe vary in several details from one producer to another. These methods were described by Boorsma (1900), Burkill (1935), Stahel (1946), Tammes (1950), van Veen and Schaefer (1950), Steinkraus et al. (1960) and Ko and Hesseltine (1961). The essential steps in the procedure are outlined in Figure 1.

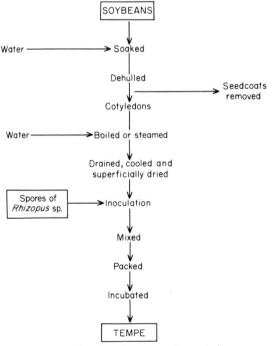

Fig. 1. Flow sheet for a tempe fermentation

Commerical air-dried soybeans are hydrated by soaking in water overnight. The hulls are removed, and the dehulled soybeans are steamed or boiled with excess water until the beany and bitter taste disappears. The cooked beans are spread on a plaited bamboo tray for draining, surface drying, and cooling. The beans are then mixed with an inoculum containing mycelium of the required *Rhizopus* species, packed in banana leaves, and left two nights at room temperature (25° to 30°C). The procedure is complete when the white mycelium of the growing *Rhizopus* mould has tightly bound the soybeans together into a white compact cake. This tempe cake is not consumed as such, but within a day it is prepared as described.

B. Raw Material

All of the more than 10 varieties of soybean which are cultivated in Indonesia (Anonymous, 1964) can be used for the manufacture of tempe. Also, other soybean varieties like Seneca (Steinkraus *et al.*, 1960), Hawkeye (Martinelli and Hesseltine, 1964) and Clark (Steinkraus *et al.*, 1965) can be used to make tempe. Unpublished work at the Northern Regional Research Center has demonstrated that any variety of soybean is satisfactory.

C. Preparation of Soybeans

Before the soybeans are fermented, they should undergo three main processes, namely hydration, dehulling and cooking.

1. Hydration

In the traditional way, hydration by soaking in water makes dehulling easier. During the soaking period, bacterial growth takes place, but this is not important. Most of the bacteria will be removed or killed during the subsequent procedures of dehulling, washing and cooking. In some studies, however, much attention was paid to this part of the process. Stahel (1946) began his laboratory procedure by boiling the dry soybeans in a quantity of water. After cooling, the soybeans were peeled and just enough fresh water was added to cover the beans. Then they were left to stand and ferment for one day, probably to produce acid. The fermentation was then stopped by boiling the beans in the fermented liquid. Actually, after the soybeans have been boiled in water and peeled, they have already been hydrated and the dehulled soybeans are ready to be inoculated. But Stahel (1946) felt it necessary to ferment for one day more. Obviously this author considered the fermentation in soakwater to be very important.

Steinkraus *et al.* (1960) added lactic acid to the soakwater and, in a later publication (Steinkraus *et al.*, 1961), announced that the soakwater had been inoculated with *Lactobacillus plantarum* (Orla-Jensen) Holland. As an alternate method of acidifying the soakwater, they also proposed the addition of acetic acid (Steinkraus *et al.*, 1965). The acidified soakwater was later used for boiling the soybeans, so that the boiled soybeans had an acid pH value. They were of the opinion that the acidity was necessary to inhibit development of spoilage bacteria during the mould fermentation.

Hesseltine *et al.* (1963) found that mould inhibitors in the soybeans, which disturb the mould fermentation, need to be washed out. This may occur during the soaking period. From this point of view, the acidification of the soakwater as done by Steinkraus and his colleagues could be advantageous if the mould inhibitors are more water-soluble at a lower pH value. However, no work has as yet been done on this aspect of the fermentation.

In our work in the laboratory as well as at a pilot production unit (Rusmin and Ko, 1974), we just soaked the soybeans in excess tapwater at ambient temperature overnight without any additional procedure. Although the soakwater became turbid due to bacterial growth, bubbles of foam arose and the pH value dropped from 7.0 to 4.5; most of the bacteria were removed or killed during the next steps of dehulling, washing and cooking. Our experience showed that this simple method did not cause any problems in the next steps of the process.

2. Dehulling

Dehulling the soybeans is necessary to enable the tempe fungus to grow on the cotyledons (Steinkraus *et al.*, 1960). The traditional tempe manufacturer places the soaked soybeans into a basket made of plaited bamboo strips, and easily frees the hulls from the cotyledons by treading with his feet. When this work takes place at the edge of a river or stream, the loosened seedcoats float away with the streaming water (Ko and Hesseltine, 1961). Similarly, the mass of dehulled soybeans is put into a tank filled with water and the floating seedcoats are skimmed away by means of a small plaited bamboo disk. Small batches of soaked soybeans can also be freed from the seedcoats by rubbing with the hands. Obviously, these traditional methods are time consuming.

Dehulling can be improved by using a simple roller mill. The distance between rollers should be adjusted to somewhat less than the size of the soaked soybeans. In such a mill, the pairs of cotyledons are slightly pressed from each other by the rotating cylinders and, at the same time, the hulls are pushed away. The hulls can then be removed by flotation with water (Ko, 1965). An abrasive vegetable peeler (Steinkraus et al., 1960) and a properly-spaced burr mill are also useful to loosen the hulls from hydrated beans (Steinkraus et al., 1965). The soybeans can also be dehulled mechanically when they are dry (Martinelli and Hesseltine, 1964). They can be preheated to loosen the hulls from the cotyledons and, after cooling, can be passed through a burr mill adjusted to loosen the hulls without cracking the cotyledons. The hulls are then removed from the cotyledons by passing the mixture over a gravity separator (Steinkraus et al., 1965).

Tempe of good quality was also made from full-fat soybean grits, which consist of cracked cotyledons broken into 10–15 pieces. Since grits absorb water very easily, soaking overnight was not necessary; soaking for three to four hours gave the same good results (Martinelli and Hesseltine, 1964). This soaking time of soybean grits could even be decreased to 30 min (Hesseltine and Wang, 1972).

Basically it is not essential to remove the seedcoats, because after dehulling and cooking the germinating mould spores can reach the surfaces of the cotyledons and grow into them. Actually, many traditional tempe producers purposely do not completely separate the seedcoats from the beans. In this way, production output is higher and unit price is minimal. However, the quality of the product will be somewhat less, but the lower price may be more attractive for people in a lower income group. However, quality-minded tempe manufacturers remove the seedcoats as much as possible.

3. Cooking

After the soybeans have been hydrated and dehulled, they are cooked or steamed. Disappearance of the bitter taste is a practical indication that heating has been sufficient and can be stopped. To obtain a uniformly moist substrate, cooking in an excess of water is easier than steaming. It was reported that material losses of dehulling, soaking and cooking of whole soybeans reached 24%, and those of dehulled soybean grits by soaking and cooking 38% (Smith *et al.*, 1964). Although data of losses during steaming are not available, material losses in cookwater may be higher than losses by steaming, but it should not be concluded that steaming is better.

Soybeans contain substances which can disturb the fermentation and should be removed during soaking and cooking of the beans. Removal of these materials is necessary, as was demonstrated when soybean grits were boiled with soakwater until excess water evaporated, but the soybeans were still moist; thus, there was no loss of solids. However, on this substrate, the mould showed poor mycelial growth and the fermented product had an unpleasant odour and astringent taste. On the other hand, if the soakwater and cookwater were discarded, a high-quality product with good mycelial growth was obtained. It was concluded that removal of a heat-stable, water-soluble factor or factors is required for fermentation of soybeans into tempe (Hesseltine *et al.*, 1963; Smith *et al.*, 1964). At least 30 min steaming at 100°C was required to destroy completely this mould-growth inhibitor (Smith *et al.*, 1964).

4. Drying and Cooling

After cooking or steaming, the soybeans should be drained, cooled and surface-dried before inoculation with the mould takes place. These precautions are necessary to prevent rapid bacterial growth in the free water on the soybeans which, in turn, disturb good growth of the mould. Technically, boiled or steamed soybeans are not sterile and may contain 10^3 to 10^4 bacterial spores per gram. Furthermore, traditional tempe inoculum is always contaminated with bacteria; even inoculum made by growing a pure culture of *R. oligosporus* on steamed rice contains per gram 10^4–10^5 bacterial spores (Rusmin and Ko, 1974). Notwithstanding the presence of this considerable amount of bacterial spores, the fermentation will not be interrupted provided enough mould spores are used (10^4–10^5 per gram substrate), the incubation temperature is favourable (30–37°C) and the soybeans are not too wet.

Under these conditions, bacterial growth is slow and the fermentation is finished before the bacteria have multiplied. During preparation of the tempe, including frying or cooking, the bacteria will be killed. If, during inoculation, the soybeans are still too wet, bacterial growth will overgrow the fungus. A slimy brown mess with a putrid smell will be the result of the fermentation, and the product must be discarded.

D. Packaging

After inoculation of the soybeans, it is important to pack them in a proper way to obtain a final product in which a white mycelium has developed abundantly and has bound the soybeans into a compact cake. To achieve this result, an equilibrium should be maintained between keeping the beans moist and restricting contact of the beans with air. Unrestricted air supply and dehydration of the beans result in poor mycelium growth and excessive sporulation (Steinkraus et al., 1965). On the other hand, too little contact with air suppresses mould growth. This happens when the soybean layer is too thick, in which case mycelium growth is poor in the centre of the cake (Martinelli and Hesseltine, 1964).

In the traditional method, inoculated soybeans are packed in folded banana leaves to make parcels measuring approximately 1 × 6 × 8 cm and then tied with dried rice stalk or bamboo string. For preparing small parcels of tempe, leaves of *Tectona grandis, Hibiscus similis* or *Erythrina* sp. are also used. After incubation, the unopened parcels are brought to market for sale. Large-size tempes are made by spreading the inoculated soybeans on a bamboo tray of approximately 40 × 100 cm, the bottom of which has previously been covered with banana leaves. The layer of soybeans, approximately 4 cm thick, is covered with more banana leaves. After incubation, the large cake is cut into convenient pieces of approximately 10 × 15 cm and offered for sale. Some forms of tempe are fermented in bamboo cylinders (Sudarmadji, 1975).

In Surinam, leaves of several *Musa* and *Maranta* species are used for making tempe packages. An attempt to substitute parchment paper, cloth impregnated with paraffin, oilcloth and tinfoil for these plant leaves was not successful (Stahel, 1946).

Martinelli and Hesseltine (1946) made a special study of several kinds of containers which could be used for tempe fermentation. They found that metal trays were better than wooden ones. Stainless-steel trays, measuring 8 × 35 × 2.5 cm, and larger aluminium trays (36 × 46 × 2.5 cm) with perforations in the bottoms of 0.6 mm on 1.3-cm centres and covered with perforated plastic sheets, gave good tempe in 20 to 22 hours. Large metal trays holding approximately 4 kg of soybean grits were incubated for only the first 15–16 h at 31°C; after this, the fermentation continued to completion at room temperature (28°C). They also got good results with plastic bags measuring 21 × 13 cm, and plastic tubing with a diameter of 10 cm, perforated manually with a sewing needle of about 0.6 mm diameter. The perforations were about 1.3 cm between centres. The same good results were obtained with cellophane tubing.

In a pilot-plant study, Steinkraus *et al.* (1965) successfully made tempe by spreading about 3 kg of inoculated beans on metal dryer trays (35 × 81 × 1.3 cm) with woven stainless-steel 3-mm mesh bottoms and covered with waxed paper. In a laboratory in Indonesia (Bandung Institute of Technology), we used covered zinc trays (31 × 11 × 1.5 cm) which were perforated at distances of 1.5 cm with 1-mm holes at the bottom as well as in the cover. For making small cakes of tempe weighing approximately 100 g, plastic bags perforated with a needle were very convenient.

E. Incubation and Mould Growth

Incubation temperature is important for a successful tempe fermentation. In a time–temperature study, it was shown (Martinelli and Hesseltine, 1964) that the range of temperatures in which tempe was produced was from 25 to 37°C (Table 2).

Table 2.

Time–temperature relationship in tempe fermentations

Incubation temperature (°C)	Fermentation time (h)	Reference
25	80	Martinelli and Hesseltine (1964)
28	26	
31	22–24	
37	22–24	
35–38	15–18	Steinkraus *et al.* (1965)

The differences observed at several incubation temperatures appeared to be related to the length of time required for fermentation. At higher temperatures, the soybeans dry out, and there is the chance that thermophilic bacteria, whose spores survive during cooking of the soybeans, will overgrow the mould and result in a bad product. Obviously, the incubation temperature not only determines the fermentation time but also favours growth of the mould over bacteria. The wide temperature range in which tempe can be produced allows traditional tempe manufacturers in Indonesia to produce tempe without an incubator, since the mean temperature in most places in Indonesia is between 20° and 30°C.

During the period of active mould growth, considerable quantities of heat are released and the temperature in the bean cake rises from 10° to 16°C above that of the incubator if it is not removed (Stahel, 1946; Steinkraus *et al.*, 1960; Martinelli and Hesseltine, 1964). If the temperature becomes too high, mould growth is suppressed and bacteria can interfere with the fermentation.

Without realizing these principles, traditional tempe manufacturers developed the following incubation procedures to prevent excessive temperature rise during fermentation. Small parcels of inoculated soybeans in banana leaves are kept in bamboo baskets, and covered by gunny sacks. When the fermentation has started and too much heat is released, the parcels are taken out of the basket and aerated for a while in the open air. The incubation is continued after the parcels have been cooled. When large-size tempe is made on trays, gunny sacks are placed over the banana leaves which cover the inoculated soybeans. Then the trays are incubated on bamboo racks in some part of the house. When the fermentation has started and the temperature rises, the gunny sacks are removed. If the temperature becomes too high, the trays are removed from the rack to cool off in the open air outside the house. In this way, the traditional tempe manufacturer adjusts the incubation temperature and takes care that he gets tempe in time, usually after approximately 40 hours.

To control the temperature during fermentation, Martinelli and Hesseltine (1964) held a 4-kg batch of soybean grits in a metal tray for the first 15–16 h at 31°C; after this the fermentation was continued to completion at room temperature (28°C). Steinkraus *et al.* (1965) reported that heat produced by the mould was dissipated about as rapidly as it formed, when they fermented 3-kg batches of soybeans on metal dryer trays (35 × 81 × 1.3 cm) with woven stainless-steel bottoms covered with waxed paper. Incubation was in a fermentation room maintained at 35–38°C and 75–85% relative humidity. A good product was obtained in 15–18 hours. For the production of tempe on an industrial scale, availability of an incubation room with air conditioning is desirable.

A successful fermentation should result in a compact, white cake in which the soybean cotyledons are completely covered, penetrated, and bound together by the white mycelium of *Rhizopus* sp. Until recently, there were contradictory reports regarding the penetration of the soybeans by the mould mycelium. Stahel (1946) found no penetration, but others (Boorsma, 1900; van Veen and Schaefer, 1950; Steinkraus *et al.*, 1960) observed some penetration, although Steinkraus *et al.* (1960) suggested that most of the action of the mould upon the beans followed the action of secreted enzymes. Sudarmadji (1975) found with a scanning electron microscope that the *Rhizopus* mould did not penetrate the soybeans more than two cell layers. He was of the opinion that enzymes must be excreted by the mould, and that it is they that facilitate the chemical changes occurring during the tempe fermentation.

Using several different fixatives and staining techniques, Jurus and Sundberg (1976) were able convincingly to demonstrate fungus hyphae in the deeper layers of the cotyledons where the hyphae became quite thin, and to measure the depth of penetration and frequency of surface hyphal invasion. The depth of mycelial growth was up to 742 μm, which is about 25% of the average width of a soybean cotyledon, and this extensive invasion of the soybean may partially explain the rapid physical and chemical changes during tempe fermentation.

IV. KEEPING QUALITIES AND PRESERVING METHODS

When fermentation is finished, the raw, white tempe cake has an attractive, bland, slightly nutty odour. These properties change within six hours when the product is kept at room temperature in the original package. Gradually, the clean white colour turns into brown, and the surface becomes wet and slimy, supposedly due to increased bacterial growth. Meanwhile, the flavour becomes disagreeable and putrid; ultimately, a strong ammonia odour will be detected. These are the reasons why tempe is sold fresh at the markets in Indonesia, and why it is prepared the same day into many kinds of dishes.

Stahel (1946) reported that, when the package was opened and the tempe had the opportunity to dry slowly, exposed to the air, the product could be eaten even two days later. Sometimes the cake is sliced into pieces approximately 2 mm thick and dried under the sun. These dried pieces remain edible for several days. However, these preservation methods are not usual. Traditionally, tempe is used when it is still fresh.

A popular kind of local preservation involves deep frying thin slices of tempe in coconut oil after it has been dipped into a water suspension of rice-flour, salt and spices. When kept in closed glass or tin containers, or in sealed plastic bags, the crisp slices remain tasty for at least several weeks. This method of preservation, which is used commercially in Indonesia, is easy and does not require special or expensive equipment.

Some methods have been tested to prolong the shelf life of fresh tempe. On a laboratory scale, Hesseltine et al. (1963) placed slices of tempe in boiling water for 5 min to destroy the mould and enzymes. It was then placed in cellophane packages and frozen in a deep freezer. After 100 days storage, it was deep-fat fried; the appearance, odour, and taste were the same as for tempe immediately after fermentation.

In a pilot-plant process, Steinkraus et al. (1965) cut fresh tempe into 2.5 cm squares and moved them directly into a hot-air dryer at 93°C. In 1.5–2.0 h, the moisture level was lowered to 2–4%. The dehydrated tempe was packaged in pliofilm bags and sealed. In this form, the dried product was stored for months at room temperature (22–26°C) without noticeable changes in colour or flavour.

Iljas et al. (1970) heated sliced tempe in boiling water for five minutes and kept it in sealed cans in three different ways. There was no significant change in acceptability of the product whether the can was sealed and immediately stored at −29°C, or whether it was filled with water, steam-vacuum sealed, heated at 155°C for 20 min and stored at room temperature. However, when it was dehydrated in a cabinet dryer at 60°C for 10 h, sealed in cans, and stored at room temperature up to 10 weeks, no significant difference in acceptability was observed, but thereafter, the acceptability tended to decrease as the storage period progressed.

V. CHANGES IN CHEMICAL COMPOSITION

Enzymic digestion and synthesis of new compounds take place during fermentation of soybeans into tempe. Proteins are broken down, which results in increases in soluble nitrogen compounds from 0.5 to 2.0% (van

Veen and Schaefer, 1950; Steinkraus *et al.*, 1960, 1965; van Buren *et al.*, 1972) and in free amino acids which increased from 1 to 85 times (Stillings and Hackler, 1965; Murata *et al.*, 1967). Hydrolysis of lipids decreased their amounts from 22.5% to 14.1% (Boorsma, 1900; van Veen and Schaefer, 1950; Wagenknecht *et al.*, 1961; Murata *et al.*, 1967; van Buren *et al.*, 1972; Iljas, 1972; Sudarmadji, 1975) and resulted in an increase in free fatty acids from 0.5 to 21% (Steinkraus *et al.*, 1965; van Buren *et al.*, 1972).

A decrease in reducing compounds was reported, due to their utilization by the mould (Steinkraus *et al.*, 1960) and in hemicellulose from 2.01 to 1.13% (van Veen and Schaefer, 1950), while fibre content increased from 3.7 to 5.8% due to mould mycelium (Boorsma, 1900; van Veen and Schaefer, 1950; Steinkraus *et al.*, 1960; Murata *et al.*, 1967; van Buren *et al.*, 1972). The soluble solids rose from 13 to 21% (Steinkraus *et al.*, 1960).

The percentage contents of a number of vitamins, including riboflavin, niacin, vitamin B_6, pantothenic acid, biotin, folate compounds and nicotinic acid increased, while the thiamin content decreased (Roelofsen and Talens, 1964; Murata *et al.*, 1967, 1970).

These chemical changes together improve the original properties of soybean; taste, flavour, appearance and edibility of tempe become more attractive than cooked soybeans. Moreover, tempe can be easily prepared and is much better digested than soybeans (van Veen and Schaefer, 1950; Wagenknecht *et al.*, 1961); even patients with dysentry and nutritional oedema were able to assimilate tempe (van Veen, 1946; Roelofsen, 1946).

Tempe made of pure soybeans has never been the cause of food poisoning. Aflatoxin is not produced by *R. oligosporus* (Hesseltine, 1967; van Veen and Graham, 1968); moreover, this mould prevented aflatoxin production by *Aspergillus flavus* (Ko, 1974).

VI. NUTRITIVE VALUE

Some studies reported that the nutritive value of tempe in rat-feeding tests was approximately equal to that of cooked soybeans (Jansen and Donath, 1924; Steinkraus *et al.*, 1961; Murata *et al.*, 1971). But, a less (Smith *et al.*, 1964; Hackler *et al.*, 1964) as well as a better nutritive value of tempe was reported (György, 1961). These differences may be due to differences in materials and conditions used in the experiments. Differences in the nutritive value of tempe in rat-feeding experiments were not serious, if one considers the other good qualities and the improved edibility of tempe for human consumption over that of plain-cooked soybeans (Smith et al., 1964; Iljas, 1972).

VII. OTHER TEMPE-TYPE PROCESSES

At least five other fermentation processes are known which resemble the production of tempe. Two are new processes: one was developed by Stanton and Wallbridge (1969) at the Tropical Products Institute (T.P.I.) in London; the other is the Tate and Lyle (Imrie, 1973) process for making single-cell protein. Three other food fermentations have been used for centuries in the Orient. These are oncom, natto and thua-noa.

A. Tropical Products Institute Process

The T.P.I. process was developed to make a tempe-like product from cassava (*Manihot esculenta* Cranz; Stanton and Wallbridge, 1969). This low-protein, staple root crop is nutritionally improved by growth within it of *Rhizopus oligosporus* and other fungi. In this process, the fungus is grown for 5 to 7 days at 27°C to produce inoculum. The sporulating mycelial pads are homogenized and blended with cassava flour which has been pasteurized for 18 h at 70°C. When inoculated, the cassava flour contains about 1–2 million propagules per millilitre of substrate. After thorough mixing of the inoculum with the flour, the resulting stiff dough is put in a mincer and extruded as spaghetti-like strands 3–5 mm in diameter. These strands are then cut into short pieces about 10 cm in length.

The extruded product has a moisture level of 45 ± 3% and a pH value from 4.5 to 6.7, depending on the type of mixture used. Other methods were also tried, such as pelletizing or cutting into thin sheets as in the preparation of flat noodles. Various types of binders were incorporated into the dough, including such materials as magnesium stearate.

The spaghetti-like pieces are placed in shallow, small-mesh aluminium trays to the depth of the tray (2.5 cm) and covered. Incubation conditions are optimal at 30°C, with relative humidity of 95–97% and a pH value of

5.5. Aeration is necessary in the incubators. The T.P.I. fermentation process takes about 40–80 h to complete; when sporulation has just started on the spaghetti-like pieces and the entire mass is held together by the mycelium, it is time to harvest.

B. Oncom (Ontjom)

This widely consumed fermented food has been produced in West Java for centuries, but very little has been published on its preparation and nutritional value (Hesseltine, 1965; Steinkraus *et al.*, 1965; Ho, 1976). Oncom is prepared from peanut press cake (material remaining after extraction of peanut oil) which is limited to the Bandung region in West Java. A second type (oncom merah or oncom tahu) is made from the residue from processing of soybeans for soybean milk. The mould used is a species of *Neurospora* originally stated to be *N. sitophila* but now claimed by Ho (1976) to be *N. intermedia* Tai. This conclusion is based on a study of 71 oncom cultures which give imperfect mating reactions when crossed with *N. sitophila*.

The product made from peanut press cake is covered with mycelium and spores of *Neurospora* sp. and is pink or orange because of sporulation. The process involves breaking up the cake and soaking in water for 24 h. The peanut mass is washed, steamed and moulded into rectangular cakes about 3 × 10 × 20 cm, which are then placed in bamboo frames and covered with banana leaves. Inoculation is carried out by sprinkling the pink koji from a previous fermentation over the cakes. Fermentation is allowed to proceed at room temperature in a shady place with good aeration. After fermentation, the cakes are cut into pieces and sold. This food is prepared in the same way as tempe, that is fried or roasted, and has an almond flavour and a light-brown colour. Strains of the mould have very active lipase and proteolytic enzymes.

C. Natto

This Japanese fermented food is prepared from whole soybeans somewhat like tempe. However, the micro-organism used is *Bacillus subtilis* (*B. natto*) rather than a mould. All of the publications on this food have been by Japanese researchers. According to Nakano (1972), annual natto production is approximately 6.10^6 kg; it is mainly produced in the northern part of Japan. In modern natto manufacture, whole soybeans are soaked in water, cooked, and inoculated with a pure culture of bacteria. The inoculated beans, with their seedcoats intact, are placed in plastic or paper-thin wood containers in a warm (40–43°C) incubator. Fermentation is completed within 14–18 h when the whole beans are covered with a strong mucilaginous material. Natto is sold in the fermentation container; the finished product is eaten in the home with cooked rice and shoyu. Hayashi and his associates recently published a series of nine articles on the nutritional value of natto, which appeared in Japanese under the title: *Experimental Studies on the Nutritive Value of Natto* (Hayashi, 1959a–d; Hayashi and Nagao, 1975a, b; Hayashi *et al.*, 1975, 1976a, b).

D. Thua-nao

Thua-nao is a fermented soybean food of northern Thailand that closely resembles natto (Sundhagul *et al.*, 1970). This fermented soybean product is used in place of fermented fish as a seasoning agent, and is added to vegetable soups and hot dishes. In some areas, it is used directly in the diet. Like natto, the micro-organism used in the fermentation is *B. subtilis*. Whole soybeans are boiled for about 3 h until they are soft and thoroughly cooked. Excess water is drained off and the soybeans are inoculated and incubated for about 48 h at 35°C. These authors (Sundhagul *et al.*, 1970) were able to isolate production cultures and to carry out satisfactory pure-culture fermentations. In the native process, about 1 to 2 kg of dry beans were used; after cooking, the beans were twice their normal size. The cooked beans were placed in bamboo baskets lined with banana leaves and then covered with additional leaves. The beans were considered properly fermented when they were covered with a sticky, viscous, colourless material with the pungent odour of ammonia. Like natto, the beany flavour disappeared and the beans changed from a light brownish-yellow to a greyish-brown colour. Unlike natto, thua-nao is lightly mashed into a paste to which salt and other flavouring agents, such as garlic, onion, or red pepper are added. The paste is wrapped in banana leaves and again cooked before selling or eating. The cooked moist paste can be kept for about two days under normal conditions. For storage, the paste is formed into small balls about 1.3 cm in diameter, pressed into thin chips, and then sun-dried. The dried chips may be kept for several months without spoiling. The flow sheet for the traditional processing of thua-nao as presented by Sundhagel *et al.* (1970) is shown in Figure 2. The chemical composition of thua-nao is given in Table 3.

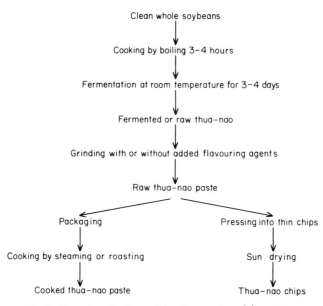

Clean whole soybeans

↓

Cooking by boiling 3-4 hours

↓

Fermentation at room temperature for 3-4 days

↓

Fermented or raw thua-nao

↓

Grinding with or without added flavouring agents

↓

Raw thua-nao paste

Packaging ← → Pressing into thin chips

↓   ↓

Cooking by steaming or roasting   Sun drying

↓   ↓

Cooked thua-nao paste   Thua-nao chips

Fig. 2. Flow sheet for the traditional processing of thua-nao

Table 3.

Chemical composition of paste and chips of thua-nao

Component	Paste (% content)	Chips (% dry weight)
Protein	16.0	36.8
Fat	7.4	14.8
Carbohydrate	11.5	19.4
Fibre	5.2	12.9
Ash	2.5	4.9
Water	56.4	11.1

E. Tate and Lyle Process

Ceratonia siliqua L. (carob tree) grows throughout the Mediterranean area and in parts of the United States, South America and Rhodesia. After removal from the pod, the beans are used to produce a commercial gum. The seeds make up only 10–12% of the pod. The pods, after the seed has been extracted, contain as much as 50% by weight of sugar which can be extracted with water. With addition of ammonium sulphate or ammonium phosphate, the resulting sugar solution makes an excellent substrate for growth of a variety of fungi. Imrie and Vlitos (1975) and Imrie (1973) studied this water solution for production of fungal protein; from over 300 organisms, they selected a wild strain of *Aspergillus niger* (M-1) which grew rapidly and produced a tannase that removed tannin from the product. They developed a batch pilot-plant process of 100 litres for cultivating this mould on carob extract medium, harvesting, and drying the mycelium. Further studies showed the dried mycelium to be nontoxic and palatable as an animal feed.

REFERENCES

Anonymous (1964). *In* 'Proceedings Rapat Kerdja Kedelai (Soybean Workshop)', (Sadikin Sumintawikarta, ed.), R.E.S. 64/21, p. 2, Bogor, Indonesia.

Boedijn, K. B. (1958). *Sydowia* **12**, 321.

Boorsma, P. A. (1900). *Geneeskundig Tijdschrift voor Nederlandsch Indie* **40**, 247.

Buren, J. P. Van Hackler, L. R. and Steinkraus, K. H. (1972) *Cereal Chemistry* **49**, 208.

Burkill, I. H. (1935). 'A Dictionary of the Economic Products of the Malay Peninsula', Volume 1, p. 1080. Crown Agents, London.

Darwis, A. and Grevenstuk, A. (1935). *Geneeskundig Tijdschrift voor Nederlandsch Indie* **75**, 104.

Diokno-Palo, N. and Palo, M. A. (1968). *The Philippine Journal of Science* **97**, 1.

Dupont, A. (1954). Ph.D. Thesis: Fakultas Ilmu Pasti dan Ilmu Alam, Bandung.

György, P. (1961). *National Academy of Science, National Research Council, Publication no. 843* p. 281.

Hackler, L. R., Steinkraus, K. H., van Buren, J. P. and Hand, D. B. (1964). *Journal of Nutrition* **82**, 452.

Hayashi, U. (1959a). *Japanese Journal of the Nation's Health* **28**, 568.

Hayashi, U. (1959b). *Japanese Journal of the Nation's Health* **28**, 574.

Hayashi, U. (1959c). *Japanese Journal of the Nation's Health* **28**, 580.

Hayashi, U. (1959d). *Japanese Journal of the Nation's Health* **28**, 588.

Hayashi, U., Naruse, A. and Oura, Y. (1975). *Bulletin of Teikoku-Gakuen, No. 1*, 7.

Hayashi, U. and Nagao, K. (1975a). *Bulletin of Teikoku-Gakuen, No. 1*, 13.

Hayashi, U. and Nagao, K. (1975b). *Bulletin of Teikoku-Gakuen, No. 1*, 21.

Hayashi, U., Nagao, K., Tosa, Y. and Yoshioka, Y. (1976a). *Bulletin of Teikoku-Gakuen*, *No. 2*, 9.

Hayashi, U., Nagao, K. Takahashi, H. and Wakabayashi, K. (1976b). *Bulletin of Teikoku-Gakuen*, *No. 2*, 19.

Hermana, and Roedjito, S. W. (1971). *Penelitian Gizi dan Makanan* **1**, 52

Hesseltine, C. W., de Camargo, R. and Rackis, J. J. (1963). *Nature, London* **200**, 1226.

Hesseltine, C. W. (1965). *Mycologia* **57**, 149.

Hesseltine, C. W., Smith, M. and Wang, H. L. (1967). *Developments in Industrial Microbiology* **8**, 179.

Hesseltine, C. W. and Wang, H. L. (1967). *Biotechnology and Bioengineering* **9**, 275.

Hesseltine, C. W. and Wang, H. L. (1972). *In* 'Soybeans: Chemistry and Technology', (A. K. Smith and S. J. Circle, eds.), Vol. 1, p. 470. Avi Publishing Company, Westport, Connecticut.

Hesseltine, C. W., Swain, E. W. and Wang, H. L. (1976). *Developments in Industrial Microbiology* **17**, 101.

Ho, C. C. (1976). *Abstracts of the Fifth International Fermentation Symposium, Berlin*, p. 364.

Iljas, N., Peng, A. C. and Gould, W. A. (1970). *Ohio Report* **55**, 22.

Iljas, N. (1972). Ph.D. Dissertation: The Ohio State University.

Imrie, F. K. E. (1973). *Journal of the Science of Food and Agriculture* **24**, 639.

Imrie, F. K. E. and Vlitos, A. J. (1975). *In* 'Single-Cell Protein', (S. R. Tannenbaum and D. I. C. Wang, eds.), p. 223. M.I.T. Press, Cambridge, Mass.

Jansen, B. C. P. (1923). *Mededeelingen van de Burgelijke Geneeskundige Dienst van Nederlandsch Indie* **I**, 68.

Jansen, B. C. P. and Donath, W. P. (1924). *Mededeelingen van de Burgelijke Geneeskundige Dienst van Nederlandsch Indie* **I**, 26.

Jurus, A. M. and Sundberg, W. J. (1976). *Applied and Environmental Microbiology* **32**, 284.

Ko, Swan Djien and Hesseltine, C. W. (1961). *Soybean Digest* **22**, 14.

Ko, Swan Djien. (1965). *In* 'Research di Indonesia', (M. Makagiansar and R. M. Soemantri, eds.), Vol. 2, p. 312. Bidang Teknologi dan Industri.

Ko, Swan Djien. (1974). *In* 'Proceedings of the IV International Congress of Food Science and Technology', (E. Portelo Maris, ed.), Volume 3, p. 244. Instituto Nacional de Ciencia y Tecnologia de Alimentos, Madrid, Spain.

Martinelli, A. and Hesseltine, C. W. (1964). *Food Technology* **18**, 167.

Mertens, W. K. and van Veen, A. G. (1933). *Geneeskundig Tijdschrift voor Nederlandsch Indie* **73**, 1223.

Murata, K., Ikehata, H. and Miyamoto, T. (1967). *Journal of Food Science* **32**, 580.

Murata, K., Miyamoto, T., Kokufu, E. and Sanke, Y. (1970). *Journal of Vitaminology* **16**, 281.

Murata, K., Ikehata, H., Edani, Y. and Koyanagi, K. (1971). *Agricultural Biological Chemistry* **35**, 233.

Nakano, M. (1972). Waste Recovery by Microorganisms, p. 27. Ministry of Education, Malaysia, Kuala Lumpur.

Oey, Kam Nio and Lie, Goan Hong (1964). Daftar Analisa Bahan Makanan, 4th ed., 1967, p. 40.

Prinsen Geerligs, H. C. (1895). *Pharmaceutisch Weekblad* **32**, no. 33.

Roelofsen, P. A. (1946). *Vakblad voor Biologen* **26**, 114.

Roelofsen, P. A. and Talens, A. (1964). *Journal of Food Science* **29**, 224.

Rusmin, S. and Ko, Swan Djien (1964). *Applied Microbiology* **28**, 347.

Saono, S., Gandjar, I., Basuki, T. and Karsono, H. (1974). *Annales Bogorienses* **5**, 187.

Smith, A. K., Rackis, J. J., Hesseltine, C. W., Smith, M. and Robbins, D. J. (1964). *Cereal Chemistry* **41**, 173.

Stahel, G. (1946). *Journal of the New York Botanical Garden* **47**, 285.

Stanton, W. R. and Wallbridge, A. (1969). *Process Biochemistry* **4**, 45.

Steinkraus, K. H., Yap, Bwee Hwa, Van Buren, J. P., Provvidenti, M. I. and Hand, D. B. (1960). *Food Research* **25**, 777.

Steinkraus, K. H., Van Buren, J. P. and Hand, D. B. (1961). *Publication 843, National Academy of Science, National Research Council*, 275.

Steinkraus, K. H., Lee, C. Y. and Burk, P. A. (1965). *Food Technology* **19**, 119.

Steinkraus, K. H., Van Buren, J. P., Hackler, L. R. and Hand, D. B. (1965). *Food Technology* **19**, 63.

Stillings, B. R. and Hackler, L. R. (1965). *Journal of Food Science* **30**, 1043.

Sudarmadji, S. (1975). Ph.D. Dissertation: Michigan State University.

Sundhagul, M., Smanmathuropoj, P. and Bhodachareon, W. (1970). *Report No. 1, Applied Science Research Corporation of Thailand*, pp. 1–14.

Tammes, P. M. L. (1950). *Landbouw* **22**, 267.

Thio, G. L. (1972). *Report of the Royal Tropical Institute, Amsterdam, The Netherlands*.

van Veen, A. G. (1932). *Geneeskundig Tijdschrift voor Nederlandsch Indie* **72**, 1379.

van Veen, A. G. (1935). *Geneeskundig Tijdschrift voor Nederlandsch Indie* **75**, 2050.

van Veen, A. G. (1946). *Voeding* **7**, 173.

van Veen, A. G. and Schaefer, G. (1950). *Documenta Neerlandica et Indonesica de Morbis Tropicis* **2**, 270.

van Veen, A. G. (1966). *In* 'Biochemistry of Some Foodborne Microbial Toxins', (R. I. Mateles and G. N. Wogan, eds.), p. 171. M.I.T. Press, Cambridge, Massachusetts, U.S.A.

van Veen, A. G. and Graham, D. C. W. (1968). *Cereal Science Today* **13**, 96.

Vorderman, A. G. (1902). *Geneeskundig Tijdschrift voor Nederlandsch Indie* **42**, 395.

Wagenknecht, A. C., Mattick, L. R., Lewin, L. M., Hand, D. B. and Steinkraus, K. H. (1961). *Journal of Food Science* **26**, 373.

Wang, H. L. and Hesseltine, C. W. (1966). *Cereal Chemistry* **43**, 563.

Wang, H. L., Swain, E. W. and Hesseltine, C. W. (1975). *Journal of Food Science* **40**, 168.

6. Edible Mushrooms

W. A. HAYES AND S. H. WRIGHT

Mushroom Science Unit, Department of Biological Sciences, University of Aston, Birmingham, England

I. INTRODUCTION

A. Species Cultivated

Mushrooms are the fleshy fruits of fungi, the most important of which belong to the class Basidiomycetes. A small number of species are sufficiently flavoured and digestible to be used as food, but exploitation on an industrial scale is confined almost entirely to one temperate species, namely *Agaricus bisporus* (Lange) Sing. The basidiomycetes display a considerable degree of variation in morphology and life cycle but, to understand the methods of artificial culture, a simplified outline of the main morphological features and life cycle of *A. bisporus* is given in Figure 1.

This species is characterized by formation of two binucleate spores on the basidia of the spore-forming tissue (hymenium or gills) of the fruit-body. In most cases, single spores are fertile, i.e. generate hypha and mycelium with the capacity to form fruitbodies. In such homothallic basidiomycetes, fertility is a consequence of nuclear fusion and meiosis. It occurs in the basidium and is controlled, firstly by the arrangement of the post-meiotic nuclei in spores and, secondly, by genetic factors of a system known as the incompatibility system. For a detailed account of these mechanisms the reader should refer to the mycological texts of Burnett (1968) and Raper (1966).

Contrary to popular belief, *A. bisporus*, the common edible mushroom, is botanically distinct from the wild field mushroom *A. campestris*, but is also to be found in the wild throughout the temperate zone of the northern hemisphere, especially on soils which are rich in nitrogen and organic matter. In artificial culture, decomposed or composted plant residues are manufactured in order to establish growth. World-wide production of *A. bisporus* is at present estimated to be in excess of 600 000 tonnes per annum.

The second most widely cultivated species is *Lentinus edodes* (Berk) Sing., known as the forest mushroom or Shiitake. This species is heterothallic, fertile dikaryotic mycelium being formed from fusion of different homokaryons. Although its natural distribution extends all over the temperate regions of Eastern Asia, its cultivation on a large scale is confined to Japan, where it is grown on wood. Present day production amounts to approximately 130 000 tonnes per annum.

Another mushroom which is cultivated and is traditional to Asian

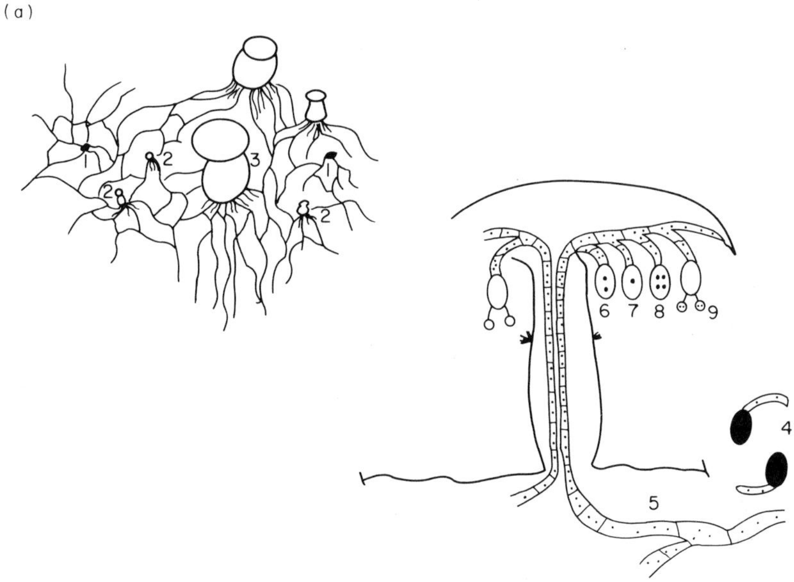

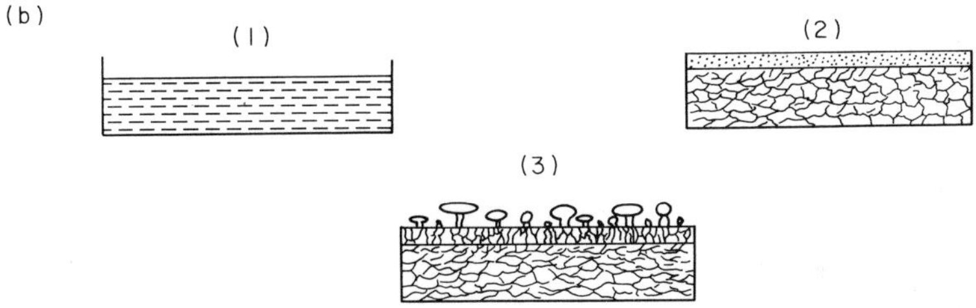

(a)

(b)

Fig. 1a. Diagrammatic representation of main morphological features of *Agaricus bisporus*.
(1) Indicates mycelium aggregates or fruitbody initials; (2) primordium or pinhead; (3) buttons, which mature to form open mushrooms bearing an hymenium, basidia and spores; (4) germinating spore; (5) vegetative tissue composed of hyphae with multinucleate cells; (6) binucleate basidium; (7) diploid cell; (8) basidium with four nuclei after meiosis; and (9) mature basidium bearing two binucleate spores.

Fig. 1b. Diagrammatic representation of main stages in artificial culture of *Agaricus bisporus*.
(1) Shows compost inoculated with grain spawn and incubated at 25°C for 10–14 days to establish vegetative phase; (2) colonized compost cased with a layer of soil, incubated for 7–10 days at 25°C followed by incubation for further 10 days in aerated room at 16°C; (3) three-dimensional growth of mycelium in compost leading to linear growth in casing layer. Initials, primordia and fruitbodies form on surface in flushes or breaks at approximately weekly intervals.

countries is *Volvariella volvacea* (Fr.) Sing., a tropical species which like *L. edodes* is heterothallic. Unlike both *L. edodes* and *A. bisporus*, this species is adapted for growth on fresh or slightly decomposed plant residues. Its natural distribution corresponds in the tropics to those areas where the growing of rice is traditional. Rice straw is the most suitable substrate,

F

which gives rise to its popular name, the padi straw mushroom. It is also referred to as the Chinese mushroom and, in recent years, improved cultivation methods have contributed to increased production, which amounts to approximately 40 000 tonnes per annum (Table 1).

Other species which are artificially cultivated but of relatively minor economic importance are *Flammulina velutipes* (Fr.) Sing., the Winter mushroom, which is adapted for growth on wood, and *Pleurotus oestreatus* (Fr.) Kummer, the Oyster mushroom, which like *V. volvacea* can be grown on fresh or partly decomposed plant residues. Although these species are traditional to Eastern Asia, small-scale cultivation has now extended to Europe.

Table 1

World production of major cultivated species of mushroom in 1975.
From Delcaire (1978)

Species	Common name	Quantity (tonnes)
Agaricus bisporus	Common edible mushroom	670 000
Lentinus edodes	Shiitake	130 000
Volvariella volvacea	Chinese or Straw mushroom	42 000
Flammulina velutipes	Winter mushroom	38 000
Pleurotus species	Oyster mushroom	12 000

In recent years, interest has been shown by researchers throughout the world in some other species which may be amenable to artificial culture, thus diversifying further the species of fungi which may be considered as a future food. Among the species known to be cultivable are *Pholioto nameko* (S. Ito) S. Ito and Imai, *Auricularia polytricha* (Mont) Sacc., *Tramella fuciformis* Berkeley, and *Stropharia rugosoannulata* Farlow apud Murr., but as yet large-scale exploitation has not been undertaken. Consideration may also be given to species which, because of their rarity, may be classed as 'exotic'. These include truffles, *Tuber melanosporum* Vitt., morells *Morchella esculenta* Fr. which belong to the class Ascomycetes, the cep, *Boletus edulis* Bull. ex Fr. and the Chanterelle, *Cantherellus cibarius* Fr. While these are not cultivated artificially, they may be regarded to be at least a minor part of the world trade in mushrooms. In France, Italy and India, they are gathered from their natural environments and exported at exhorbitant cost.

B. Production and Consumption

It is thought that the origins of artificial culture relate back to the experiences of seventeenth century French gardeners who observed that horse manure used for cultivating melons often produced mushrooms when moistened with the water used for cleaning and washing wild mushrooms. Early attempts at cultivating were done in the open air, but during the nineteenth century the underground quarries, which were formed through the removal of stone for the construction of Paris, were found to be good locations for cultivation in all seasons of the year. These quarries or caves are still in use today in France, but, according to Abercrombie (1817), in England sheds, barns, glasshouses or 'tent like' protectors made from canvas were used. These are the first recorded example of purpose-built structures for mushroom cultivation.

The art of artificial culture was introduced to the United States of America by English, French and Scandinavian gardeners employed by the affluent inhabitants of Philadelphia and New York, but up to the early part of this century, mushroom cultivation was mainly practised in France. Significant developments in the United States greatly improved techniques, particularly the development of a reliable method for preparing inoculum, by Sinden, in 1932. The relevance of aeration and carbon dioxide concentrations in the atmosphere were appreciated by Lambert (1938) who also specified requirements for the preparation of composts. However, mushroom production was still confined to France, U.S.A. and England up to the outbreak of the Second World War, but production levels were increasing, total production being approximately 50 000 tons per annum.

Immediately following the Second World War, production was quickly re-established to its pre-war level and cultivation extended to other countries in Europe, notably The Netherlands, Italy and Scandinavia. From 1950–60, production of *A. bisporus* increased at an average rate of 8% per annum. During this decade, mushroom cultivation was introduced to Japan, Australia, New Zealand and Taiwan (Formosa or Republic of China). During the next decade (1960–70), therefore, mushroom production continued to increase at an average rate of 12% per annum. The contribution of Taiwan to this increase was significant, since at the beginning of the decade production was small, and by the end of the decade Taiwan was the world's third largest producer, following the U.S.A. and France.

During the present decade, world production of *A. bisporus* continues to increase and production is being established at significant levels in India, Pakistan, some African and South American countries, South Korea and China. Production is at present estimated to be in excess of 900 000 tonnes per annum (Table 2).

Table 2

World-wide mushroom consumption in kilograms per head
of population in 1974.
From Edwards (1976)

West Germany	2.020
France	1.220
Holland	0.810
Italy	0.710
United States of America	0.800
United Kingdom	1.030
Canada	1.300
Taiwan	0.580
South Korea	0.030
China	0.001

Two main factors have contributed to increased production of *A. bisporus* world-wide: (1) extension of cultivation to different geographic zones, and (2) an increase in the productive efficiency of cultivation techniques, although this growth in the production and availability of mushrooms has been matched by an increase in the consumption of mushrooms, particularly evident in West European countries, Canada and U.S.A. Consumption of *Lentinus edodes* and *Volvariella volvacea* is confined to the Orient, but *Agaricus bisporus*, preserved in cans, is exported from Taiwan, South Korea and China to the major importing countries, namely U.S.A., Canada and West Germany. France and The Netherlands are also major exporters of canned mushrooms. Only a limited international trade exists for fresh mushrooms, West Germany, England, Belgium and Spain being importers, and exporting countries being France, The Netherlands, Eire, Denmark and Poland.

The increase in consumption of mushrooms reflects some change in outlook to this food, which traditionally was considered to be at best a somewhat expensive luxury. In England, for example, in the eighteenth and nineteenth centuries, mushrooms were only available to Royalty and the aristocracy, the consumer having to rely on those gathered from meadows and forests. Since some wild species are known to be poisonous, this may be regarded as somewhat dangerous. The introduction and extension of artificial mushroom cultivation throughout the world has not only provided a safe means by which mushrooms can be obtained for food purposes, but has also led to improvements in the technology of production methods, which have resulted in mushrooms being available at relatively low cost to almost every household.

However, the full impact of artificial culture has yet to be realized. The recent introduction of cultivation to low-income developing countries will obviously contribute to the continued growth of both production and consumption. Also, more recent developments in the methodology of culture offer greater scope for a more industrialized approach to cultivation in the high-income Western countries, where consumption is well established. It is likely, therefore, that mushrooms will figure more prominently in future diets.

II. PROCEDURES IN THE PRODUCTION OF *AGARICUS BISPORUS*

Cultivation of this mushroom accounts for about 75% of the mushrooms produced world-wide and, since the first discovery of artificial culture, most of the improvements to methods of cultivation have resulted from empirical trials, the experience of amateur growers and gardeners, as well as, in this century, professional cultivators. Also, during this century, many scientific contributions have provided a basis for a more complete understanding of procedures.

A. Pure-Culture Inoculum

It was shown by Ferguson (1902) that single basidiospores of *A. bisporus* do not germinate in the absence of actively growing mycelium. Later studies by Lösel (1964, 1967) showed that germination was inhibited by low and high pH values and in the presence of high concentrations of carbon dioxide. Isovaleric acid was also shown to stimulate spore germination and these earlier observations led Rast *et al.* (1976) to study the biochemistry of dormancy and germination. Four conditions were identified as necessary to allow germination to proceed: (i) removal of excess carbon dioxide which functions as a metabolic block of some enzymes in the tricarboxylic-acid cycle; (ii) acceleration of glycogenesis from catabolism of acetyl-CoA until there is a depletion of lipid reserves in the spore; (iii) breakdown of trehalose and formation of mannitol; and (iv) an increase in the carbon flow through the hexose monophosphate pathway.

A method of preparing grain spawn described by Sinden (1932) was an important contribution to the improvement of culture practice. Hitherto, inoculum was prepared from mycelium taken from soil where wild mushrooms grew; this served as the inoculum. The various methods devised were unreliable, due largely to the range of contaminants introduced by these procedures.

In the production of grain spawn, pure cultures are first established from single or multispore cultures obtained from selected mature mushrooms. In the methods employed, aseptic techniques are adopted and cultures are maintained on sterilized agar-containing media or composted horse manure and straw mixtures. Selected cultures are propagated by transferring mycelium to grain (wheat, rye or millet) previously soaked in water and sterilized.

Manufacture of grain spawn is now a highly mechanized and specialized process, and application of modern aseptic techniques has eliminated all sources of contamination. Spawn so prepared can be distributed over large distances. Production of a pure-culture inoculum (comparable to clean reliable seed in plant cultivation) is basic not only to the success of artificial culture but also provides a basis for the 'guarantee of edibility' that is provided by artificial culture techniques, an assurance not given by the alternative option of gathering mushrooms from their natural locations.

B. Preparation of Compost—a Primary Substrate

The major nutrients for mushroom growth are obtained from composts, which, by tradition, are prepared from horse manure and wheat-straw mixtures. Spawn will develop mycelium in unfermented horse manure, but satisfactory development is prevented by the self-generation of heat and competition from other micro-organisms. It is therefore necessary to compost the manure in order to produce a medium that will remain stable and in which the quantity of readily available nutrients for competing organisms is decreased.

Modern techniques of composting are modelled on systems devised by Sinden and Hauser (1950, 1953) in which mixtures were prepared in two stages, Stage I being over a period of seven days, and Stage II, which is done in an aerated room in trays or on shelves, over a minimum of five days, and includes a steam pasteurization phase. Emphasis is given to the size and shape of the Stage I stack in order to maximize microbial activity, a process which is continued through to Stage II by provision of control over aeration and temperature. A 'Long Method' was developed by Rasmussen (1962) in which Stage I is extended to 16 days, and the ingredients and stack shape and size are adjusted in order to produce an inactive compost, before filling into trays or shelves for a shortened Stage II procedure which primarily involves a pasteurization with steam. These procedures introduced for the first time the concept, first shown by Lambert (1941), of pasteurization into substrate preparation and thus removed the danger of pests and pathogens being introduced into the culture system. Methyl bromide fumigation was found by Hayes and Randle (1970) to be a satisfactory alternative to steam and this has been successfully applied by some cultivators. Its use has also made it possible to prepare high-yielding composts in remote regions of the world where steam is costly and supplies not readily available.

More recently, considerable commercial interest has been shown in a procedure referred to as '3-phase-1 system', bulk pasteurization or tunnel composting (Denham, 1977). This involves a conventional Stage I procedure but, rather than pasteurize in trays or shelves, the compost is filled into a large insulated tunnel and each end is closed by double section insulated doors. A centrifugal fan provides air to a system of underground ducts. Heat exchangers in the duct provide dry heat which is also equipped with a steam injector for pasteurization purposes. In this closed system, accurate monitoring of temperature and aeration can be done and it is claimed that a more homogeneous substrate is produced. An important difference, however, is that compost is filled into the growing containers or beds after pasteurizing and this is more vulnerable to contamination than *in situ* methods.

Although standard methods and procedures are proving to be an invaluable guide to composters throughout the world, in practice modifications are necessary in order to allow for variables which

Fig. 2. Preparation of compost substrates in England (above) and Taiwan (facing page).

inevitably exist in the materials, facilities, and systems of growing at different locations (Fig. 2).

C. Casing Soil—a Secondary Substrate

It was known from the beginnings of artificial culture that it was necessary to cover a compost with a layer of soil in order to induce formation of fruitbodies. This procedure is known as 'casing'. A sub soil was known to be more reliable than top soil but, in the early part of this century, with the introduction of sterilization procedures, a steam-sterilized top soil proved satisfactory. It was also appreciated that a neutral or slightly alkaline soil was more satisfactory.

In about 1950, a sphagnum peat neutralized with limestone entirely replaced the use of soil in England, and its use soon extended to Europe and some other parts of the world. However, where suitable peat is not available, soil continues to be used, but sterilization treatments have now been refined by the use of steam–air mixtures and where steam is not available formaldehyde is a satisfactory chemical alternative. Peat is relatively free of pests and pathogens and is not normally sterilized.

The casing layer serves as a mechanical support for the developing

Fig. 2 *(continued)*

fruits and functions in maintaining the correct water relations for the culture as a whole. More recently, the significance of micro-organisms in the casing have been appreciated, following the studies of Eger (1961), Hayes *et al.* (1969) and Hayes (1972). The casing layer should therefore be considered as a substrate which supports not only the mushroom but also an associated microflora, which apparently benefits the mushroom and supports its growth through the important transition from mycelial growth to a fruitbody. It is normal practice to apply the casing layer from 10–14 days following inoculation with spawn, by which time the primary compost substrate is fully colonized with the mycelium of *A. bisporus*.

D. Incubation, Environmental Conditions and Protected Cropping

The phases of growth, namely vegetative (mycelium in compost) and reproductive (fruitbodies in casing), develop optimally at two different temperatures, 25°C and 16°C, respectively (Flegg 1968, 1970). In addition, the gaseous environment radically influences the normal growth processes, carbon dioxide in the atmosphere being stimulatory in the vegetative phase, but inhibitory to the formation of fruits (Tschierpe,

1959; Tschierpe and Sinden, 1964; Long and Jacobs, 1968). Relative humidity of the air immediately above the casing is also known to be an important environmental variable. Flegg and Gandy (1962), for example, showed that at relatively high humidities (80–90%) the onset of fruiting tended to be earlier and the weight of mushrooms greater than at low humidities (40–50%). According to Storey (1972), at very high humidities and when evaporation from the casing layer is decreased, water may condense from the air onto the casing layer, cropping is delayed and yields lowered.

In artificial culture, these special environments are created and controlled in special structures or growing houses. As more knowledge was gained of the relationship between environmental factors and growth, simple systems of ventilation, which were a feature of growing in caves and the early purpose-built structures, have given way to the concept of air conditioning. On modern mushroom farms, this allows more precise control of aeration requirements, temperature and humidity throughout the year.

While the advantages of sophisticated automatic controls have not been established in economic terms, they contrast sharply with some of the

Fig. 3. Growing house in Taiwan.

newer concepts being applied in some countries new to cultivation. In Taiwan, for example, mushroom cultivation is done in structures made from bamboo and dried leaves of the local vegetation and employing natural systems of ventilation (Fig. 3).

Growing in underground caves and quarries, again with natural systems of ventilation, has been common practice in France since the origins of *A. bisporus* culture. Other structures which provide a degree of insulation can be readily adapted for mushroom cultivation. These include disused poultry and pig houses, glasshouses, aircraft hangers, barns, basements, out of use railway tunnels, or simple comparatively cheap buildings constructed of polythene, recently pioneered at the Lee Valley Experimental Horticulture Station in England. In such structures, the need for maximum utilization of space is not an important economic factor, and a balance is maintained between the quality of the substrate and the total capacity of the room to ensure satisfactory growing conditions.

E. Evolution of Growing Systems

The substrates required for mushroom cultivation are of necessity bulky, and dictate to a large extent the various systems that are adopted in the artificial culture of *A. bisporus*. In addition, the different conditions of aeration, temperature and humidity required for the normal growth process have also influenced the evolution of growing systems.

1. *Ridge beds.* Even up to about 40 years ago, and before the introduction of mechanization, the beds of compost were arranged on the floors of the caves, glasshouses or even on the ground in the open. In order to maximize the surface area of the bed and for ease of harvesting, beds were arranged in ridges, giving rise to the term 'ridge bed'. In this system, the materials were transported to the growing area by hand and obviously imposed a limit to the scale of a production unit. As a method of growing it is now obsolete.

2. *The shelf system.* This has evolved from a 'Standard House' for mushroom growing developed by growers in Pennsylvania, U.S.A. The same basic design of growing house is still widely adopted in the U.S.A. and Canada. In these, the beds are arranged in shelves along the length of the house and in tiers of up to seven or eight high. Wooden catwalks are incorporated for the purposes of harvesting (Fig. 4). Modified versions of

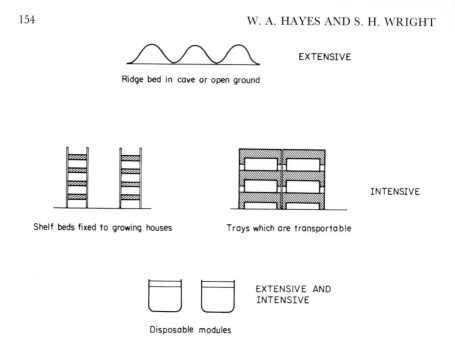

Fig. 4. Alternative growing systems for *Agaricus bisporus* culture.

the shelf system were developed later this century in the Netherlands and Taiwan.

In the shelf system, pasteurization, incubation and cropping are done in one room. The labour requirement for filling substrates onto the fixed shelves imposes limits to scale, and about 4.5 crops per annum are normal for this system. Better utilization of space is, however, possible with the introduction of bulk pasteurization techniques in tunnels, but the risks of infection are at present thought to be high.

3. *The tray system.* This is used on large and medium sized units in France, England, Italy, U.S.A., Canada, Europe, Australia and South Africa. Small trays, as containers for growing, were first introduced in the U.S.A. in 1930, but, after the Second World War, development of mushroom production saw the introduction of large trays with a cropping area of 1.5 to 2.5 m². These are easily arranged in stacks and moved mechanically, and therefore made way for a highly mechanized system involving all the important stages of culture. It also introduced a high degree of flexibility into the methods of culture, and rooms which are specifically equipped for a particular stage of the process are a feature of a modern tray-production unit. These rooms are: (i) for substrate

pasteurization; (ii) incubation; (iii) fructification; and (iv) cropping rooms. This has also allowed greater intensification in growing. Normally, about six crops per annum from a given cropping house are achievable, but there is a tendency for this to be increased by shortening cropping time.

4. *Disposable containers.* This is a comparatively recent development in mushroom growing and, instead of a wooden tray, the pasteurized compost is put in a low-cost disposable container. In its simplest form, a plastic sack is a convenient and practical container, or in some instances substrates are moulded in rectangular blocks before being covered with plastic film.

For this system, pasteurization may be done in trays or in the new method of bulk pasteurization. It is particularly applicable to extensive systems of growing where low-cost buildings are used for cultivation. As with trays, a high degree of mechanization is possible but, compared to the tray system, a relatively small capital investment is required. The main advantage of systems based on disposable containers is in disease control. The possibility of transmitting disease from one crop to another is eliminated.

Of interest also is the successful application of disposable containers for provision of 'kits' for growing by amateur enthusiasts. Pasteurized substrates are inoculated with spawn before being overwrapped with polythene. A casing soil is also included, leaving only a simple procedure of maintaining the correct temperatures and watering to the amateur grower.

F. Crop Management

Following the application of the casing soil, it is usual to maintain a temperature which is optimal for vegetative growth (25°C) until from six to ten days later, when the temperature is adjusted to 15°C to 18°C and fresh air is introduced to the growing environment. Since this climate change initiates fructification, its timing is critical. If adjusted too early, fruits will form suboptimally below the surface and, if at a later stage the mycelium will grow to the surface and form an impermeable mycelial barrier, which will greatly reduce the number of fruits. Tschierpe (1972) emphasizes the importance of maintaining a high relative humidity in order to prevent drying out of the very fine hyphae as they reach the surface.

From the developed mycelial aggregates or 'initials', primordia or pinheads develop, a proportion of which develop further into the characteristic mushroom fruits. The beds 'flush' every six to nine days depending on temperature, humidity and strain used. Fruitbodies are harvested by hand, at three stages of fruit development, referred to as 'buttons' (veil remaining unbroken), 'cups' (veil partially broken) and 'flats' (gills visible and forming spores). Beds continue to yield flushes for up to twelve weeks, but the total harvest decreases rapidly following the third flush. After the sixth flush, it is usually considered to be uneconomical to continue the crop further and the culture is discarded.

Despite the considerable control that has now been introduced into cultivation of mushrooms, in order to maximize the full growth potential of the crop some considerable individual skill and judgement are required in management. This not only relates to maintaining the right climatic conditions for the various stages of growth, but also to the art of maintaining the correct water relations during the various stages of cropping, and the difficult task of maintaining the culture free from harmful competitors, pests and pathogens.

G. Occurrence of Competitors, Pests and Disease

The objective in artificial culture is to maintain, as far as possible, a pure culture of *A. bisporus* under semi-agricultural conditions. Later in this chapter, the importance of other associated micro-organisms in the nutrition, growth and development of *A. bisporus* will be discussed, but of interest here are those life forms which can be regarded as harmful or deleterious to the crop.

These organisms can be categorized into two groups, namely competitors and true disease-causing organisms. In the case of the former, their presence is the result of imperfections in one or more of the stages of culture. The extent of damage is related directly to the extent or degree of imperfection, and will not spread to infect other normal crops. On the other hand, the pest and disease-causing organisms are progressive and, once established, increase in numbers. If not controlled, their effects decrease yields to uneconomic levels and in some instances can lead to complete eradication of the crop. However, since the introduction of precisely controlled pasteurization procedures in substrate preparation, and the application of control and manipulation of environmental conditions, the exclusion and control of contaminating organisms are achieved with greater ease than with other plant crops.

1. Competitors

A large number of micro-organisms have been found colonizing the substrates of mushroom culture, but only a few are known to interfere with growth. The mechanisms of this inhibition are not known, but toxic excretions appear to be a likely cause. In many instances, the occurrence of competitors is indicative of a particular defect in the procedures of artificial culture and they are therefore sometimes referred to as 'indicator' fungi. Others appear to develop without causing any apparent damage; these may be conveniently termed 'weed' fungi.

Establishment of the common soil fungi in a compost substrate is indicative of post-pasteurization contamination with soil or dust particles. Common soil contaminants include *Diehliomyces microspores* (Diehl and Lamb) Gil., *Sporendonema purpurescens* (Ben) Mason and Hughes, and *Chrysosporium luteum* (Cost) Carm. Growth of *Papulospora byssina* Hotson and *Scopulariopsis fimicola* (Cost and Matr.) Vuill., known as brown and white plaster mould respectively, reflects a prolongation of composting particularly during Stage I. The presence of what are termed 'Ink Caps', namely *Coprinus fimaturius*, is the result of insufficient composting time, while the damaging olive-green mould *Chaetomium olivaceum* is the result of insufficient available carbohydrate to allow complete microbial transformation of ammonia and amines or inadequate aeration during Stage II.

The range of fungi found both in the compost and the casing substrate, and regarded as abnormal inhabitants, has been adequately reviewed by Atkins (1974), but since the adoption of high standards of hygiene on mushroom farms their occurrence is now relatively rare.

2. Disease Organisms

Some bacteria, fungi and viruses infect mushrooms and cause disease. A common disease is 'bacterial blotch' caused by *Pseudomonas tolaasii* Paine, which inhabits most soils used for casing mushroom beds. Pseudomonads are also responsible for a rapidly spreading disease known as 'mummy'.

Fig. 5. *Agaricus bisporus* fruitbodies infected with *Mycogone perniciosa* Magnus.

Schisler *et al.* (1968) reported on the intracellular occurrence of pseudomonads in diseased material, but no reliable information is available on its etiology. However, nowadays its occurrence is rare.

Damaging fungal pathogens include *V. fungicola* (syn = *V. malthousei* Ware) and *Mycogone pernicosa* Magnus (Fig. 5). Infections cause malformations of the stipe and fruitbody and, in the case of the former, a characteristic brown spotting on the surface of the pileus. Disease severity is proportional to the level of infection and, being soil borne, attention to hygiene is essential in order to avoid rapid spread and severe infections.

Viruses are now known to be the cause of disorders variously called La France (Sinden and Hauser, 1950); Watery stipe (Gandy, 1960); X-disease (Kneebone *et al.*, 1962) and Dieback disease (Gandy and Hollings, 1962). A common characteristic of the disease is the formation of drumstick-like fruitbodies and premature opening of the veils. Six virus particles have been isolated from diseased mushrooms; five of these are polyhedral, being 19, 25, 29, 35 and 50 nm diameter, and one bacilliform measuring 19 × 15 nm (Hollings and Stone, 1969; Hollings, 1972; Nair, 1972). Mycelium (Last *et al.*, 1967), spores (Schisler *et al.*, 1967) and spawn (Nair, 1972) are known to transmit virus particles and, with this knowledge, control of virus disorders is relatively easy.

3. Pests

Many of the arthropod pests associated with the mushroom crop are confined to four distinct groups, namely: sciarids, cecids, phorids and mites (see Hussey *et al.*, 1969; Hussey, 1972).

The larvae of the sciarid species *Lycoriella solani*, *Lycoriella auripila* and *Bradysia brunnipes* burrow into the stipe and caps and interfere with normal development of mycelium in the casing soil. Species of the genera *Heteropeza* and *Mycophila* are the common cecids, while the phorid *Megaselia nigra* is attracted by volatiles evolved from growing mushroom mycelium. Invading mites include the tarsonemid mite *Tarsonemas myceliophagus*, and the red-pepper mite *Pygmephorus* spp. Mite invasion is often associated with the occurrence of other weed fungi in composts. Inferior preparation and pasteurization of substrates often results in the occurrence of nematodes both saprophagous *Rhabditis* species and pathogenic mycelial feeders, such as *Ditylenchus myceliophagus* Goodey and *Aphenolocoides composticola* Franklin.

H. Disease Control and Terminal Disinfection

Notwithstanding the importance of substrate pasteurization, the early recognition and adoption of correct disease-control measures was perhaps one of the factors that has contributed most to the establishment of mushroom cultivation as a predictable semi-industrial process. It is now recognized that the most vulnerable time for contamination is in the early stages of the culture process, particularly up to and including the time of the application of the casing substrate. Thus, it is imperative for successful culture to adopt rigorous hygiene measures in compost preparation and pasteurization, incubation and in pasteurization of the casing layer. Also, during these operations, a relatively high degree of sophistication is applied to the construction of rooms in order to exclude potential contaminants. A simple but effective barrier is provided by filters located in the air ducts and maintaining positive pressures throughout the working area. Although not comparable to the axenic procedures used in preparation of pure culture spawn, nevertheless they represent a relatively sophisticated technology which is unique to mushroom cultivation in what is usually regarded as an horticultural activity.

Most of the major diseases are prevented in this way, but some diseases, particularly blotch, can be effectively controlled by maintaining the correct temperature and relative humidity regimes during the remaining cropping stage. However, it is also recognized that, given sound spawn and substrates, the cropping stage is less demanding and relatively simple systems are at least equal to the more elaborate systems, with the capability of a high degree of manipulative control and environment.

In the event of disease occurring, a range of chemical agents toxic to the pests and pathogens may be applied during the process. The use of the more generally available fungicides, however, poses special difficulties, since both host and pathogen are fungi (hyperparasitism). However, in recent years, the use of the fungicides, methyl benzimadazol, carbanic acid (Benomyl) and thiabenzadole, are selective in their action and overcome this difficulty. Biological methods of disease control have not yet been applied to commercial growing situations, but active research in this area may prove fruitful. Its application to a commercial process such as mushroom cultivation seems more promising than its application to conventional food crops (see Nair, 1973).

In view of the possibility of transmitting disease from one crop to the next, an important final procedure in the culture process is to disinfect not only the crop but also the entire structure. This terminal disinfection is achieved by the use of steam in order to maintain temperatures of about 60°C for 12 hours. Since this procedure is costly and damaging to the growing rooms, fumigation with methyl bromide is now known to be satisfactory and, in many situations, a better alternative. However, the use of disposable containers for growing obviates this need, and is an important advantage over the traditional tray and shelf systems of growing.

III. PRODUCTION OF OTHER EDIBLE SPECIES

In recent years, some attention has been given to the production of edible species with different nutritional and environmental requirements from *A. bisporus*. Unlike the conventional cultivated mushroom, they can be grown on substrates which do not require a prolonged composting process. Also, a second substrate, a casing soil, is not required to initiate fruitbody formation, but there is a requirement for light to allow normal development and production of fruitbodies. While commercial exploitation of these mushrooms is at present on a relatively small scale, their acceptance as a food and the simplicity of the substrate preparation stage, suggest that they may figure more prominently in the future of artificial cultivation.

A. Cultivation of the Chinese and Oyster Mushroom on Straw or Partially Decomposed Substrates

Outdoor growing of *Volvariella volvacea*, the Chinese mushroom, has been practised for many centuries in China (Chang, 1972). In the traditional method, bundles of previously wetted straw are arranged in a deliberate manner into stacks, usually of a specific size and shape, on an elevated soil base. This is often done in a rice field, between successive rice crops. Spawn, prepared on chopped rice straw, is placed between the bundles during the preparation of the stack, which is maintained moist throughout the cultivation process. It is therefore usually located in a shady position, near an irrigation channel or stream.

Once arranged, the stack temperatures rapidly increase to 40–50°C and fruits of *V. volvacea* form in 12 days, on the sides and top of the stack. Normally, fruits are harvested at an immature (button) stage and, like *A. bisporus*, successive flushes develop over a 40–50 day harvesting time, after which the exhausted beds are discarded.

The productivity of the beds is dependent on outdoor climatic conditions; heavy rains, low temperatures and excessive direct sunlight decrease yields. Chang (1965) records yields of between 6.8 to 8.5 kg per 100 kg of dry straw.

Some progress towards a more industrialized approach to cultivation of this species has been made during the last decade. Indoor cultivation on straw beds is now practised in Hong Kong, Taiwan and the Philippines. Ho (1972) describes a simple growing house prepared from a bamboo framework, covered with polythene film. Straw or sugar-cane leaves are used for the roof in order to block out solar radiation. Ventilation is provided by a small fan in an air duct.

Unlike *A. bisporus*, there is a low light requirement and this is provided artificially. Additional nutrients applied as supplements to the rice straw also improve yields. Rice bran and industrial wastes such as distillery wastes are beneficial. Chang and Yau (1970) describe the benefits of using a mixture of cotton waste and rice straw as a substrate. On the other hand, Ho (1972) demonstrated a substantial benefit from using a casing layer akin to those used in *A. bisporus* culture.

Species of *Pleurotus*, collectively known as oyster mushrooms, are also

Fig. 6. Cultivation of (a) *Pleurotus* sp. on sawdust and (b) Shiitake on logs.

primary colonizers of decaying vegetation. In nature, they are colonizers of dead wood, especially the stumps and trunks of trees. Artificial culture on logs prepared from deciduous trees is possible and is practised to a small extent in Japan. Following the early experiments of Block *et al.* (1958), who demonstrated growth of *Pleurotus* spp. on sawdust, more successful methods applicable to large-scale cultivation have recently been developed (Fig. 6). A wide range of vegetable wastes are suitable substrates for growth.

Edible species include *Pleurotus oestreatus*, *P. florida*, *P. eryngii*, *P. cornucopiae* and *P. sajor-caju*, but cultivation to date is confined to the temperature species *P. oestreatus* which forms fruitbodies at temperatures of up to 15°C. Strains of *P. florida* form fruits at temperatures of 25°C.

Sterilized crushed corn cobs were shown by Toth (1970) to be a good substrate, but a range of materials are known to be suitable. These include cereal straw (Bano and Srivastava, 1961; Zadrazil, 1976), newspaper (Zadrazil, 1976), banana pseudo-stems (Jandaik and Kapoor, 1976) and grass meal (Kalberer, 1976). According to the methods of Zadrazil (1976), a prolonged fermentation is not necessary, but a preliminary heat (80°–100°C) treatment of straw is beneficial through the release of soluble products such as phenolics and saccharides. Substrates containing high amounts of protein depress yields. Normally, the substrates are prepared on polythene film and, in order to establish the vegetative stage, incubation is done in unventilated rooms for ten days at a temperature of 20°C and a relative humidity of 75–90%. Spawn rapidly colonizes substrates to form a dense block. The polythene film is removed, and the growing room ventilated and illuminated for 12 hours in every day. After a further ten days, fruitbodies form in flushes. Yield is dependent on the environmental conditions, especially light intensity, humidity, temperature and the degree of ventilation. A yield of 250 kg of fruitbody caps per tonne of substrate is regarded as a satisfactory yield.

B. Cultivation of Shiitake and the Winter Mushroom on Wood

Many mushroom species colonize dead wood and, because of their natural lignicolous habit, methods have been developed by which the accepted edible species are cultivated. The edible species *Lentinus edodes* Shiitake, which accounts for about 25% of the world's total mushroom production, is cultivated extensively in Japan and in some other Asian countries.

Shiitake is usually cultured on logs cut from *Quercus acutissima, Q. serrata, Castanopsis cuspidata* and *Carpinus* species. Holes are drilled in the log substrate and spawn, prepared on pieces of wood or sawdust, is inserted into the log at several points. Care is taken to maintain the correct moisture contents during the incubation stage; the logs are arranged in an oblique upright direction in order to allow the natural movement of air. Incubation is done out of doors in laying yards but, after eight months, they are removed to raising yards which are usually shaded and have a relatively high humidity. Shiitake fruits develop at temperatures between 12°C and 20°C flushing in the spring and autumn. The logs are kept moist by frequent watering and fruitbodies develop over a period of six years.

According to Ando (1976), there is a light requirement during the vegetative stage for fruitbody formation. Protected cultivation has been tried experimentally, but the relatively slow growth rate limits the intensification of cultivation. However, in recent trials conducted by Akiyama *et al.* (1976), the time taken to produce the first fruitbodies can be greatly decreased if newly cut wood is used rather than wood which has been air dried and seasoned. Also, protecting the bed logs with straw and plastic mats during incubation contributes to shortening the time for first fruitbody formation to six months. A shortened substrate-colonization stage also results in a substantial reduction in the wastage of logs due to the invasion of weed fungi.

Flammulina velutipes, the winter mushroom, is also a wood-destroying species, which is cultivated in the Orient. This species produces small (3 cm diameter) fruits on long slender stipes (up to 9 cm in length) and is amenable to what is termed 'bottle cultivation' on sawdust (Mori, 1968). A medium of sawdust and rice bran is made moist, mixed and filled into one-litre polypropylene bottles, appropriately capped before sterilization. After sterilization, the medium is inoculated with spawn prepared from sawdust. In order to maintain an optimum growing temperature of 22–25°C within the bottle, incubation is done in a room maintained at 18–20°C. Bottle caps are removed after 20 days, and the surface ruffled before placing in darkness at a temperature 10–12°C, at a relative humidity of 80–85%. Primordia form after the ten days of cold temperature treatment, but the quality of the fruitbodies is improved by lowering the temperature to 3–5°C with simultaneous air movement, followed by a gradual raising of the temperature to 8°C and maintaining a relative humidity of 75–80%. Light is required for maturation of the fruitbodies. To support the long and slender stipes, a plastic support is placed on the neck of the bottle. Two flushes are obtained from each bottle to yield about 200 g fresh fruitbodies per bottle.

Bottle cultivation on sawdust has completely replaced cultivation on logs with the result that considerable improvements in yield and quality have been achieved. Production in Japan amounts to 30 000 tonnes per annum.

IV. UTILIZATION AND VALUE OF MUSHROOMS

The numerous myths and superstitions associated with growth of wild mushrooms are also reflected in some attitudes to the consumption of cultivated mushrooms. But, with the world-wide growth of artificial culture, attitudes and outlooks are changing. In Western countries, they have traditionally been regarded as a luxury food, eaten with other foods (especially meat) to provide flavour and zest. Even up to the beginning of this century, their availability was restricted to the high-income category and the aristocracy. However, mushrooms are now consumed by people in the lower income groups and increasing attention is being given to their value as a food.

They are used in modern cuisine in a variety of ways, but a number of unexplained and contrasting preferences are apparent in different countries. In the U.S.A., West Germany and Canada, there is a preference for canned mushrooms, while in Britain fresh mushrooms are preferred. There is also an increasing interest in the use of preserved mushrooms as chutneys, ketchups, or in dried or pickled form. Preservation of wild edible mushrooms is an art which has developed from the rural communities of some Eastern European countries, Italy, India and other countries of the Orient.

A. Composition and Food Value

In the published literature on the composition and nutritional properties of edible mushrooms, many conflicting facts and viewpoints are evident. Much of this stems from the wide range of analytical techniques used by various investigators to determine composition. Also, it is frequently not clear whether figures on composition relate to mycelium or the fruitbody

Table 3

Data relating to the composition and food value of mushroom fruitbodies.
From Hayes (1976)

Composition (%)		Essential amino acids (% dry protein)	
Water	91	Isoleucine	1.28
Solids	9	Leucine	2.16
Carbohydrate	2.4	Lysine	1.62
Protein (nitrogen × 6.25)	3.3	Methionine	0.39
Fat	0.4	Phenylalanine	1.55
Ash	0.9	Threonine	1.48
		Tryptophan	3.94
		Valine	1.63

Mineral content (% dry matter)		Vitamin content (mg/100 g fresh weight)		
Calcium	0.17	Nicotinamide	4.1	Trace amounts of
Potassium	4.12	Riboflavin	0.5	pyridoxine,
Sodium	0.64	Ascorbic acid	2.5	pantothenic acid,
Phosphorous	1.50	Thiamin	0.1	choline, biotin,
Iron	27.00			vitamin B_{12}
				and folic acid.

and, in some instances, different species are described under the common name 'mushroom'. In such cases, it is not possible to distinguish between wild and cultivated species. The following summary relates to *A. bisporus* fruitbodies (Table 3).

1. *Carbohydrate.* Much of the carbohydrate consists of mannitol, normally regarded as being of no nutritional significance, and therefore excluded from calculations to estimate calorie value of foods. Other carbohydrates are present, however, and according to Holtz (1971), fruit bodies of *A. bisporus* contain 0.48 mg fructose, 2.20 mg glucose, 11.5 mg of mannitol and 0.52 mg sucrose per gram of fresh weight. It was suggested by Holtz (1971) that mannitol functioned in maintaining an osmotic potential to maintain a relatively high level of water in the fruitbody.

2. *Fat.* According to the analyses of Hughes (1962), ether-soluble lipids in the *A. bisporus* fruit accounted for 1.3% of the dry weight of the white variety and 1.8% of the cream. Fatty acids present were found to be linoleic, palmitic, stearic, arachidic, oleic and lauric acids. Holtz and

Schisler (1971) claim that the lipid contains one of the highest proportions of linoleic acid known, and accounts for 63–74% of the total fatty acids present.

3. *Protein.* In the literature, there is a considerable variation in the values for protein, varying between two extremes of a zero value recorded by Chatfield and Adams (1940) to 4.88% of the freshweight (Chang, 1972). It is not possible satisfactorily to explain the discrepancies in the protein values given by various authorities, but there is little doubt that the general practice of expressing protein as crude protein (i.e. total nitrogen multiplied by 6.25) does not give a true value for protein, since some nitrogenous constituents of the cell include non-protein nucleic acids, urea, chitin and chitosans. According to the standard tables on the composition of foods which apply to Britain (McCance and Widdowson, 1969; Manual of Nutrition, H.M.S.O., 1970), fresh mushrooms contain 1.8% protein.

In recent studies, Hayes and Haddad (1976) have shown that protein values, as determined by amino-acid analysis, vary not only according to the developmental stage at which mushrooms are harvested, but also the time of harvesting. Culture practices, however, play an important role, especially watering regimes during the cropping cycle. During storage, 'true protein' content (as distinct from crude protein) of fresh mushrooms is gradually decreased over six days (Beelman *et al.*, 1976; N. Haddad and W. A. Hayes, unpublished results).

A value of 35% true protein, of the dry matter, was given by Hayes and Haddad (1976) for a typical freshly harvested mushroom. It was shown to contain all the essential amino acids required by a human adult, but it is deficient in methionine, while tryptophan and lysine were present in relatively high proportions. Using *in vitro* enzymic techniques, a digestibility value of 82% was obtained.

4. *Minerals and vitamins.* Mushroom fruitbodies contain relatively high levels of potassium, phosphorous, iron and copper but are deficient in calcium (Anderson and Fellers, 1942). These workers noted that approximately one third of the iron could be considered utilizable by the human body. Cheldelin and Williams (1942) and Anderson and Fellers (1942) concluded that fruitbodies of cultivated *A. bisporus* are an excellent source of nicotinic acid and riboflavin, and also contain significant quantities of pantothenic acid and vitamins B, C and K. They also concluded that non of the B-group vitamins was lost in processing.

Using *A. bisporus* as an example, it is evident that mushrooms may be classed as a highly-flavoured, low-calorie proteinaceous food, supplying also some vitamins and minerals. As such, mushrooms are seen to be a valuable complement to other foods, especially to the staple plant foods.

B. Medicinal Properties

In the Orient, mushrooms are frequently consumed for their healthful and therapeutic properties. Ailments claimed to be alleviated or cured by *Lentinus edodes* have their origins in the remedies practised during the Ming dynasty (1368–1644) and include illnesses now known to be caused by viruses, high blood pressure and heart diseases.

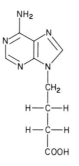

Fig. 7. Structure of eritadenine (2(R),3(R)-dihydroxy-4-(9-adenyl)-butyric acid), a plasma cholesterol-lowering substance extracted from Shiitake.

Compounds from mushrooms which show antitumour activity have received comparatively little attention since the discovery by Lucas *et al.* (1959) of a potential antitumour agent, calvacin, found in the wild edible fungus *Calvatia gigantea*, the giant puffball. The polysaccharide, lentinan, a β1, 3-glucan isolated from *Lentinus edodes*, has been shown to have antitumour activity in mice (Chihara *et al.*, 1969; Hamuro *et al.*, 1976). Dennert and Tucker (1973) showed that lentinan altered immunoresponsiveness through its action as a T-cell adjuvant.

More interest has been shown in the anticholesterol activity of eritadenine (2(R),3(R)–dihydroxy–4–(9 adenyl)butyric acid; Fig. 7) and also by the Shiitake mushroom (Chibata *et al.*, 1969; Tokita *et al.*, 1972; Tokuda *et al.*, 1976; Suzuki and Ohshima, 1976). Tokuda and his coworkers conclude that hypocholesteremic activity is due to acceleration of cholesterol metabolism by stimulation of cholesterol esterification and the promotion of cholesterol transport from α-lipoprotein to β-liprotein. Activity is also associated with an increase in the excretion of cholesterol.

Extracts from several mushrooms were shown by Cochran and Lucas (1959) and Goulet *et al.* (1960) to possess antiviral activity. An interferon-inducing RNA was reported by Tsunoda and Ishida (1970), while Yamamura and Cochrane (1976) isolated another inhibitor of myxovirus multiplication, AC2P, a high molecular-weight polysaccharide composed mainly of pentoses.

Although many natural foods are being investigated as potential sources of substances which may bear some relevance to modern drug therapy, searches in the higher fungi are comparatively recent. To date, only a few potentially useful substances have been discovered, but none has proved to be clinically useful. It should also be stressed that there is no scientific evidence to advocate incorporation of mushrooms into diets as therapeutic or prophylactic agents. Nevertheless, mushrooms have in the past figured in folklore remedies for a wide range of ailments and further research may prove more fruitful.

V. CONCLUSIONS AND FUTURE PROSPECTS

The technology associated with artificial culture of mushrooms can readily be identified with microbiology, fermentation and bio-engineering. In many respects, commercial methods of cultivation can be viewed as the application of routine laboratory methods in the propagation of micro-organisms to a large-scale agricultural process. In both, the objective is to maintain a pure culture of the organism being cultured, free from harmful contaminants.

Pure-culture spawn production is common to all species cultivated and, in the artificial culture of *Flammulina* sp. in bottles, every stage in the culture process is akin to laboratory methods. In cultures of species of *Pleurotus* and *Volvariella*, some of the sophistications of substrate manufacture, involving the use of autoclaves in order to establish pure-culture substrates, are replaced by either a short fermentation or a low-temperature heat treatment. However, for the most extensively cultivated species, namely *A. bisporus*, substrate manufacture is complicated, prolonged, and includes a precisely controlled pasteurization stage.

The two basic procedures which are concerned with the establishment of the vegetative stage of growth, spawn and substrate manufacture, are technically the most demanding, requiring sophisticated equipment and trained personnel. Subsequent procedures, which are primarily concerned with development of the reproductive or fruitbody stage of the life cycle, provide the crop for harvest and are by comparison less demanding on technical skills.

Even so, in the intensive production units, so much a feature of production in the industrialized advanced high-income countries, considerable effort is made to exclude the entry of contaminants during the reproductive or cropping phase. Air filtration and maintenance of positive pressure in work areas where cultures are exposed, and automated systems of environmental control, are examples of the degree of sophistication. Those contrast markedly with the outdoor systems adopted for the cultivation of *Lentinus edodes* and *Volvariella volvacea* in the Orient.

Indeed, relatively simple systems have been applied in Taiwan and in some parts of India for production of *A. bisporus*. In the remote regions of the Western Himalayas, cultivation is being developed as a cottage industry, where it is practical to grow *A. bisporus* throughout the year without heating or cooling the growing atmosphere. In some localities, schemes based on the separation of the two levels of technology in mushroom culture are being applied in rural development programmes. By centralizing spawn and compost manufacture, the specialist expertise and required high technology are centralized in what is termed a 'mother unit', which provides the cultures, fully prepared for cropping, to a large number of 'satellite' growing units located in nearby villages for growing in mud huts or even in village dwellings. This same concept has been applied in England for the marketing of modules for amateur growers and for the general public for cultivation in the home.

The spread of *A. bisporus* cultivation to the low-income emerging countries is indicative not only of changing attitudes to mushrooms as a food, but also of a more general scientific understanding of the factors which govern growth and development of this fungus. Definition of the nutritional requirements and their application to composting methods and techniques has, above all, contributed to the extension of growing world-wide (Gerrits *et al.*, 1967; Hayes, 1969, 1972). For laboratory culture, dextrin and a mixture of simple sugars, hydrolysed casein, mineral salts, biotin, thiamin and a source of acetate are minimal requirements. In composts, the carbohydrates are provided by lignin, cellulose and hemicelluloses, and it has been shown by Hayes (1969) and Stanek (1972) that protein is obtained primarily from the biomass of aerobic thermophilic micro-organisms built up during composting.

This understanding has provided a basis for an extension to the range of materials that can be utilized for mushroom cultivation. An increasing proportion of substrates are used which incorporate a wide variety of vegetable materials. In addition to the traditionally used wheat straw used for bedding horses, barley and oat straw and hay prepared from a wide range of grass species can be utilized. In the Orient, where rice straw is plentiful, substrates are prepared at low cost by incorporating almost any locally available animal manure. Other waste products, such as maize leaves, sugar cane bagasse and a variety of agricultural, municipal and industrial waste products, are utilized in mushroom substrates.

In recent years also, the significance of bacteria in the soils used as a casing layer has been demonstrated (Eger, 1962; Park and Agnihotri, 1969; Hayes et al., 1969; Hayes, 1972, 1974; Nair and Hayes, 1974; Hayes and Nair, 1976). Strains of *Ps. putida* and closely related types are now known to be associated with the production of fruits. Soils can therefore be screened for, and deficient soils avoided for culture purposes. This alone has resulted in the successful establishment of mushroom production in some locations.

Besides the ability of a fungus to grow on a wide range of substrates, a prerequisite for the industrial exploitation is an ability to grow quickly. Unlike the lower fungi, mycelial growth rate of the higher fungi is relatively slow. Consequently, attempts at production of mycelial biomass have not been developed beyond the experimental stage (Humfeld and Sugihara, 1952; Litchfield, 1968; Heinemann, 1963). For the production of mushroom fruitbodies a further extension of time is required, not only to initiate development of the first fruits, but because a prolonged incubation is required in order to maximize yields. This varies from about forty days for *A. bisporus* culture to the extreme of seven years in the growing of *L. edodes* on logs.

Of necessity, a solid surface is required for production of fruitbodies and the evolved systems contrast markedly with the concept of fungal biomass production using liquid nutrients and fermentation techniques (Solomons, 1975). Even so, for production of *A. bisporus*, cultures are arranged in layers and, being a batch process, up to six crops per annum can be harvested. The relative yields per unit area in agricultural terms

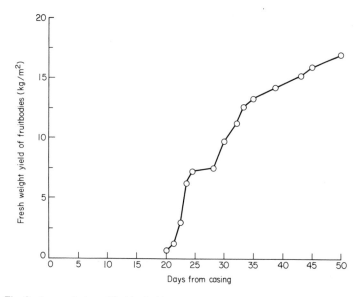

Fig. 8. Accumulation of fruitbody biomass during typical *Agaricus bisporus* culture.

are impressive. For example, a modern production unit cultivating *A. bisporus* would expect to yield about 540 tonnes of dry matter per hectare per annum, or 160 tonnes of dry protein per hectare per annum. It should be noted also that these figures do not include the biomass accumulated in the compost and that which is normally discarded in accepted harvesting practice, e.g. stalks and blemished fruits.

However, there are a number of factors inherent in the evolved processes of artificial culture which impose major constraints on a more extensive exploitation of these fungi as significant producers of biomass in a truly industrial sense. These are: (1) substrates are bulky and difficult to manoeuvre, requiring a high degree of mechanization and energy input in the collection of raw materials; (2) different environmental conditions for growth require insulated and soundly constructed growing structures in most regions of the World; (3) low biological efficiency in the conversion of substrates; fresh weight yield relative to fresh weight of substrate rarely surpasses the 20% level in any of the species cultivated; and (4) the sporadic production of fruit and the differential maturation of fruits on harvest requires hand harvesting. These constraints are prominently reflected in the economics of mushroom production. Generally, in Western countries, cultivation on a large scale is recognized as being capital and labour intensive and, despite the high degree of mechanization, the labour requirement contributes 50% to the total production costs. Material costs for substrates (compost and casing soil), spawn and biocides, represent by comparison about 35% of total costs. Fuel inputs also represent an important component of the total costs of production.

In contrast, in the developing countries, simple and unsophisticated systems of growing are adopted. Mechanization is minimal and buildings made from locally available wood, thatch or mud provide sufficient protection. Production costs are low in comparison with Western countries, largely because of the availability of low-cost labour. Substrates therefore assume the greatest share of production costs.

Fig. 9. Mechanical harvesting of *Agaricus bisporus*.

Progress in research and application of mushroom science to the artificial culture of mushrooms has resulted in diversification of methods applicable to a range of locations and circumstances. Current research and development are being directed to both intensive and extensive systems of culture. Especially relevant to the economics of production is the development of mechanical harvesters (Persson, 1972; Fig. 9).

Machines have been developed in The Netherlands which have increased cutting rates from the accepted rate for hand harvesting of 20 kg per hour to 1000 kg per hour. Although this may be more compatible with production of biomass, it has at present limited application, since the consumer demands a fruit of a given size and shape free from blemishes.

Long-term research in genetics, breeding and physiology of *A. bisporus* aimed at improving the biological efficiency of artificial culture is likely to lead to continued improvements to existing systems, but industrialization may depend on some new concepts (Hayes, 1976). These are based on utilization of the residual substrates after cropping for growth of other edible species, thus also allowing the application of 'crop rotation', an established practice in the cultivation of crop plants, as a precaution against the build-up of disease organisms.

Although a wide range of species can be cultivated by artificial techniques, their future role in mushroom cultivation will depend on acceptance of novel types by the consuming public. However, if biomass generated by micro-organisms is to contribute to future human foods, it may be argued that consumers may be more prepared to accept a natural food of proven edibility, such as mushrooms, rather than other synthetic foods of fungal origin (Solomons, 1975). The fact that basidiomycetes convert waste materials into a highly flavoured proteinaceous fruit, which complements many of the plant staple foods, is clearly relevant to the needs both of the emerging and industrially advanced countries. It is likely that plant foods will figure more prominently in future diets but, as with other biomass foods, the future role of mushrooms will be governed by the economics of production methods and costs relative to other animal and vegetable foods.

REFERENCES

Abercrombie, J. (1817). 'Abercrombie's practical gardener'. 2nd Edition. Cordell and Davies, London.
Akiyama, H., Akiyama, R., Akiyama, I., Kato, A. and Nakazawa, W. (1976). *Mushroom Science* **9**, 423.
Anderson, E. E. and Fellers, C. R. (1942). *Proceedings of the American Society of Horticultural Science* **41**, 301.
Ando, M. (1976). *Mushroom Science* **9**, 415.
Atkins, F. C. (1974). 'Guide to Mushroom Growing'. Faber, London.

Bano, Z. and Srivastava, H. C. (1962). *Food Science* **12**, 363.
Beelman, R. B., McArdle, F. J. and Parrish, G. K. (1976). *Mushroom Science* **9**, 333.
Block, S. S., Taso, G. and Han, L. (1958). *Journal of Agriculture and Food Chemistry* **6**, 923.
Burnett, J. H. (1968). 'Fundamentals of Mycology'. Edward Arnold Ltd., London.
Chang, S. T. (1965). *World Crops* **17**, 47.
Chang, S. T. (1972). 'The Chinese Mushroom'. Chinese University, Hong Kong.
Chang, S. T. and Yau C. K. (1970). *Mushroom News*, U.S.A. **18**, 9.
Chatfield, C. and Adams, G. (1940). *United States Department of Agriculture Circular 549*.
Cheldelin, V. H. and Williams, R. J. (1942). *University of Texas Publication* No. 4337, 105.
Chibata, I., Okumura, K., Takeyama, S. and Kotera, K. (1969). *Experientia* **25**, 1237.
Chihara, G., Maeda, Y., Hamuro, G., Sasaki, T. and Fukuoka, F. (1969). *Nature, London* **222**, 687.
Cochran, K. W. and Lucas, E. H. (1959). *Antibiotics Annual* 1958–1959, 104.
Delcaire, J. R. (1978). *In* 'Biology and Cultivation of Edible Mushrooms', (S. T. Chang and W. A. Hayes, eds.). Academic Press, New York.
Denham, T. G. (1977). *In* 'Composting', (W. A. Hayes, ed.). M.G.A., London.
Dennert, G. and Tucker, D. (1973). *Journal of the National Cancer Institute* **51**, 1727.
Edwards, R. L. (1976). *The Mushroom Journal* **45**, 282.
Eger, G. (1961). *Archiv für Mikrobiologie* **39**, 313.
Ferguson, M. (1902). *Bulletin of the United States Bureau of Plant Industries* **16**, 72.
Flegg, P. B. (1968). *Journal of Horticultural Science* **43**, 441.
Flegg, P. B. (1970). *National Agricultural Advisory Service, Quarterly Review* **90**, 61.
Flegg, P. B. and Gandy, D. G. (1962). *Journal of Horticultural Science* **37**, 124.
Gandy, D. G. (1960). *Nature, London* **185**, 482.
Gandy, D. G. and Hollings, M. (1962). *Annual Report, Glasshouse Crops Research Institute* 1961, 103.
Gerrits, J. P. G., Muller, F. M. and Bels-Koning, H. C. (1967). *Mushroom Science* **6**, 225.
Goulet, N. R., Cochrane, K. W. and Brown, G. C. (1960). *Proceedings of the Society of Experimental Biology and Medicine* **103**, 96.
Hamuro, J., Maeda, Y., Fukuoka, F. and Chihara, G. (1972). *Mushroom Science* **9**, 477.
Hayes, W. A. (1969). *Mushroom Science* **7**, 173.
Hayes, W. A. (1972). *Mushroom Science* **8**, 663.
Hayes, W. A. (1974). *In* 'The Casing Layer', (W. A. Hayes, ed.), M.G.A., London.
Hayes, W. A. (1976). *Nutrition and Food Science* **42**, 2.
Hayes, W. A. and Haddad, N. (1976). *The Mushroom Journal* **40**, 2.
Hayes, W. A. and Nair, N. G. (1976). *Mushroom Science* **9**, 259.
Hayes, W. A. and Randle, P. E. (1970). *Annual Report, Glasshouse Crops Research Institute* 1969, 166.
Hayes, W. A., Randle, P. E. and Last, F. T. (1969). *Annals of Applied Biology* **64**, 177.
Heinemann, B. (1963). United States Patent 3 086 320.
Ho, M. S. (1972). *Mushroom Science* **8**, 257.
Hollings, M. (1972). *Mushroom Science* **8**, 733.
Hollings, M. and Stone, O. M. (1969). *Science Progress, Oxford* **57**, 371.
Holtz, R. B. (1971). *Journal of Agriculture and Food Chemistry* **19**(6), 1272.
Holtz, R. B. and Schisler, L. C. (1971). *Lipids* **6**(3), 176.
Hughes, D. H. (1962). *Mushroom Science* **5**, 540.
Humfeld, H. and Sugihara, T. F. (1952). *Mycologia* **44**, 605.
Hussey, N. W. (1972). *Mushroom Science* **8**, 183.
Hussey, N. W., Read, W. H. and Hesling, J. J. (1969). 'The Pests of Protected Cultivation'. Edward Arnold, London.
Jandaik, C. L. and Kapoor, J. N. (1976). *Mushroom Science* **9**, 6671.

G

Kalberer, P. P. (1976). *Mushroom Science* **9**, 653.
Kneebone, L. R., Lockard, J. D. and Hagar, R. A. (1962). *Mushroom Science* **5**, 461.
Lambert, E. B. (1938). *Botanical Review* **4**, 397.
Lambert, E. B. (1941). *Journal of Agricultural Research* **62**, 415.
Last, F. T., Hollings, M. and Stone, O. M. (1967). *Annals of Applied Biology* **59**, 451.
Litchfield, J. H. (1968). *In* 'Single Cell Protein', (R. I. Mateles and S. R. Tannenbaum, eds.), M.I.T. Press, London and Cambridge.
Long, P. E. and Jacobs, L. (1968). *Mushroom Science* **7**, 373.
Lösel, D. M. (1964). *Annals of Botany* **28**, 541.
Lösel, D. M. (1967). *Annals of Botany* **31**, 417.
Lucas, E. H., Byerrum, R. U., Clarke, D. A., Reilly, H. C., Stevens, J. A. and Stock, C. C. (1959). *Antibiotics Annual* 1958–1959, 493.
Manual of Nutrition (1970). Her Majesty's Stationery Office.
McCance, R. A. and Widdowson, E. H. (1960). *Medical Research Council Special Report*, Series No. 297. H.M.S.O. London.
Mori, K. (1968). *Mushroom Science* **7**, 577.
Nair, N. G. (1972). *Mushroom Science* **8**, 155.
Nair, N. G. (1973). *The Mushroom Journal* **16**, 140.
Nair, N. G. and Hayes, W. A. (1974). *In* 'The Casing Layer', (W. A. Hayes, ed.), M.G.A., London.
Park, J. Y. and Agnihotri, V. P. (1969). *Nature, London* **222**, 984.
Persson, S. P. E. (1972). *Mushroom Science* **8**, 115.
Raper, J. R. (1966). 'Genetics of Sexuality in Higher Fungi'. Ronald, New York.
Rasmussen, C. R. (1962). *Mushroom Science* **5**, 91.
Rast, D., Greuter, B., Lendenman, J. and Zobrist, P. (1976). *Mushroom Science* **9**, 59.
Schisler, L. C., Sinden, J. W. and Sigel, E. M. (1967). *Phytopathology* **57**, 519.
Schisler, L. C., Sinden, J. W. and Sigel, E. M. (1968). *Phytopathology* **58**, 944.
Sinden, J. W. (1932). United States Patent 1 869 517.
Sinden, J. W. and Hauser, E. (1950). *Mushroom Science* **1**, 52.
Sinden, J. W. and Hauser, E. (1953). *Mushroom Science* **2**, 123.
Solomons, G. L. (1975). *In* 'The Filamentous Fungi', (J. E. Smith and D. R. Berry, eds.), Vol. 1, pp. 249–264. Edward Arnold, London.
Stanek, M. (1972). *Mushroom Science* **8**, 797.
Storey, I. F. (1972). *Mushroom Science* **8**, 95.
Suzuki, S. and Olshima, S. (1976). *Mushroom Science* **9**, 463.
Tokita, F., Shibukawa, N., Yasumoto, T. and Kaneda, T. (1972). *Mushroom Science* **8**, 784.
Tokuda, S., Tagiri, A., Kano, E., Sugawara, Y., Suzuki, S., Sato, H. and Kaneda, T. (1976). *Mushroom Science* **9**, 445.
Toth, E. (1970). *Gradinarstvo* **6**, 42.
Tschierpe, H. J. (1959). *Gartenbauwissenschaft* **24**, 18.
Tschierpe, H. J. (1972). *Mushroom Science* **8**, 553.
Tschierpe, H. J. and Sinden, J. W. (1964). *Archiv für Mikrobiologie* **49**, 405.
Tsunoda, A. and Ishida, N. (1970). *Annals of the New York Academy of Sciences* **173**, 719.
Yamamura, Y. and Cochran, K. W. (1976). *Mushroom Science* **9**, 495.
Zadrazil, F. (1976). *Mushroom Science* **9**, 621.

7. Algal Biomass

JOHN R. BENEMANN, JOSEPH C. WEISSMAN AND
WILLIAM J. OSWALD

*Sanitary Engineering Research Laboratory, University of California,
Berkeley, California, U.S.A.*

I. INTRODUCTION

Micro-algal single-cell protein production is, in many ways, more akin to conventional agricultural processes than to the normal industrial single-cell protein (S.C.P.) production processes. Some of the distinctive characteristics of industrial S.C.P. production are low land requirements, independence of climatic factors, controlled environments, pure microbial cultures and high continuous productivity. These do not apply to micro-algal S.C.P. production. Mass cultivation of algae in outdoor pond systems has many of the general attributes of agriculture, namely extensive land use, dependence on sunlight and climate, invasion by pests, weeds and herbivores, and large requirements for water and fertilizers. However, some of the properties of algal mass culture resemble

177

those of other industrial microbiological processes and might confer to algal S.C.P. production some advantages over conventional agriculture. These properties include the hydraulic nature of the culture (allowing easy handling), the ability at all times to provide nutrients at optimal concentrations, the year-round optimal ground cover resulting in maximal utilization of sunlight, absence of non-productive unusable tissues (roots, stems, etc.), relatively high productivities, and a high protein content of the micro-algal biomass (up to 50% of dry weight) which is similar to that of other micro-organisms. In theory, algal S.C.P. production could incorporate the advantages of both agriculture and microbial biomass production, resulting in an efficient, practical and economic source of protein. In practice, the technology of micro-algal cultivation is still limited by the problems of micro-algal species control, harvesting, and design and operation of engineered cultivation systems. Basically, such systems must consist of shallow ponds continuously or intermittently mixed and diluted with fresh or recirculated growth media. Dewatering and drying, algal cell-wall and membrane digestibility, and high pigment content are processing problems in preparation of micro-algal S.C.P. from harvested biomass.

Besides large land requirements and exposure to climate, resulting in variable insolation and temperature, the major difference between large-scale algal S.C.P. production and most other large-scale S.C.P. production processes is the open nature of algal systems, which allows ready invasion by competing algae and other colonizing types of microscopic pond life. Maintenance of the desired species or types of micro-algae is a limiting technological problem. Definition of desirable micro-algae involves those specific properties which allow economical operations and an acceptable product. Nutritional acceptability, based on extensive feeding data, is the primary requirement for S.C.P. The required specificity of the cultivation technology depends on the ultimate end use of the algal product (e.g. feed or food) as well as on intermediate processing steps. In general, cultivation of green algae need not be as species-specific as that of the blue-green algae, since production of toxic substances is well known to be associated with the latter but not with the former (Gorham, 1964; Shilo, 1967).

Strategies for selective micro-algal cultivation technologies might resemble higher plant agriculture as well as include novel methods which are uniquely applicable to micro-algal production in ponds. Of course, aseptic cultivation of pure cultures is possible and has been carried out with algae in industrial production under heterotrophic conditions. However, this is either not economically feasible or would have little advantage over other S.C.P. production processes. Cultivation under non-sterile conditions in controlled environment greenhouse or clear plastic-covered ponds is possible, but is not sufficient to prevent invasion by competing algae and undesirable micro-organisms (Tamiya, 1957). Covered greenhouse agriculture is, at present, a feasible process only where a speciality crop of high value is cultivated (Bivort, 1977). Even if the required advances in plastics and construction technology are achieved, covered cultivation of micro-algae offers few advantages and several drawbacks. Aside from high cost, these include overheating problems, decreased light intensity, restriction of potential sites, and increased maintenance requirements. This review will emphasize micro-algal cultivation in open ponds. However, covered algal ponds might be attractive, and economically feasible, in some situations.

The second requirement for species control in micro-algal S.C.P. production, after nutritional acceptabilty of the algae, is easy harvesting of the cultivated algal species. Harvesting has been in the past a limiting economic factor in micro-algal biomass production. The dilute nature of the standing crop in micro-algal cultures (150–700 mg per litre), the microscopic size of the plants, the large volume that must be processed due to continuous operation of the ponds, and the large differences between micro-algal types, have made development of a universal and inexpensive harvesting process difficult (Golueke and Oswald, 1965). Centrifugation and chemical flocculation (using either lime or alum) have proved to be most universal, but also very expensive (above U.S. $100/10^6$ litres), harvesting methods. Filtration or screening techniques are not applicable to small single-cell, or the smaller colonial, micro-algae since they either clogg or pass through the filters or screens. Settling or flotation of algae is a well-known phenomenon in the laboratory and field, but has not yet been adapted to harvesting micro-algae from mass culture. In theory, this could be a very inexpensive harvesting method, costing less than U.S. $5/10^6$ litres. Table 1 gives a comparative evaluation of various algae harvesting methods. Cost estimates are subject to large uncertainties, depending on assumptions and proponents. In general, Table 1 illustrates the fact that the more expensive harvesting methods do not require algae species control, while inexpensive methods normally do depend on it.

The filamentous or large colonial micro-algae can be removed

Table 1

Comparison of available micro-algae harvesting processes

Process	Estimated cost (U.S. $/$10^6$ litres)	Approximate concentration factor	Relative energy requirement	Dependence on micro-algae type	Quality of algae produced
Centrifugation	140	40–100	Very high	Minor	Good
Coagulation-flotation	120	85	High	Minor	Poor
Coagulation-sedimentation	110	50	High	Minor	Poor
Ultrafiltration (non-fouling)	100	50	High	None	Good
Intermittent sand filtration	70	—	Low	Some	Poor
In-pond settling with chemicals	40	—	Low	Some	None
Microstraining	15	30	Low	Strong	Good
Settling without chemicals	5	20–200	Very low	Unknown	Good

From Benemann et al. (1977a). Costs are estimated for a 40 million litre/day harvesting facility and are given in 1976 U.S. dollars. Settling without chemicals is calculated on the basis of continuous operation of a sedimentation tank operating on a 3–4 hour detention time.

economically from pond effluents, and concentrated about 30-fold by simple fine-mesh screens (pores above 25 μm). The algae that collect on the screens are removed by a water spray. Either the water spray or the screen may move, one in relation to the other. If the water spray is fixed and a slowly rotating screen is used, the device is known as a microstrainer. Microstrainers are well known devices for removing blue-green algae from water reservoirs. Cost of harvesting algae using microstrainers is estimated at about U.S. $15 per million litres for an approximately 40 hectare pond system (Table 1). Cultivation of micro-algal species that can be harvested by such processes, such as the filamentous blue-green algae *Spirulina* spp., is one approach to the problem of algae separation and biomass production.

A final requirement for species control is for micro-algae which can be cultivated continuously with high productivity. Yields above 40 dry tonnes/hectare/year are expected from algal cultures (see Section VII, p. 200). These three requirements—suitability as an S.C.P. source, easy harvestability, and high yields—must be optimized in a production system that will be able to compete economically with alternative plant or S.C.P. sources. Algal mass-cultivation technology has not yet developed to the point where all of these requirements can be met in a single system; however, recent advances suggest this may be possible in the near term.

The study and development of algal mass cultures for S.C.P. production has been pursued for over 30 years. Micro-algae were first suggested by Sphoer and Molner (1949) as a protein source while, in Germany, they were viewed as a potential source of fats during World War II (Harder and Witsch, 1942). These early suggestions were translated into a pilot project for production of species of the unicellular green alga *Chlorella* sp, carried out in 1951 by the Arthur D. Little Company under sponsorship of the Carnegie Institution (Little, 1962). Technical difficulties and poor economics discouraged further development of this system. Commercial cultivation of *Chlorella* spp. in open and covered ponds was achieved a few years later in Japan (Tamiya, 1957; Krauss, 1962), where the algal product was used as a prized food-colouring agent, thus defraying high production costs. The high content of pigment in micro-algae (including xanthophylls, chlorophylls and carotenoids) and its appeal as a speciality ('health') food is the basis for present commercial interest of micro-algal production. This is the case for the world's largest currently operating micro-algal production facility, namely, the one tonne per day Spirulina plant of the Sosa Texacoco Co., near Mexico City, and the many, smaller, green algae production systems in Japan and the Far East.

Inexpensive production of micro-algae would be possible using waste-treatment ponds. In such 'oxidation' ponds, the algae function as oxygen producers, and production costs would be totally or partially covered by waste treatment credits. When dealing with municipal sewage product quality, questions limit application of this concept to S.C.P. production at present. However, the micro-algae produced on animal wastes would be more acceptable as a feed source.

The state of the art of these various methods of micro-algal cultivation and present or projected economics are reviewed in this chapter. A brief description of relevant physiological and ecological information is also included. Excluded from this review are the utilization of marine macro-algae (seaweeds) as food sources (since they fall, by size alone, outside the S.C.P. definition), cultivation of micro-algae for agricultural fertilizer (Benemann *et al.*, 1977a, 1977b) and the sustaining of food-chains for higher organisms (in aquaculture). Aquaculture is an important application of micro-algal cultivation, and many of the problems of aquaculture can be traced to difficulties in producing a continuous, dependable microalgae supply (Rhyther and Goldman, 1975). Thus development of species-specific micro-algal mass cultivation technology would have an important impact on aquaculture, and thereby increase the potential for food production by this technology. A recent review of photosynthetic S.C.P. has been published (Waslien *et al.*, 1977).

II. MICRO-ALGAE IN LABORATORY AND FIELD

Micro-algae have been the subject of intensive ecological, physiological and biochemical studies for over one hundred years. They are the most important primary producers of organic matter in most aquatic systems, and are the basis for often complex foodchains, many of which are of commercial significance. Eutrophication is another aspect of micro-algal ecology subject to intensive study because of its practical impact. Despite the large body of knowledge about algal ecology (Fogg, 1975), it provides few guidelines to micro-algal cultivation. Ecological studies do not, in general, attempt to achieve specific results, and experimental studies of micro-algal species competitions have usually been passive or neutral regarding outcome. Nevertheless, much information has been developed regarding environmental factors which affect specific algal species and result in their dominance or replacement by competing algae. Although the predictive value of current ecological theory is not high, the general features of phytoplankton ecology are well developed, and many specific cases have been analysed in detail. Application of the techniques and insights of phytoplankton ecology to large-scale algal cultivation could solve many of the outstanding problems.

Mass cultures of algae in continuously or intermittently diluted and mixed ponds present a much simpler ecosystem than those normally studied by ecologists. Problems of vertical stratification, horizontal 'patchiness' and diurnal variations in phytoplankton distribution are minimized in culture ponds. Sampling for populations and nutrient concentrations can be readily instituted; if the hydraulic characteristics of the ponds are known to be adequate, effluent sampling alone will suffice. This renders the study of pond-culture ecology much simpler than that of natural systems, although few detailed studies have yet been undertaken. Adaptations of micro-algae to survival in the natural environment are going to determine their relative competitiveness in mass cultures, and the operational variables (mixing, dilution rate, nutrient concentrations, depth and pH value), which can and must be adjusted to optimize the requirements of high productivity and species control. Although the problems can be reasonably well defined, their resolution is not yet possible without development of new theoretical insight and extensive empirical investigation.

In the laboratory, micro-algae may be cultivated on a number of different inorganic (autotrophic) media; in general, each algal species has a wide range of tolerated nutrient compositions, as well as physical or chemical conditions (such as temperature, light intensity and salinity; Nichols, 1973). Many synthetic nutrient solutions have been formulated, usually with nutrient concentrations far in excess of what most algae experience in nature: a situation also true in algae cultivation ponds. The nutrients required are known algal and plant constituents, including phosphorus, nitrogen, sulphur, iron, magnesium and manganese, as well as trace metals and ions. Vitamins and some cofactors are required by some green algae, particularly flagellates, but seldom by blue-green algae. The limitations of the laboratory environment require care in extrapolating to outdoor cultures. Nevertheless, laboratory cultures have provided practically all biochemical knowledge, most physiological insight, and considerable theoretical understanding of many ecological phenomena of interest in algal culture.

The growth kinetics of algae are of crucial importance for applications involving large-scale, mixed, continuous or batch cultures. Any theory which would predict algal productivity or algal species competition is complicated by the fact that large-scale ponds have a plug-flow component (e.g. they are not perfectly mixed) and are exposed to transients of temperature and light intensity (which should be the limiting nutrient). This non steady-state component in algae mass culture, a combination of batch and continuous cultures under constantly changing conditions, presents a severe difficulty in developing an appropriate mathematical theory for outdoor algal mass cultures. Such a theory would need to describe and predict productivity as well as species competition. Neither the computer programme describing growth and yield of *Chlorella* spp. in chemostats as a function of photosynthetic intracellular parameters (Dabes *et al.*, 1970), nor the stoicheiometric approach to primary productivity in lakes (Platt *et al.*, 1977), is directly applicable to the requirements of algal mass culture in outdoor systems.

As an approximation, which is good enough for most present purposes, growth can be represented by continuous-culture theory as formulated by Monod (1950) and developed by Herbert *et al.* (1956). The relative growth constant μ (which represents the instantaneous growth rate,

$$\mu = \frac{1}{N}\frac{dN}{dt},$$

where t indicates time, and N cell concentration) is related to the doubling time G of the algal cells (which, in the absence of recycling, is the same as the hydraulic detention time of the culture) by the equation $\mu = 0.69/G$. With everything being constant, larger algae would be expected to grow slower due to a smaller surface: volume ratio and this is, indeed, often found (Fogg, 1975). Indeed, in recent studies with sewage oxidation ponds, a correlation between detention time of the culture and size of the algae was established (Benemann *et al.*, 1977a).

In a chemostat, one nutrient often becomes the limiting factor for cell growth in determining cell concentration (X) and thereby productivity $(p = \mu X)$. The relationship between substrate concentration and cell growth is normally expressed by Monod kinetics. However, intracellular nutrient concentrations are more immediately responsible for observed growth rates than extracellular nutrients (Droop, 1974). Since light is the key nutrient of photosynthetic microbial S.C.P. production, it is the one of interest. Sunlight, being a combination of different wavelengths absorbed by pigments of different absorption bands, must be considered a multiple nutrient. Indeed, it is possible to find two algae (for example, a green and a blue-green) co-existing in a light-limited chemostat because each alga uses a portion of the spectrum preferentially (Weissman and Benemann, 1978). Co-existence of two species on a single limiting nutrient is theoretically excluded in chemostat theory (Stewart and Levin, 1973; for a qualification, see p. 196).

In nature and in algal ponds, many species can co-exist because of multiple limiting nutrients (Titman, 1976), e.g. different nutrients limit different species, due to heterogeneity of the environment and environmentally caused transient oscillations. Less species diversity, often uni-algal populations, are observed in rapidly diluted cultures. The optimal dilution rate is determined by species-control requirements and maximum productivity, which in turn are determined by the algal concentration response to dilution rate (growth rate in a stable continuous culture). Algal concentration is limited by the effects of increased self-shading (resulting in increased pigment production and even more self-shading) and by cell respiration and maintenance. Productivity, which in light-limited pond cultures is directly related to solar light conversion efficiency, is optimal at moderate growth rates (often half maximal; Oswald, 1970). The light-absorption characteristics of an algal culture follow the Beer-Lambert Law allowing derivation of a relationship between light intensity and algal density, productivity and percent of light conversion efficiency (Shelef et al., 1968). The photosynthetic machinery which allows algae to convert sunlight, carbon dioxide, and water into biomass is the same as in higher plants, and subject to the same limitations in theoretical light conversion efficiency. Therefore, micro-algae have the same upper achievable productivity rate, namely about 6% of total solar energy into biomass heat of combustion under light-limited conditions. In practice, only about half of this can be achieved as a sustained year-round yield.

Besides metabolic adjustments (e.g. changes in pigment concentration) sinking, buoyancy and motility are used by micro-algae to optimize light utilization and nutrient uptake. In ponds, mechanical mixing and shallow depth minimize the need for active relative motion of micro-algae, perhaps accounting for the predominance of some micro-algal species in out-door cultures which are not often found to dominate in natural systems.

In unmixed or undermixed ponds exhibiting thermal stratification on a diurnal basis, flagellated algae appear to have an advantage since they can migrate easily through the water column and avoid temperature extremes. Loss of algal biomass due to sinking to the pond bottom can be an important factor in the yields obtained. Control of temperature fluctuations in shallow culture ponds is not feasible, and heating of ponds may be practised in only a few special situations, such as the utilization of power-plant waste heat. Thus, temperature, like light intensity, must be considered an unavoidable variable which cannot be controlled. Pond operations must be continuously adjusted to maintain desirable populations in face of these changing and uncontrollable environmental conditions (see Section III, p. 187).

Effects on algal growth rates, which determine both yields and population dynamics of different concentrations of nutrients, salts, micronutrients (particularly trace metals and organic growth factors), have been extensively documented (Healy, 1973). Interspecies effects, through extracellular products excreted by algae, have been suspected for a long time, but remain to be demonstrated in algal ponds. In mass cultures consisting of high densities of algae growing well below maximal growth rates, extracellular products could become a critical factor and must be studied. The high nutrient and algal concentrations make outdoor mass cultures a different ecological situation for most natural ecosystems.

Zooplankton grazing is another ecological phenomenon which is of crucial importance to algal mass culture. Zooplankton grazing is undesirable since it often results in rapid and drastic decreases in yields. The relatively short hydraulic detention times of algal cultures, shorter than the maximum growth rate of zooplankton, is not always effective in removing them because of the large number of offspring involved and the existence of relatively unmixed regions in large-scale systems. The grazing organisms may be controlled by various chemicals; however, this alternative may be unappealing economically and for product quality reasons. Another alternative is the use of coarse mesh screens (with pores greater than 100 μm in diameter) for rapid (intermittent or continuous) removal of zooplankton (Benemann et al., 1978a). Control of zooplankton infestations and algal pathogens is of crucial importance in the development of algae cultivation technology.

III. CULTIVATION TECHNOLOGY OF ALGAE IN LARGE-SCALE PONDS

Controlled cultivation of micro-algal species must be accomplished within limits imposed by engineering feasibility and economic reality. This prevents the use of sterile growth units and media. As already discussed, species-specific cultivation technology will need to be developed which allows continuous maintenance of particular inoculated strains for prolonged periods of time. The inocula themselves can be built up under successively less rigorously controlled conditions. The inoculation level and degree of control over its production will be parameters determining the economics of such systems. The minimum engineering and operational characteristics of large-scale pond systems designed for low-cost, high productivity algal cultivation are reasonably well known (Oswald and Golueke, 1960). The basic design is called a 'high-rate pond', a large, shallow compacted dirt pond bordered by a low levee (about 1–1.5 m high), divided into a long, continuous 10 to 30 m wide channel by means of baffles. The operational pond depth is 20 to 50 cm, depending on the engineering requirements of levelling and mixing, and the operational optimization of temperature fluctuations, algal concentrations and harvesting costs. Mixing is provided by one or more mixing stations using very low head-high capacity pumps or, preferably, paddlewheels. In general, constant low-mixing speed of 10–30 cm/sec are used; however, a variable mixing schedule might allow minimizing power requirements while preventing algal settling. Power requirements for mixing are relatively minor as long as mixing speeds do not exceed about 30 cm/sec.

The costs of high-rate pond construction can, in principle, be very low; it is estimated at U.S. $10 000–20 000 per hectare for the basic earthworks, baffles, paddlewheels and influent and effluent structures (Benemann et al., 1977a, 1978b). Except for a concrete apron next to the mixing stations, the ponds can be unlined, with sealing provided by a clay layer in high-porosity soils. Spray sealing of ponds with a thin impermeable asphalt or plastic layer might be economically feasible. The costs of ponds are considerably higher than preparation of agricultural crop lands; higher and more sustained yields as well as decreased operational costs must justify the capital investment. Nutrient supply, including carbon dioxide injection, would not be a significant expense; but the cost of the nutrients can be of critical importance. Water utilization, on an area basis, would be higher than that of conventional irrigated agriculture—however it would be lower on a protein basis. Assuming a rate of inorganic nutrient recovery similar to that in agriculture, inorganic fertilizers would be utilized for micro-algal S.C.P. production at an equivalent economic cost. Since micro-algae are effective in decreasing nutrient concentrations in natural, eutrophic, and even highly fertilized bodies of water, micro-algae might utilize nutrients more effectively than higher plants. The minor nutrients and micro-elements should not provide any special difficulties; they may even be provided from sea salts.

Provision of carbon dioxide to ponds is a critical economic parameter which must be minimized by siting the system near an available source of waste (free) carbon dioxide, such as a power plant or a naturally-enriched carbon dioxide source. In this respect, algae production differs from conventional plant cultivation, in which carbon dioxide is provided from the air. The diffusivity of carbon dioxide across the air–water interface would severely limit algal productivity, and would require both an enriched carbon dioxide source and a mechanical process for its introduction. Although pond carbonation is not difficult, it still requires engineering development.

Only a limited number of interrelated operational variables can be adjusted during pond operations. These include hydraulic dilution and loading rate, mixing velocity and schedule, inorganic nutrient concentrations, depth and pH value. Table 2 shows normal limits of various pond parameters and the economically feasible control methods (Benemann *et al.*, 1977a). Many of these are, of course, interacting. It is possible to vary detention times of various types and sizes of organisms independently, allowing some control over algae and zooplankton populations. Insolation and temperature cannot be controlled, and must consequently be compensated for by changing pond operations. This technology is not yet sufficiently advanced. However, in recent experiments, small-scale ($10\ m^2$) high-rate oxidation ponds were operated under various regimens of detention times, mixing, and selective biomass-recycle to determine the conditions under which large, filterable, colonial or filamentous algae could be cultivated (Benemann *et al.*, 1977a, 1978a). Pond detention times were shown to be an important factor in determining the morphology and size of the pond algae, and thereby their harvestability by microstrainers. In Fig. 1 are results obtained with two ponds operated at different detention times.

Table 2

Economically feasible pond operations in production of algal biomass

Parameter	Control method(s)	Normal limits
Algal concentration	Harvesting, dilution, recycle, inoculation	150–700 mg/litre
Depth	Dilution, harvesting	20–50 cm
Hydraulic detention time	Dilution	1.5–6.0 days
Phytoplankton detention time	Biomass recycle, dilution	1.5–6.0 days
Zooplankton detention time	Harvesting-recycling with DSM screen	0.5–6.0 days
Hydraulic loading	Dilution (applicable to oxidation ponds)	2–20 cm/day
pH Value	Carbon dioxide, dilution	6.0–10.5
Nutrient additions	Add with dilution water or independently	Should not be limiting
Oxygen tension	Mixing, carbonation	0–25 mg/litre
Light absorption	First four parameters, mixing	Absorbtion of 99–99.9% of incident light

The long-detention time pond maintained its harvestability, while the short-detention time pond could not be harvested by microstraining after a few days. The longer detention time ponds produced less algal biomass; an optimization of harvestability and yield has yet to be achieved. In recent experiments, the parameters which determine algal settling (with-

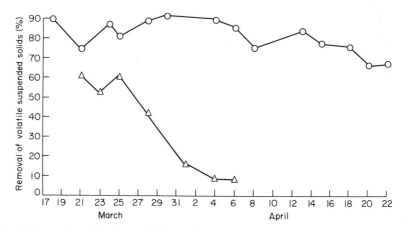

Fig. 1. Comparison of the harvestability of a long detention-time pond (circles: detention time was eight days, and 33% of the harvested algae were recycled) with a short detention-time pond (triangles: detention time was four days, and none of the harvested algae was recycled). The data show that the long detention time-recycle pond was much more harvestable than the short detention-time pond.

out flocculating chemicals) have been investigated (Benemann *et al.*, 1978b). In high-rate sewage oxidation ponds, a flocculated, rapidly settling algal culture could be maintained through a combination of correct detention times and mixing speeds. One key parameter affecting flocculation of the algae is their nutritional status. The low cost of such 'auto-flocculation' as a harvesting method (Table 1, p. 180) warrants its further development.

IV. CULTIVATION OF GREEN ALGAE

As noted previously, the first large-scale outdoor cultivation of micro-algae was carried out by Arthur D. Little Co. (Little, 1955). They used a plastic tube about 1.3 m wide and filled to about 10 cm depth with culture (leaving another 5 cm of air space). The culture was installed on a rooftop as a 150 foot-long continuous tube with two 180° turns. A mineral medium was used with carbon dioxide-enriched air (5%, v/v) and the algal crop was harvested by centrifugation. There were numerous operational problems, which included puncturing of the plastic, overheating of the culture, infestation by rotifers, suspected nutritional deficiencies, and a poor climate during most of the experiments. Yields were generally poor, although for a few days about 10 g/m²/day were produced. Under favourable conditions, 20 g/m²/day were thought possible. Costs were estimated at 25¢ U.S./lb in 1953, but involved a rather large number of optimistic assumptions (Fisher, 1955).

Work on *Chlorella* spp. was successfully continued in Japan, where the high commercial value of dried Chlorella (about $5/kg even 20 years ago; Krauss, 1962) encouraged a small industry which later spread to Taiwan. A highly mechanized system for cultivation was developed, which consisted of large circular ponds with centrally-pivoted mixing arms. Difficulties in maintaining a culture of *Chlorella* sp. and obtaining satisfactory yields resulted in the adoption of greenhouses for covering the ponds, and cultivation of the *Chlorella* sp. on acetate as a carbon source in controlled dark cultures. The algae were separated by centrifugation and then dried. The economics of the process are marginal, even for the small, high-priced, specialty food (or feed) market, and it is obviously not competitive in S.C.P. production. At present, about 1500 tonnes per year are produced in about a dozen installations in Japan and Taiwan (Tsukada *et al.*, 1977).

Research carried out for over 20 years in West Germany has concentrated on cultivation of *Scenedesmus* sp. (Soeder and Pabst, 1970; Stengel, 1970). This work was a great advance over the Japanese cultivation technology in using the basic high-rate pond design, and in generally being able to maintain the *Scenedesmus* sp. culture. The project, however, has not yet resulted in development of a practical industrial process, either in West Germany or in the underdeveloped countries where it was introduced (Venkateraman *et al.*, 1977). A key limiting factor is certainly the high costs of harvesting by centrifugation. The desirability of replacing the *Scenedesmus* strain utilized by these researchers with a larger species capable of continuous filtration was recognized (Soeder *et al.*, 1967).

Utilization of algal ponds for treatment of wastewaters is a wide-spread and inexpensive method of municipal sewage disposal. In such ponds (termed oxidation ponds), raw or settled sewage or industrial wastewaters are introduced into a series of normally relatively deep (1 to 3 m) ponds, where aerobic and anaerobic bacterial action decomposes organic material. Micro-algae, most commonly small single-cell or colonial green, or flagellated types, flourish in the ponds and maintain a high tension of dissolved oxygen in the upper layers of the ponds, growing at the expense of nutrients released by the bacteria. Engineered oxidation ponds are an elaboration of the ancient city moat and the popular fish ponds of the Far East. Algal productivity is normally low because algae settle out and, anyway, they are not harvested from the ponds. Most oxidation ponds are actually designed to minimize algal concentrations in effluents as discharge is considered undesirable. It was recognized early that, by adopting a high-rate pond design (Oswald, 1960), oxidation ponds could be much more efficient and effective in waste treatment, while at the same time producing large quantities of algal biomass. Several projects have been carried out to investigate the feasibility of producing waste-grown algae (Dugan *et al.*, 1970; North American Aviation Co., 1967) with a major ongoing project in Israel developing an integrated wastewater treatment, water reclamation S.C.P. production system (Shelef *et al.*, 1976).

Wastewaters are large and inexpensive, and dependable supplies of nutrients and water are required for micro-algal production. The economics of the process could be very favourable since a considerable subsidy is available from the waste treatment function of the system. However, for S.C.P. production, it must be free from pathogens, pesticides, other organic chemicals, and, most importantly, heavy metals. Although, according to Hintz *et al.* (1966), there is no evidence of toxicity in animal-feeding experiments, until the issue of quality is satisfactorily and completely answered, algae grown on municipal sewage should not be extensively utilized for feed or S.C.P. production. More immediate application would be generation of methane which, together with primary sludge, would partially satisfy local demands (Benemann *et al.*, 1977a). Liquid food-processing or husbandry wastes could be much more readily utilized for S.C.P. since their composition is predictable and greater control of algal species and product quality is possible. In the future, waste-grown micro-algae could well be the cheapest protein source; however, considerable work on establishing the quality of any such product is required.

V. CULTIVATION OF FILAMENTOUS BLUE-GREEN ALGAE

Blue-green algae are well known for their toxicity, having caused many cattle deaths (Gorham, 1964). Nevertheless, some blue-green algae have been used as traditional food sources in many parts of the world. Species of *Nostoc*, heterocystous nitrogen-fixing organisms which occur as small balls growing in river beds, have been reported to have been eaten by people both in South America and in Southeast Asia (Ortega, 1972). Most remarkable is the story of *Spirulina* which at present ranks as the most promising micro-alga for S.C.P. production.

In the early 1960s, a French anthropological expedition to Chad sent back samples of a local food 'DIHE' which was the dried green scum collected by fine nets from the shores of a local lake (Leonard, 1966). The alga involved was identified as *Spirulina platensis*, a filamentous helical blue-green organism. This shape makes the alga particularly easy to harvest with fine mesh screens. The alkaline waters of Lake Chad, and other African lakes, appear to be specifically suitable for growth of *Spirulia* spp. since often they are the predominating organism. Historical research reveals that the Aztecs ate *Spirulina* spp. before Cortez, and possibly used it as a food staple (Farrar, 1966). The Institut Français du Pétrole carried out an active research programme into cultivation of this species of alga, resulting in a production process utilizing bicarbonate-rich media and carbon dioxide enrichment using gas syphons (Durand-Chastel and Clement, 1975). In Mexico, the Sosa Texacoco Co. operates a very large solar evaporator which concentrates alkaline subsurface waters for bicarbonate production. *Spirulina geitleri* (or *maxima*) is found in areas around the solar evaporators where bicarbonate concentrations are suitable.

The Sosa Texacoco Co. (Santillan and Durand-Chastel, 1977) developed a one tonne per day cultivation harvesting and processing pilot plant which has been fully operational for over two years and which may double in capacity in 1980. The cultivation unit consists of a single, ten hectare pond with a longitudinal baffle. Mixing is provided by only one small pump, and the depth is two to three feet. This pond can just, but only just, be considered a high-rate pond. Some operational problems exist, particularly in the winter when production is apparently interrupted for about two months. The pond is fertilized with nitrate and iron. Apparently carbon dioxide or other nutrients are not added, suggesting some exchange with the surrounding solar evaporator waters.

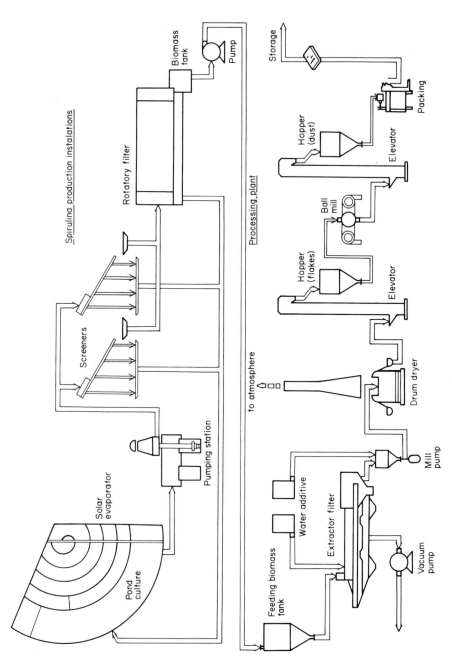

Fig. 2. Flow diagram of the Sosa Texacoco pilot plant for producing *Spirulina* sp.

Algae are harvested by a two-stage filtration process, first with an inclined screen and then a microstrainer. The harvested biomass is further concentrated by vacuum filtration, and then dried by drum drying or spray drying (Fig. 2). The operation is a private venture, and detailed process information or costs are not available. Yields are given as 10 g/m²/day averaged over the year, with 18–20 g/m²/day being peak productivities (Durand-Chastel and Silve, 1977). The product is sold, mainly to Japan, at about U.S. $5/kg, which is considerably less than *Chlorella* spp. sells for in that market. Recently, however, the Japanese market has not absorbed foreign imports and new uses for this product must be developed.

The alga *Spirulina* sp. appears to be well adapted for rapid growth at high pH values, thus allowing maintenance of a relatively pure culture in alkaline media. Appearance of contaminating algae is, however, still a problem. In the laboratory, it grows well on a low-salt and low-alkalinity medium; but, under these conditions, it loses its particular advantage and cannot be maintained in open culture. It can be grown on sewage with addition of bicarbonate (Kosaric *et al.*, 1974) or salt (Benemann *et al.*, 1977a). Outdoor cultivation of *Spirulina* spp. away from alkaline lakes requires re-utilization of the bicarbonate-rich medium (otherwise it would be prohibitively expensive) and re-supply of used nutrients. This can be done continuously for several months. The minimum number of times the media will have to be recycled depends on the cost and concentration of the bicarbonate. Probably the bicarbonate concentration being used (8–16 g/litre) can be decreased drastically without loss of species control by maintaining a high pH value in the cultures with a simple pH-stat controller, which regulates the rate of carbon dioxide addition. *Spirulina* sp. grows well on seawater at various dilutions (Benemann *et al.*, 1977a) and its cultivation on brackish, saline or irrigation drainage waters, which are unsuitable for conventional agriculture, is a distinct possibility. However, the temperature requirements for cultivation of *spirulina* spp.—it needs a day-time temperature of at least 20°C—restrict its production to tropical and semitropical climates.

One technique that can be used to help maintain cultures of *Spirulina* sp. that are being invaded by flagellated or single-cell green algae is the method of selective biomass recycle (Weissman and Benemann, 1977). In brief, a portion of the harvested biomass (which preferentially concentrates the *Spirulina* sp.) is returned to the growth pond. This artificially increases the detention time (which is the reciprocal of the growth rate) of the alga which thus obtains a competitive advantage. In Fig. 3 is shown the application of this principle in a chemostat culture in which species of *Chlorella* and *Spirulina* are competing. Without biomass recycle, the *Chlorella* sp. displaces the *Spirulina* sp. and, with a 67% recycle, the *Spirulina* sp. displaces the *Chlorella* sp. Selective biomass recycle is only one limited technique for species control; but, together

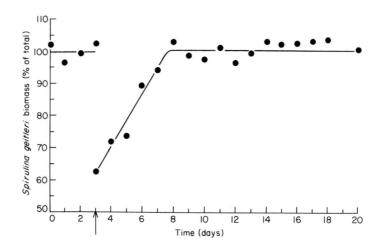

Fig. 3. Competition between *Spirulina* sp. and *Chlorella* sp. in a chemostat. A 3-litre chemostat was operated at a dilution rate of 0.0344 h^{-1}, with a modified blue-green algal medium and light as the limiting factor. Under these conditions, *Chlorella* sp. will outcompete *Spirulina geitleri*. However, if the filamentous *Spirulina geitleri* is harvested and partially recycled, it outcompetes the unicellular *Chlorella* sp. In the experiment, at day 3 (indicated by arrow), about 40% of *Chlorella* sp. was mixed in the culture. At day 10, less than 1% *Chlorella* sp. was present in the culture. From Weissman and Benemann (1978).

with control of pH value and medium optimization, it should be possible to maintain the culture and obtain good yields (10 g/m²/day) using inexpensive pond operations. One technique for maintaining the culture is the maintenance of a high concentration of *Spirulina* sp. and operation on a semi-batch basis. The present high value of *Spirulina* sp. and the extremely large market for it as a feed additive (at prices above conventional protein sources) assure a bright commercial future once technology is matured by experience.

VI. NUTRITIONAL ASPECTS OF
MICRO-ALGAL SINGLE-CELL PROTEIN

Perhaps more effort has been expended on nutritional studies of micro-algae than on development of production processes. However, out of thousands of known species of algae, only a few are traditional foods or have been subjected to scientific nutrition studies. Most of the algal types and species found in nature are not of interest in S.C.P. production due to indigestibility or toxicity, presence of silicone or carbonate shells, and lack of economic cultivation technology. Species of *Chlorella* and *Scenedesmus* have been traditionally the most commonly investigated micro-algae, since they are the most frequent to appear and persist under cultivation conditions, normally utilized with freshwater, using either sewage or mineral nutrients. Species of *Spirulina* have received more attention recently. Mixed algal populations from sewage-oxidation ponds have also been extensively studied. A recent review has detailed most algal feeding trials (Waslien, 1975).

Micro-algae must be evaluated both quantitatively (percent protein by dry weight) and qualitatively (nutritional acceptability). The high content of pigments in micro-algae is of great interest at present, and has value in animal-feed production, but is not a requirement in S.C.P. production and is undesirable for human consumption. Thus, a detailed discussion of micro-algal pigments is not given here. The composition of micro-algae with respect to their content of protein, carbohydrates and lipids can be very variable. Contents of protein decrease and those of lipids increase under conditions of nitrogen limitation. Other nutritional limitations also have strong effects on the composition of micro-algae. Protein content may vary from 8 to 50%, lipid from <1 to 86%, carbohydrate from 4 to 40%, and ash from 4 to 45%. Sewage-grown algae or seawater-grown algae have generally higher ash and lower protein contents (Waslien, 1975). In general, desirable strains for S.C.P. production may be defined as those that contain 50% crude protein, determined as the total nitrogen content multiplied by 6.25. This exaggerates actual protein content because of non-protein nitrogen; actual protein contents of about 35% are probably the highest that may be expected. Table 3 gives the composition of the three micro-algae most commonly studied and a mixed sewage-grown algae culture. Amino-acid compositions of micro-algal S.C.P. show a generally good distribution but, as with many plant proteins, there is a general deficiency in sulphur-

Table 3

Chemical composition of selected micro-algae

Component	Micro-algae			
(% dry weight)	Chlorella sp. 7–11–05[1]	Scenedesmus sp.[2]	Chlorella-Scenedesmus[3] (10:1, sewage-grown)	Spirulina maxima
Protein[a]	55.5	53.0	41.8	55.5
Fat	7.5	13.0	7.2	12.7
Carbohydrate	17.8	13.5	27.4	17.4
Ash	8.25	6.5	19.1	7.4
Moisture	7.0	6.0	4.5	7.0
Crude fibre	3.1	8.0	—	—
Ascorbic acid	1.46	—	3.96	1.03
β-Carotene	5.02	—	6.02	2.25
Pantothenic acid	1.12	1.2	0.46	—
Pyridoxin	0.3	—	0.11	0.043
Thiamin	0.77	0.17	0.115	0.138
Riboflavin	—	0.42	0.269	0.285

[a] Protein contents were calculated by multiplying the value for the total nitrogen content by 6.25. True protein contents are about two-thirds of the values quoted.
[1] Lubitz (1963).
[2] Dam et al. (1965).
[3] Cook (1962).
[4] Clement et al. (1967).

containing amino acids, in particular methionine. The lack of a large content of storage proteins (as in corn or wheat) makes changing the amino-acid composition of micro-algae a difficult genetic task, although some improvements should be achievable with modifications in culture media and strain selection. The vitamin content of micro-algae (Table 3) is generally high.

Of primary importance in the utilization of micro-algal S.C.P. is the digestibility of the cell wall. Cell walls of green algae are difficult to break down. Those of blue-green algae contain unusual fatty acids which are not digestible. Micro-algae also contain various undigestible storage compounds, including poly β-hydroxybutyrate. The presence of possibly toxic compounds is neglected here, as such algae would not be cultivated, or possible toxins would be removed during processing. In general, the walls of both green and blue-green algae are not readily digested in the mammalian digestive system, with ruminants generally being better capable in this context. The processing technology for micro-algae that results in cell wall breakdown has not received sufficient attention. Spray drying will rupture the cell walls of blue-green, but not green, algae. Other methods of cell rupture are possible including mechanical disintegration (Hedensborg et al., 1970). Direct feeding to swine of settled concentrated pond-grown algae is reported as a common practice in the Far East, and is apparently successful.

Animal-feeding studies have been carried out with rats, chickens (Grau and Klein, 1957; Blum and Calet, 1975), swine and cattle (Hintz et al., 1966), fish (Tan and Hwang, 1966) and sheep. In most cases, when algae constituted a large (above 60%) fraction of the diet, growth disturbances and histopathological changes, particularly of the liver, were observed. Where algae represented only a smaller fraction of the diet, or where purified protein obtained from an ethanol extract was used, no overt pathological changes were reported (Waslien, 1975). Remarkably, no adverse effects related to sewage contamination were noted in any of several tests of sewage-grown micro-algae when used as additives to less than 20% of the diet in animals. The nutritional quality of micro-algal proteins, as estimated from the biological value of absorbed protein and the protein efficiency ratio, is about as good as that of soybean and wheat protein, but is not comparable to milk or egg protein (Waslien, 1975). The nutritional value of micro-algae is dependent not only on algal species and strain but also on the processing (e.g. sun drying as compared with spray drying). In general, micro-algae can be considered a suitable animal-feed additive if sufficient quality control in cultivation and processing is achieved.

Results of micro-algae testing trials in human nutrition followed the general pattern observed in animals. High ingestion rates of algae decreased protein digestibility except where purified solvent-extracted proteins were used. However, when algae were used as a supplemental nitrogen source, as when algae are added to a rice-protein diet, its value (measured as the nitrogen balance) was as high as that of animal protein added to the same diet (Lee *et al.*, 1967). This is the critical test of a S.C.P. source, namely whether it can replace part of a high-quality (and cost) protein in a diet or improve a poor-quality protein diet. Use of micro-algae as a human food source was also extensively investigated as part of the United States and Russian space projects (Calloway and Waslien, 1971).

Use of algae, whether for astronauts or inhabitants of Indian villages, is limited by acceptance and palatability. Sewage-grown micro-algae S.C.P. will probably find difficult acceptance directly for human nutrition, even if safety could be guaranteed. In general, micro-algae have a strong taste and even stronger colour. These undesirable properties can be minimized by restricting the algal content to a small fraction of the diet. The taste of *Spirulina* sp., for example, is not unpleasant in small amounts. Pressed into pills, micro-algae have found a large health-food market in the Far East and, to some extent, in the U.S.A. In the future, should micro-algal production, processing technology and economics progress beyond the level of a speciality market, algal production as a feed supplement and human-food additive should be feasible.

VII. ECONOMICS OF MICRO-ALGAL SINGLE-CELL PROTEIN

The present costs of cultivating, harvesting, and preparing micro-algae for S.C.P. and related uses is too high to justify commercial production. The commercial production systems used in Japan and Taiwan have a minimum production cost of about U.S. $5/kg, the market price of *Chlorella* sp. being about twice that. The Spirulina product of Sosa Texacoco sells for approximately U.S. $5/kg, which is probably well above production costs, discounting process development. Labour and drying costs are the largest factors in the Spirulina manufacture; however, exact cost figures are not available. Costs for sewage-grown micro-algae are competitive with alternative protein sources if low-cost algal harvesting processes are used or if algal harvesting costs are charged to the waste-treatment function of the ponds. If chemicals are used to separate micro-algae, these must be removed prior to utilization, although alum is not harmful.

The economics of micro-algae S.C.P. production is dictated by capital and operational costs as well as the yield of algae. Yield data are at present available in insufficient amounts for justifying large-scale S.C.P. production systems. Yields of over 20 g/m²/day, which extrapolates to 73 tonnes/hectare/year, have been reported for small-scale systems operated at high dilution rates for short periods of time (Soeder, 1977) and productivities exceeding 100 tonnes/hectare/year have been reported from high-rate sewage oxidation ponds (Shelef *et al.*, 1976; McGarry *et al.*, 1974. Solar conversion efficiencies of 2% of total solar energy into algal biomass can often be calculated. However, production of *Spirulina* sp. in Mexico averages 10 g/m²/day, with peak values of 18 to 20 g/m²/day during the most favourable season. As the Sosa Texacoco system does not appear to be optimized for high-yield production, because the pond is too deep and undermixed and nutrient control is not rigorous, it should be possible to sustain production rates of 50 tonnes/hectare/year for algal cultivation on synthetic media. This, of course, presupposes no significant interference by pests, algal contamination and herbivorous zooplankton. If 50 tonnes/hectare/year can be achieved, with a 35% protein content, this would be the most efficient plant-protein producing system at present known.

This output should allow considerable investment and operational expenses for micro-algal production and processing. These costs can be divided into capital costs for the pond systems, costs for the nutrients required to cultivate the algae (including water), mixing and pumping power, harvesting and processing costs, and labour and management expenses. At present all of these costs are, to a considerable extent, speculative. Processing costs for dewatering and drying of the algae are reasonably well established, as they are similar to those of other S.C.P. sources. It may be roughly calculated that the costs for spray drying are about U.S. 25¢/kg starting from a paste with 5% solids content. These costs could still be decreased by alternative drying or protein-extraction methods. Harvesting costs are given in Table 1 (p. 180); a suitable primary harvesting and secondary concentration process must be operated in series. For purposes of S.C.P. production, the expensive physical or chemical concentration (Table 1) methods are not suitable; screening methods (as used in harvesting of *Spirulina* sp.) and settling or flotation are most applicable. These would cost about U.S. 4–10¢/kg of algae. Pumping and mixing power inputs can be considered minimal in operation of such systems. Micro-algae should utilize nutrients about equally well as higher plants, and costs of fertilizer on a unit protein basis should thus be similar to those in conventional agriculture. Nutrients could be supplied by animal wastes, lowering costs and allowing a waste-treatment credit.

The capital cost of pond systems can be quite low if only a shallow earthen pond with wide channels and a few mixing stations are required. Costs of such ponds are probably about U.S. $10 000/hectare, excluding land costs. The requirement for pond lining, to prevent erosion at high mixing speeds, will restrict the speed allowed. Various additional charges, including nutrient and water-supply lines, access and influent–effluent boxes, could increase costs to U.S. $20 000/hectare, particularly if accurate flow control and pond operations are requirements. Labour costs for the basic operation and maintenance should be quite low; the hydraulic nature of the system allows large-scale operations. It has been calculated that, in principle, assuming yields of about 50 tonnes/hectare/year, the cost of micro-algal production could be as low as U.S. 2–4¢/kg (before drying). However, these calculations do not allow any significant expenses for pond management: These include the necessary sampling and analysis of pond conditions, production of inoculum, elimination of weeds, diseases and herbivores, cost of water and special chemicals and land charges and profits. These must be considered the critical unknowns in production of algal S.C.P. At present, experience is too limited to allow prediction of when, or whether, algal S.C.P. can be produced commercially for human foods or animal feeds in competition with grains, soybeans or fishmeal. Commercial production of micro-algae, such as *Spirulina* sp., as a speciality feed additive should be feasible applying present experience, and a large market could be developed in the near future. Such a commercial development would rapidly establish the technical and economic limits of micro-algal production. Specific predictions of where these limits may be, or what production costs may be, are premature.

Twenty-five years ago, the technology of micro-algal production took its first tentative step from laboratory to pilot plant (Little, 1953). The second step to full-scale commercialization is yet to be taken. Early hopes and optimistic visions have not been realized as fast as many expected. Exaggerated claims about expected yields have led to accusations that a 'cruel illusion was proffered by laboratory scientists and writers who proposed that we could feed the world on algae' (Odum, 1965). Such accusations were based on extrapolation of laboratory or pilot-scale management inputs to large-scale systems, which lead to the conclusion that practical algal cultivation is not feasible (Odum, 1965). The arguments here presented indicate that, if the ecology of an algal production system is sufficiently understood, inputs and management could be minimized and low-cost algal production become a reality.

REFERENCES

Benemann, J. R., Koopman, B. L., Weissman, J. C., Eisenberg, D. M., Murry, M. and Oswald, W. J. (1977a). *Sanitary Engineering Research Laboratory Report No. 77–5.* University of California, Berkeley.

Benemann, J. R., Murry, M., Weissman, J. C., Hallenbeck, P. C. and Oswald, W. J. (1977b). *Sanitary Engineering Research Laboratory Report No. 78–3.* University of California, Berkeley.

Benemann, J. R., Weissmann, J. C., Eisenberg, D. M., Goebel, R. P. and Oswald, W. J. (1978a). *Sanitary Engineering Research Laboratory Report No. 78–79.* University of California, Berkeley.

Benemann, J. R., Pursoff, D. and Oswald, W. J. (1978b). *Report on Department of Energy Contract No. EX-78-X-C1-1605.* University of California, Berkeley.

Benemann, J. R., Weissman, J. C., Eisenberg, D. M. and Oswald, W. J. (1978c). *Sanitary Engineering Research Laboratory Report No. XXX.* University of California, Berkeley.

Bivort, A. (1977). *National Science Foundation Report No. XX.*

Blum, J. C. and Calet, C. (1975). *Annals du Nutricion e Alimentacion* **29,** 651.

Calloway, D. H. and Waslien, C. I. (1971). *Environmental Biology and Medicine* **1,** 299.

Clement, G., Giddey, C. and Menzi, R. (1967). *Journal of the Science of Food and Agriculture* **18,** 497.

Cook, B. B. (1962). *Journal of Nutrition* **52,** 243.

Dabes, J. N., Wilke, C. R. and Sauer, K. H. (1970). *Lawrence Radiation Laboratory Report No. 19958.* University of California, Berkeley.

Dam, R., Lee, S., Fry, P. C. and Fox, H. (1965). *Journal of Nutrition* **86,** 376.

Droop, M. R. (1974). *Journal of the Marine Biological Association of the United Kingdom* **54,** 825.

Dugan, G. L., Golueke, C. G., Oswald, W. J. and Rixford, C. E. (1970). *Sanitary Engineering Research Laboratory and School of Public Health. Second Progress Report No. 70–1.* University of California, Berkeley.

Durand-Chastel, H. and Silve, M. D. (1977). *Proceedings of the European Seminar on Biological Solar Energy Conversion,* p. 57. Grenoble-Antrans, France.

Durand-Chastel, H. and Clement, G. (1975). *Proceedings of the 9th International Congress on Nutrition, Mexico, 1972,* Vol. 2, p. 85. Kruger, Basel.

204 JOHN R. BENEMANN *ET AL.*

Farrar, W. V. (1966). *Nature, London* **211,** 341.
Fisher, A. W. (1955). *Proceedings of the World Symposium on Applied Solar Energy*, p. 24. Phoenix, Arizona.
Fogg, G. E. (1975). 'Algal Cultures and Phytoplankton Ecology'. University of Wisconsin Press.
Gorham, P. R. (1964). *In* 'Algae and Man', (C. D. F. Deschusmed, ed.), p. 307. Academic Press, New York.
Golueke, C. G. and Oswald, W. J. (1965). *Journal of the Water Pollution Control Federation* **37,** 471.
Grau, C. R. and Klein, N. W. (1957). *Poultry Science* **36,** 1046.
Harder, R. and Witsch, H. von (1942). *Forschungdienst* **16,** 270.
Healy, F. P. (1973). *Critical Reviews in Microbiology* **2,** 69.
Hedensborg, G., Mogren, H. and Enebo, L. (1970). *Biotechnology and Bioengineering* **12,** 947.
Herbert, D., Elsworth, R. and Telling, R. C. (1956). *Journal of General Microbiology* **14,** 601.
Hintz, H. F., Heitman, H., Weir, W. C., Torell, D. T. and Meyer, J. H. (1966). *Journal of Animal Science* **25,** 675.
Kosaric, N., Nguyen, H. T. and Bergougnou, M. A. (1974). *Biotechnology and Bioengineering* **26,** 881.
Krauss, R. (1962). *American Journal of Botany* **49,** 425.
Lee, S. K., Fox, H. M., Kies, C. and Dam, R. (1967). *Journal of Nutrition* **92,** 281.
Leonard, J. (1966). *Nature, London* **209,** 126.
Little, A. D., Inc. (1953). *In* 'Algal Culture from Laboratory to Pilot Plant', (J. S. Burlew, ed.), p. 600. Carnegie Institute of Washington Publication.
Lubitz, J. A. (1963). *Journal of Food Science* **28,** 229.
McGarry, M. G., Ng, K. S., Leving, N. H. and Lee, T. L. (1974). *Process Biochemistry* **(9),** 14.
Monod, J. (1950). *Annales de l'Institut Pasteur, Paris* **79,** 390.
Nichols, H. W. (1973). *In* 'Handbook of Phycological Methods', (J. R. Stein, ed.), pp. 79-84. Cambridge University Press.
North American Aviation Inc. (1967). *Final Report No. SID-76-461. PH 86-65-31.*
Odum, H. T. (1965). 'Environment, Power and Society'. Wiley Interscience, New York.
Ortega, M. M. (1972). *Botanica Marina* **15,** 162.
Oswald, W. J. (1976). *In* 'Ponds as a Wastewater Treatment Alternative', (E. Gloyna, ed.), p. 257. College of Engineering, University of Texas.
Oswald, W. J. (1970). *In* 'Prediction and Measurement of Photosynthetic Productivity', (C.D. Jones, ed.), p. 473. Centre for Agricultural Publishing and Documentation, Wageningen, Netherlands.
Oswald, W. J. and Golueke, C. G. (1968). *In* 'Single-Cell Protein', (R. I. Mateles and S. R. Tannenbaum, eds.). M.I.T. Press, Cambridge, Massachussetts, U.S.A.
Oswald, W. J. and Golueke, C. G. (1960). *Advances in Applied Microbiology* **2,** 223.
Oswald, W. J. (1960). *Journal of the Sanitary Engineering Division SA.4,* 71.
Platt, D., Dennan, K. L. and Jassby, A. D. (1977). *In* 'The Sea, Ideas and Observations on Progress in the Study of the Seas', (E. D. Goldberg, ed.), Vol. VI, pp. 807-837. John Wiley, New York.
Ryther, J. H. and Goldman, J. C. (1975). *Annual Review of Microbiology* **29,** 429.
Santillan, C. and Durand-Chastel, H. (1977). *Proceedings of the International Seaweed Conference*, Santa Barbara, California.
Shelef, G., Moraine, R., Meyden, A. and Sandbank, E. (1976). *In* 'Microbial Energy Conversion', (H. G. Schlegel and J. Berneen, eds.), p. 427. Erich Goltz., Gottingen.

Shelef, G., Goldman, J. C., Sobsey, M., Harrison, J., Gee, H., Halperin, R. and Oswald, W. J. (1970). *Sanitary Engineering Research Laboratory Report No. 68–4.* University of California, Berkeley.

Shelef, G., Oswald, W. J. and Golueke, C. G. (1968). *Sanitary Engineering Research Laboratory Report No. 68–3.* University of California, Berkeley.

Soeder, C. J. (1977). *In* 'Microbial Energy Conversion', (H. G. Schlegel and J. Barnea, eds.), p. 59. Erich Goltz, Gottingen.

Shilo, M. (1967). *Bacteriological Reviews* **31**, 180.

Soeder, C. J., Hegewald, E., Pabst, W., Payer, H. D., Rolle, I. and Stengel, E. (1967). *In* 'Jahrbuch Landesamt für Forschung', p. 295. Nordrhein-Westphalen, West Germany.

Soeder, C. J. and Pabst, W. (1970). *Berichte der Deutschen Botanischen Gesellschaft* **83**, 607.

Stengel, E. (1970). *Berichte der Deutschen Botanische Gesellschaft* **85**, 589.

Spoehr, H. A. and Milner, H. W. (1949). *Plant Physiology* **24**, 120.

Stewart, F. M. and Levin, B. R. (1973). *American Naturalist* **107**, 171.

Tamiya, H. (1957). *Annual Review of Plant Physiology* **8**, 390.

Tan, C. and Hwang, F. (1966). *Proceedings of the Food and Agriculture Organization Symposium on Warm-Water Pond Fish Culture*, Rome.

Titman, D. (1976). *Science, New York* **192**, 463.

Tsukada, O., Kawahra, T. and Miyachi, S. (1977). *In* 'Biological Solar Energy Conversion', (A. Mitsui, S. Miyachi, A. San Pietro and S. Timura, eds.), p. 363. Pergamon Press.

Venkateraman, L. V., Becker, W. E. and Shamala, T. R. (1977). *Life Sciences* **20**, 63.

Waslien, C. I. (1975). *Critical Reviews in Food Science and Nutrition* **6**, 77.

Waslien, C. I., Kok, B., Myers, J. and Oswald, W. J. (1977). *In* 'Protein Resources Study', (N. Scrimshaw and N. Milner, eds.), p. 522. M.I.T. Press, Cambridge, Massachusetts, U.S.A.

Weissman, J. C. and Benemann, J. R. (1978). *Biotechnology and Bioengineering* in press.

Note added in proof

Since completion of this review, two excellent articles reviewing the historical experience and theoretical basis of microalgal mass cultures have appeared (Goldman, 1979a, b). The author tabulated the available productivity data on microalgae mass cultures which revealed that relatively little data on sustained large-scale cultivation exist. Most reported data was from short-term (one–three months) small-scale (below 100 m²) experiments. For larger scale operations (e.g. *Spirulina* production in Mexico or *Chlorella* production in the Far East), only second-hand sources are available. However, both the available data and theoretical analysis indicate that a production rate of 500 tonnes/hectare/year is currently achievable. An international meeting on algal biomass entitled 'The Production and Use of Micro-Algae Biomass' was held September 17–22, 1978 in Acre, Israel, which brought together many of the experts working in this field around the world. The proceedings of this meeting (Shelef and Soedev, (1979) form an excellent record of the current state of the art in this area. The basic conclusion of this meeting was that microalgal production is a technology well past the pilot plant stage. However, production economies are currently limited by the high cost of harvesting by centrifugation or chemical flocculation and by drying of the algal biomass. The author's work on low-cost harvesting of microalgae by bioflocculation and settling presented at this meeting has recently been demonstrated to be successful at the pilot scale with 0.1 hectare high rate sewage oxidation ponds. Evidence presented by several authors at this meeting strongly support previous research indicating the high nutritional value and safety of microalgal single cell protein. Production of *Spirulina* has become of great world wide interest and several projects are attempting to develop low-cost production technologies. The key problem is to develop media which do not require high concentrations of bicarbonate and cultivation processes that maintain a pure culture and allow effective media recycling. This would probably require a periodic destruction of contamination in the recycled media, either by disinfection, pasturization, or specificlabile algicides. The high temperature requirement of *Spirulina* spp. restricts its cultivation to seasons or areas of high temperature. The drying problem can be overcome either by solar drying processes, wet feeding of the algae (e.g. to pigs) or fractionation of the biomass into various components–protein, pigments, chemicals–prior to drying. The production of *Dunaliella* spp. for the integrated production of these algal components is proceeding in Israel. Production of fin-fish or shellfish is another important application of microalgal cultures; it avoids both the problems of harvesting and drying and can produce, with high efficiency, a valuable animal protein. The problems of algal culture and species control become critical in these applications, and the foodchains are dependent on the continuous availability of consumable algal species. The development of species-specific microalgal cultivation technologies thus remains the biggest challenge in this field.

References

Goldman, J. C. (1979a). *Water Research*, **13,** 1.
Goldman, J. C. (1979b). *Water Research*, **13,** 119.
Shelef, G. and Soeder, C. J. (1979). *The Production and Use of Micro-Algae Biomass, Proceedings of the Akko Introductory Symposium.* Published by NCRD (Jerusalem) and GSF (Munchen).

8. Biomass from Whey

J. MEYRATH AND K. BAYER

Institute of Applied Microbiology, University of Agriculture, Vienna, Austria

H

I. INTRODUCTION

A. Composition of Whey

Whey is the residual liquid after removal of fat and casein from whole milk. The major component is lactose accompanied by smaller concentrations of non-coagulable protein in casein recovery, plus citric acid, minerals, vitamins and traces of fat; lactic acid in variable amounts is present depending upon the length of the period of milk storage before processing, particularly if lactic-acid bacteria were used to accomplish casein precipitation. A novel method of protein precipitation for some types of cheeses, developed in Australia (see Schulz, 1969), permits an easy and almost complete coagulation of milk proteins, including those normally remaining in the whey.

Gross compositions of whey derived from various types of cheese manufacture, which are typical for the low-acid or sweet types of whey, are shown in Table 1. Whey from milk for production of cottage-type cheese by precultivation of lactic-acid bacteria is naturally higher in lactic acid content. Besides, there are small changes in nitrogenous compounds in view of the (slight) proteolytic activity of these organisms (Table 2). According to Wasserman (1960a), whey nitrogen is composed of almost equal parts of non-coagulable and hot-acid precipitable organic nitrogen. This observation coincides with the data in Table 2. Furthermore, the hot-acid precipitable nitrogen fraction (lactalbumin and lactoglobulin) is not easily degradable by lactic-acid bacteria according to Orla-Jensen (see Wasserman, 1960a). Yeasts grown for biomass production behave similarly. As described (see p. 229) it is mainly the inorganic nitrogenous compounds and amino-acids that are assimilated.

The amounts of whey derived from curds of the various types of cheese vary as follows (kg of whey per kg of cheese): hard cheeses, 11.3; soft cheeses, 7.5–9.5; cottage-type cheeses, 4.0–9.0; casein, 30; (Drews, 1975).

Table 1

Gross composition of various types of whey. From Blanc (1969)

Composition (%)	Range of composition in whey from production of				
	Cheddar[1]	Gouda[2]	Emmental[3]	Gruyère[3]	Tilsit[3]
Dry matter	6.6–7.1	6.2–6.5	6.41–6.60	6.85–6.96	7.02–7.26
Total protein	0.82–0.95	0.8–0.9	0.65–0.73	0.67–0.73	0.67–0.79
Fat	0.12–0.36	0.04–0.08	0.35–0.56	0.41–0.59	0.39–0.55
Lactose	4.62–5.01	4.8–5.2	4.59–4.79	4.82–5.05	4.82–5.06
Ash	0.37–0.65	0.48–0.55	0.48–0.59	0.51–0.56	0.55–0.59
Calcium	0.047	0.03–0.05	0.033–0.037	0.034–0.038	0.040–0.045

[1] Berry (1923).
[2] Olling (1963).
[3] Blanc (1969).

The pH value of sweet whey varies between 4.5 and 6.7; that of sour whey between 3.9 and 4.5. It is evident (Tables 1 and 2) that lactose is the major component of whey. If dried, there will be in the order of 70% lactose and 9–14% crude protein, with a comparatively high ash content (about 9%). Because of the low protein content, whey powder is not a high-grade food or feeding material, despite its good composition in amino-acids and its content in vitamins. A detailed list of mineral

Table 2

Comparison of sweet and sour-whey composition. From Drews (1975) according to Waeser (1944)

Components		Composition in	
		sweet whey (%)	sour whey (%)
Dry matter		6–7	5–6
Ash		0.5–0.7	0.7–0.8
Crude protein of this:		0.8–1.0	0.8–1.0
genuine protein nitrogen	52.5		43.9
peptide nitrogen	31.3		33.1
amino-acid nitrogen	2.5		6.1
creatin nitrogen	2.6		2.5
ammonia nitrogen	1.0		2.3
urea nitrogen	9.1		10.3
purine nitrogen	1.0		1.8
Lactose		4.5–5.0	3.8–4.2
Lactic acid		traces	up to 0.8
Citric acid		0.1	0.1

Table 3

Detailed composition of whey from Emmental, Gruyère and Tilsit cheeses, in relation to the original milk. From Blanc (1969)

Composition	Emmental		Gruyère		Tilsit	
	Milk	Whey	Milk	Whey	Milk	Whey
Minerals (mg %)						
Ash	690 ± 0	540 ± 60	750 ± 40	530 ± 30	710 ± 20	570 ± 20
Calcium	110 ± 4	35 ± 2	120 ± 5	36 ± 2	119 ± 6	43 ± 3
Magnesium	15 ± 4	9 ± 1	13 ± 2	10 ± 2	18 ± 4	10 ± 2
Sodium	41 ± 3	45 ± 1	42 ± 2	44 ± 3	42 ± 2	46 ± 2
Potassium	138 ± 2	143 ± 11	143 ± 7	147 ± 12	143 ± 12	151 ± 7
Phosphorus	01 ± 3	45 ± 3	100 ± 2	45 ± 2	95 ± 4	47 ± 2
Chlorine	95 ± 2	99 ± 4	101 ± 4	106 ± 4	102 ± 6	112 ± 1
Water-soluble vitamins						
Thiamin	47	38	41	39	42	38
Riboflavin	186	137	233	186	262	157
Pyridoxin	34	44	60	43	44	39
Cobalamin	0.38	0.29	0.33	0.27	0.32	0.27
Calcium pantothenate	281	385	426	461	280	432
Biotin	1.7	1.8	1.1	1.7	1.4	1.6
Vitamin C	440	203	358	230	600	260
Fat-soluble vitamins (I.U./100g)						
Vitamin A	102	87	139	91	104	75

Values quoted are averages ± maximum and minimum.

Table 4

Amino-acid composition of whey protein

Amino acid (g/16 g nitrogen)	Milk albumin[1]	Whey protein[2]	β-Lactoglobulin[3]	α-Lactalbumin[3]
Lysine	10.5	7.85	11.8	11.5
Histidine	2.3	1.10	1.6	2.9
Threonine	8.0	5.93	5.0	5.5
Cystine/cysteine	4.0	1.21	3.4	6.4
Valine	6.6	5.77	6.1	4.7
Methionine	2.6	2.40	3.2	0.95
Isoleucine	7.5	—	6.9	6.8
Leucine	12.1	13.49	15.5	11.5
Tyrosine	5.3	2.53	3.7	5.4
Phenylalanine	5.0	2.97	3.5	4.5
Tryptophan	2.5	1.25	2.7	7.0
Proline	—	—	5.1	1.5
Glycine	—	—	1.4	3.2
Arginine	—	—	2.8	1.2
Aspartic acid	—	—	11.4	18.7
Serine	—	—	4.0	4.8
Glutamic acid	—	—	19.3	12.9
Alanine	—	—	7.0	2.1

[1] Amundson (1967).
[2] Surazynski et al. (1967a).
[3] Gordon and Whittier (1965).

components and vitamins is shown in Table 3, and the amino-acid composition of the proteins is shown in Table 4.

B. Problems of Usages of Whey

The volume of whey production in the countries handling this commodity is listed in Table 5. The sum of the figures in Table 5 coincides near enough with the value given by Vananuvat and Kinsella (1975) of 100 billion pounds world-wide production.

As indicated earlier, the low protein content precludes whey from being accepted as a high-grade food or feeding material. In addition, there exists widespread lactose malabsorption among adult humans, resulting in abdominal bloating, rumbling and diarrhoea due to β-galactosidase deficiency in the small intestine. This deficiency exists in 2–15% West-Europeans and Americans and 70–90% Asians, Africans, American Negros and Eskimos (Edelsten, 1974). There is also occasional incidence of failure in galactose transformation in the liver of infants, i.e. deficiency of either hexose 1-phosphate uridyl-transferase or galactoki-nase, leading to the so-called galactosaemia characterized by the occurrence of cataract of the eye lens as well as mental disorders (Lehinger, 1975). The popularity of whey as a drink or food supplement is not therefore favoured.

Of the 25 billion pounds (corresponding to 1.5 billion pounds of solids) of whey produced in the U.S.A., one third to one half is actually utilized. According to Vananuvat and Kinsella (1975), 7 billion pounds of native whey are utilized world-wide as human food (sweet whey only), of which 319 million pounds are used in the U.S.A. (Bernstein and Everson, 1973); in that country, about an equal amount is used for animal feeds. In some countries, like the U.S.A., about a quarter of the whey is processed into lactose (Amundson, 1967) and protein. Poor solubility and the laxative effects of lactose limit its applicability; the largest proportion of lactose is still probably used for penicillin production. Small quantities of whey are processed for soft drinks, whey blends, and protein fractions. The newer developments of ultrafiltration and reverse osmosis for recovery of whey proteins and lactose are not yet used extensively, but further market developments will certainly bring about a change.

It will be clear that the problems of redistributing whey to the farmers are increasing continuously. The tendency to process milk in units of

Table 5

Volume of whey production per annum in various countries. From Blanc (1974)

Country	Tonnes per annum
Australia	1459.000
Austria	680.000
Belgium	249.000
Brasil	480
Bulgaria	273.000
Canada	1208.592
Denmark	1084.000
Finland	392.000
France	4800.000
German Federal Republic	630.000
Great Britain	1469.000
Ireland	364.000
Israel	188.000
Italy	3880.000
Japan	90.000
Netherlands	2662.000
New Zealand	1830.000
Poland	1373.006
Sweden	558.000
Switzerland	1048.945
South Africa	167.000
Spain	300.000
U.S.A.	11 804.000
U.S.S.R.	3500.000

steadily increasing size, combined with concomitant increase in relative and absolute transport costs, precludes extensive utilization of the native low-solids whey on the farm. Whether it is worthwhile to evaporate a material containing some 6.5 to 7.0% dry matter for use as feeding material depends on the value of the product obtained. The feeding value of whey powder is not sufficiently high to sell it at a price that would off-set the drying costs. This is the situation in western and central Europe whereas, in the U.S.A., Bernstein and Everson (1973) considered that costs of production and sale value were about equal. A compromise solution is the partial evaporation of whey. Scheurer (1968) has calculated the economies for maximal transport distance versus degree of evaporation. This scheme is practised to some extent in Germany and Austria.

In conclusion, it has to be realized that, despite all efforts made, about one half of the 50 million tonnes of whey produced world-wide is wasted. Even if the promising processes of recovering whey proteins are going to be applied extensively, the major component, lactose, will remain unused. To destroy this material by any of the conventional means of waste treatment is prohibitive. The value for the Biochemical Oxygen Demand (BOD: 5 days) of whey is in the order of 50 000 ppm. Thorough investigations by Negaard (1975) have shown that, under Swiss conditions, investment costs of U.S. $2000 have to be made to remove one kilogram BOD/day; in addition operation costs per annum are about U.S. $62 50 to remove 1 kg BOD/day. Thus, a medium-size cheese factory producing 100 000 litres of whey per day, and therefore 5000 kg BOD/day, would have to invest U.S. $10 million, possibly as much as the cheese plant itself costs.

It goes without saying that further efforts have to be made for adequate utilization or upgrading of whey. In view of its suitable composition and its relative purity, microbiological transformation would now appear to be the method of choice. Whey can be utilized to propagate lactic-acid bacteria to be utilized as starter cultures in the manufacture of silage and Sauerkraut or to regulate the intestinal flora. Majchrzak (1976) and others (e.g. Pedziwilk, *et al.*, 1970; Janicka, *et al.*, 1976) were able to propagate propionic-acid bacteria for production of vitamin B_{12}. While promising, this process has not yet been carried out on an industrial scale. Johnstone and Pfeffer (1959) have attempted to cultivate a nitrogen-fixing strain of *Aerobacter aerogenes* on whey. With a cultivation time of about seven days, industrial prospects are not very promising. For a review of the earlier literature on microbial utilization of whey, including bacterial fermentations, alcohol, biomass and fat production, see Wix and Woodbine (1958a, b). Extensive studies on mycological synthesis of fat from whey were done by Wix and Woodbine (1959a, b). Biomass production, with high protein content under aerobic conditions, following addition of inorganic nitrogenous compounds, would now appear to be the method better suited both for obtaining valuable products and lowering drastically the load of the effluent.

C. History of Biomass Production from Whey

Considering the early development of feed yeast in Germany,

particularly during the Second World War, it is understandable that in this area whey too has been considered as raw material. In Austria, yeast was produced from whey in 1940 by Messrs. Harmer KG in Spillern. Yields of 35% dry matter based on available lactose were obtained. On a larger scale, it was produced in Linz (Austria) (Mietke and Dubrow, 1944). Whey proteins were precipitated prior to yeast propagation by the hot-acid process, removed by separation and recovered separately. Yeast yields were 30% dry matter, based on available lactose.

The Polyvit process (Waeser, 1944) was characterized by the addition of nitric acid 'to mobilize phosphate' and of ammonia. A 'torula'-type strain of yeast was used. After the yeast propagation, all of the culture liquid was dried. In Mannheim, the Waldhof process (Demmler, 1950) was in operation until a couple of years ago. The pitching yeast was derived from the sulphite liquor process of yeast production, and was said to be predominantly *Candida utilis*. In batch culture, yields of 50% dry matter, based on available lactose, were obtained but, in continuous operation, the yield could be as low as half this value. An average of 35% yield was quoted.

Fabel (1949) reported the large-scale production of *Oidium lactis* on whey. Diluted acid-hydrolysed whey (3 parts of water plus 1 part of deproteinized whey) was used to propagate the organism under submerged conditions. About 40 g dry matter of mycelium were obtained from one litre of undiluted whey. Small amounts of *Oidium lactis* mycelium were already produced in surface culture during the First World War.

In the patent of the Sav process (Société des Alcohols du Vexin; Naiditch and Dikansky, 1960), two fermenters were used to propagate *Candida kefir* (syn. *Torulopsis cremoris*) or *Kluyveromyces fragilis*. In continuous operation, a residence time of 6–9 hours was maintained in the first fermenter, and the pH value dropped to about 4. One sixth of the effluent of the first fermenter was usually transferred to the second fermenter, where fresh whey could be added with a residence time of 2 hours, the pH value rose to about pH 5 due to consumption of residual lactic acid. From whey with an initial content of 6% dry matter, a liquid of 4.3–4.5% dry matter was obtained, of which 35–40% was due to yeast. Not all of the available lactose was necessarily utilized by yeast. The whole culture fluid was evaporated to 30% and spray-dried. The normal composition of this powder was as follows:

Crude protein   32%

Fat	5%
Lactose	25%
Nitrogen-free extract	32%
Lactic acid	4%
Ash	18%
Water	8%

In Czechoslovakia, Tomíšek and Grégr (1961) attempted to overcome the major difficulties of yeast propagation on whey observed so far, i.e. inadequate aeration in commercial-scale fermenters resulting in long residence times, most likely combined with problems of contamination by lactic-acid bacteria, and therefore unreliability of the process in terms of yield and growth rate of yeast. Self-priming turbines were developed. A semi-continuous process was adopted by which 150 m³ of whey were fermented in three 20 m³-fermenters per day. A mixed culture of *Candida utilis* and *Candida pseudotropicalis* (syn. *C. casei, C. cremoris*) was used. Yields of 51–54% dry matter, based on lactose, were quoted. It is obvious that here, too, comparatively long residence times were necessary. It is not clear whether this was due to oxygen deficiency or inadequate nutrient supply.

Simek *et al.* (1964), using *Kluyveromyces fragilis*, used yeast extract and, later, small concentrations of molasses to supplement whey with growth factors. In sparger-type fermenters under continuous operation, dilution rates of 0.1–0.15 h^{-1} (1.0–1.5% molasses added) were maintained. The whole of the culture liquid was evaporated (25–28% dry matter) and then roller-dried. One m³ of whey supplemented with 10 kg of molasses resulted in a product of 40–50 kg dry matter per m³. The dried material was of the following composition:

Crude protein	54%
Fat	4%
Nitrogen-free extract	19%
Ash	22%
Lactose	0

In the 'Wheast' process (Robe, 1964), at the Knudsen Milk Products Co., California, U.S.A., it is characteristic that, prior to the yeasting process, the whey protein is precipitated but not separated. At the end of the process, this protein is recovered by centrifuges, together with the yeast (*Kluyveromyces fragilis*) produced. Yields of 33% dry matter based on lactose were quoted. The total yield (including precipitated proteins) was 22–23 g/l with a crude protein content of 58–60%.

It is apparent from this summary of processes up to the early 1960s that either, or both, yields of yeast and rate of production were not to the degree that might reasonably be expected. It has to be realized that a sugar concentration of, say, 45 g/l and a dilution rate of 0.25 h^{-1} would result in a productivity of 5.6 g $l^{-1}h^{-1}$ with an approximately equal oxygen demand. Up to the late 1960s, commercial-scale fermenters for biomass production were not usually more efficient than the equivalent of 2 to 2.5 g $l^{-1}h^{-1}$. It is obvious that fast-growing yeast in undiluted whey would be starved for oxygen, the oxygen-transfer rate of the fermenter limiting the growth rate. The consequence is that either a strongly diluted whey has to be used, or biological complications ensue, in particular the development of facultatively anaerobic bacteria such as lactic-acid bacteria can be considered to be. Probably, also, insufficient attention has been devoted to selection of appropriate strains. As will be shown later, lactose-utilizing yeasts tend to grow irregularly in continuous culture.

While, in most of the processes here quoted, nitrogen in the form of ammonium ions has been added, there were possibly deficiencies in other respects. It is the object of this chapter to analyse nutrient requirement and other environmental conditions, as well as methods of propagation in the laboratory and on an industrial scale, and discuss more fully the modern commercial processes.

II. ORGANISMS USED FOR BIOMASS PRODUCTION; ENRICHMENT, ISOLATION

A. Organisms

Considering their reputed acceptability for food and feeds as well as comparative ease of cultivation, it is not surprising that yeasts and (occasionally) yeast-like micro-organisms have been the organisms almost exclusively tested for their suitability to transform whey sugar into proteins for foods or feeds. Table 6 is a compilation of articles indicating the organisms utilized. It is apparent from Table 6 that, in most cases, *Saccharomyces fragilis* (now *Kluyveromyces fragilis*) was used for biomass production from whey. This was the species usually considered as the best lactose utilizer. Most often the selection of the strain(s) used was not the result of a screening programme, rather was it one of several strains taken from an existing culture collection.

Table 6

Organisms used for biomass production from whey

Author	Organism[a]	Source
Amundson (1967)	*Saccharomyces fragilis*	Unknown
Anvar (Ullman and Schwartz, 1973)	*Enterobacter liquefaciens*	Unknown
	Enterobacter cloacae	Unknown
	Serratia atypica	Unknown
Atkin *et al.* (1967)	*Trichosporon cutaneum* 54–169	University of California (H. J. Phaff)
	Saccharomyces fragilis	N.R.R.L.–Y1109
Bayer (1977)	*Candida intermedia*	Screening—dairy products
Bechtle and Claydon (1971)	*Saccharomyces fragilis*	N.R.R.L.–Y1109
	Candida utilis	N.R.R.L.–Y900
	Lactobacillus delbrueckii	N.R.R.L.–B445
	Leuconostoc dextranicum	N.R.R.L.–B1145
	Leuconostoc citrovorum	N.R.R.L.–B1232
	Torulopsis cremoris	N.R.R.L.–Y1172
	Torulopsis sphaerica	Kansas State University
	Candida pseudotropicalis	Kansas State University
	Lactobacillus casei	Kansas State University
	Lactobacillus bulgaricus	Kansas State University
	Streptococcus lactis	Kansas State University
Bernstein and Everson (1973)	*Torula lactosa*	Screening
	Saccharomyces lactis	Screening
	Saccharomyces fragilis	Screening
Blanchet and Biju-Duval (1969)	*Saccharomyces fragilis*	Unknown
Chapman (1966)	*Saccharomyces fragilis*	Unknown
Davidov *et al.* (1963)	*Candida tropicalis*	Unknown
Edelsten (1974)	*Saccharomyces fragilis*	A/S Alfred Jörgensen Lab;—skim milk.
Fabel (1949)	*Oidium lactis*	Screening—dairy products

TABLE 6 (cont.)

Organisms used for biomass production from whey

Author	Organism[a]	Source
Hanson et al. (1947)	*Saccharomyces fragilis*	Unknown
	Torula kefir	Unknown
	Mycotorula lactis	Unknown
	Saccharomyces anamensis	Unknown
	Candida arborea	Unknown
	Candida pulcherrima	Unknown
	Torulopsis utilis	Unknown
	Monilia candida	Unknown
	Oidium sp.	Unknown
Lembke et al. (1975)	*Candida krusei*	Enrichment
	Lactic-acid bacteria	Unknown
Porges et al. (1951)	*Saccharomyces fragilis*	N.R.R.L.
	Torulopsis cremoris	N.R.R.L.
	Candida lipolytica	N.R.R.L.
	Torulopsis utilis	N.R.R.L.
Sav (Naiditch and Dikansky, 1960)	*Saccharomyces fragilis*	Unknown
	Torula cremoris	Unknown
Šiman and Mergl (1961 a, b)	*Torula* spp.	Unknown
	Torulopsis spp.	Unknown
	Candida spp.	Unknown
Simek et al. (1964)	*Saccharomyces fragilis*	Központi Elelmiszeripari Kutatointezet
Tomíšek and Grégr (1961)	Mixed culture	Katedre kvasne chemie a technologie Vscht
	Torula utilis	
	Torula casei	
	Torula cremoris	
	Candida tropicalis	
Vananuvat and Kinsella (1975)	*Saccharomyces fragilis*	Cornell University
Wasserman (1960a)	*Saccharomyces fragilis*	Unknown
Wix and Woodbine (1959 a, b)	Filamentous fungi	Baarn Culture Collection Commonwealth Mycological Institute

[a] Species names in this table are the original ones used by the authors; some of them have since been changed (see Lodder, 1970).

Some authors have used mixed cultures. Lembke *et al.* (1975); Moebus
and Lembke (1977) used a strain of *Candida krusei* able to utilize lactate,
the latter being produced by cultivating separately or simultaneously
different types of lactic-acid bacteria on whey. Lembke *et al.* (1975)
reported that their original yeast strain, *Candida pseudotropicalis*, used in
pilot plant-scale to utilize lactate after a bacterial fermentation, was
gradually replaced by a chance contamination of *C. krusei*, obviously
better suited for the particular environmental conditions. The *Candida
utilis* strains reported in Table 6 were said to be able to utilize lactose.

A screening programme for industrial biomass production cannot be
based merely on the ability to utilize lactose. In practice enrichment,
isolation and selection of the desired organisms will have to be carried out
with the expected performances in mind. One important feature of a
biomass production process is that it will have to be cheap, and
industrialists, universally, agree that it will be too expensive to use a fully
aseptic process. Preferably, any kind of heat treatment, involving capital
costs, energy and cooling water, should be omitted. Deproteinization of
whey should not be a necessity, but an option. Thus, the organisms to be
used have to be perfectly adapted to the process, without undue risk of
being affected by undesirable contaminating organisms. The property of
high growth rate, therefore, is not only important to keep the fermenter
volume as small as possible, but also to have a high degree of certainty
that they will not be overrun by other types of organisms. They should not
be negatively affected by whey proteins, which is apparently not always
the case, as indicated by Vrignaud (1976) and as can be inferred from
some publications (Bechtle and Claydon, 1971). High growth rate is not
a property of many filamentous fungi. Cultivation periods of several days,
reported by Wix and Woodbine (1959a, b) for a large number of
filamentous fungi grown on whey in surface and shaken culture, are too
long to be considered for economic production of biomass. While bacteria
can grow fast, they are undesirable for other reasons. Thus, yeasts appear
to be the organisms of choice. The organisms have to be suitable for
continuous cultivation, preferably in one stage only, in order to keep
capital and operation costs as low as possible. While this condition would
seem to be easily met, in practice lactose-utilizing yeasts tend to show
considerable variation (Demmler, 1950; Bayer, 1977; Klein, 1975;
Mikota, 1973; Wagner, 1973).

Acid resistance is a desirable property. By either carrying out acid
washing of the yeast cream at regular intervals or maintaining

permanently a low pH during cultivation, bacterial development can be largely avoided. Bacterial types which would be able to resist a cultivation pH of 3.5 to 4.0, at a residence time of 4–5 hours in steady-state culture, could be found among acetic-acid bacteria, but lactose adds an additional selection effect. *Oidium lactis* (Fabel, 1949), an interesting organism from the viewpoint of acceptability, growth rate, yield and ease of recovery, has to be excluded in view of its low acid resistance and therefore a proneness to bacterial contamination.

In terms of recovery, cell size is important, since yeasts are normally recovered by centrifugation. Flocculence, if it is quantitative and if sufficiently dense aggregates are formed, saves on centrifuge time. If, however, the whole culture liquid is evaporated after fermentation, flocculence tends to produce non-homogeneous conditions in storage tanks prior to thermolysis and evaporation, thus turning into a disadvantage.

Uniform morphology of the yeast cells is also desirable if separate recovery of the biomass is envisaged. In particular, pseudomycelia should be absent; if not, two different recovery processes would be required, the one filtration to recover mycelium, the other centrifugation to recover residual single and budding cells, since filamentous-type organisms would choke the jet-inlets to separators. Pseudomycelium would also increase the viscosity of the culture considerably, again involving higher energy requirements. After having isolated an organism fulfilling all these properties, it will have to satisfy in terms of protein content and will have to stand the tests in feeding trials.

B. Enrichment and Isolation of Lactose-utilizing Yeasts

According to our experience, the principle of enrichment cultures is well worthwhile applying. This can be done on both mixtures of culture-collection strains and natural sources. From cultures tested for growth rate and yield on a large number of culture-collection strains, both in mixture and singly, we were unable to select one satisfying our goal.

Wagner (1973), in his survey of lactose-utilizing yeasts occurring in dairy products, dairy waste and centrifuge sludge, proposed the following procedure for enrichment and isolation. The material, in particular cottage-type cheese, was usually incubated from 1–2 days at 30°C and suspended in sterile Ringer (physiological salt) solution. After short

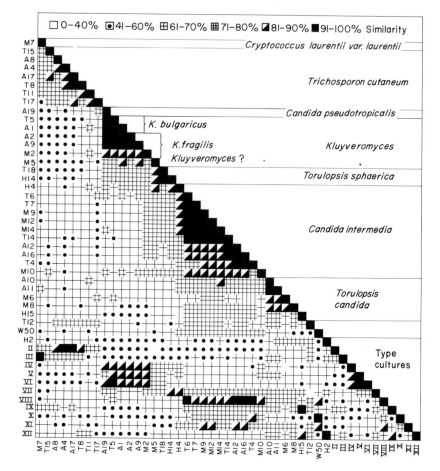

Fig. 1. Numerical classification of lactose-utilizing yeasts isolated from dairy products. From Wagner (1973).

sedimentation, it was filtered through paper filter. Portions (about 100 ml) were then passed through a membrane filter of 3 μm pore size. This filter, less rich in bacteria than the filtrate, was incubated in shake flasks (30 ml in 300 ml conical flasks), with a medium of the following composition (g/l of tap water, pH = 4.5): D (+) lactose, 20; $NH_4H_2PO_4$, 6.12; $(NH_4)_2SO_4$, 1.2; KH_2PO_4, 1.0; $MgSO_4$. $7H_2O$, 0.25; yeast extract, 0.5; antifoam oil, 1.0 ml. Incubation was at 30°C, 48 h, 250 rev./min. Starting with crude milk, the pH value was lowered with phosphoric acid to about 4.5 and processed as above. A microscopic check after shake-culture incubation usually showed yeasts in predominance.

Plating on an agar medium of the above composition was done by serial dilutions or by streak plates. After three days incubation, colonies distinguished by their large size, elevated habit and mostly glossy appearance, were checked microscopically. Yeast colonies were purified again by replating on streak plates. This repeated purification step was necessary as there was always a significant proportion of mixed cultures in the first isolation. Some yeasts, attractive because of their large size, were able to utilize lactate or glucose only when accompanied by bacteria or lactose-decomposing yeasts. Pure cultures were maintained on agar slants with a medium of the above composition, and lyophilized.

C. Identification of Lactose-utilizing Yeasts

This is usually done according to Lodder (1970). Wagner (1973) successfully used numerical taxonomy in addition. A survey of strains isolated from various dairy materials and classified numerically is shown in Figure 1. The two major groups, namely *Kluyveromyces* spp. and *Candida intermedia*, cluster quite clearly; nevertheless, there is a similarity coefficient of up to 70–80% for intermediate types. From Fig. 1 one wonders whether division of *Kluyveromyces fragilis* and *K. bulgaricus* into two different species is justified. *Candida intermedia*, so far not described (except by Tilbury *et al.*, 1974), as an abundantly occurring yeast in dairy products, was the one predominantly isolated by the above methods under Austrian conditions.

III. ENVIRONMENTAL CONDITIONS FOR YEAST PRODUCTION FROM WHEY

The species of yeast used for biomass production on whey differ (see Section II) from those most often used for other raw materials. It is worthwhile to examine to what extent environmental conditions for optimal growth rate and yield differ from the conventional process. Increasing interest in microbial biomass production also brought us to the more and more widely accepted view that the processes to be used in practice, with all possible simplification, have to dictate the isolation of the ideal strains, and not *vice versa*. This principle does not apply only to the choice of a particular species but also to selection of strains. Numerous examples could be cited to this effect. In feed- and food-yeast production, it has been universally considered that a cultivation temperature of 30–32°C is the one to be used. This has caused considerable expense in terms of cooling the fermenter, particularly in areas of high average ambient temperature and/or water scarcity. The logical way of selecting a yeast strain able to grow at 40–42°C has not been attempted for a very long time. Truly, it involved the switch-over from *Candida utilis*, the one and only species considered usable, to other species of *Candida*. Similarly, a pH of 5, or at most 4.5, was thought to be the lowest at which a yeast could be cultured at sufficiently high growth rate and yield. Adequate (expensive) precautions had then to be taken to prevent contamination. With the availability of stainless steel at reasonable prices, cultivation at far lower pH values became possible without suffering from corrosion effects. However, it took a comparatively long time to realize that it is worthwhile in many cases to select new acid-resistant strains. It is now almost common knowledge that, at a pH of 4 or below, a yeast strain selected for the particular substrate growing in steady-state culture at a specific growth rate of about 0.2 to 0.3 h^{-1} will meet few competitors.

This experience, valid as it is from the biological point of view, cannot always be made use of in practice. A biomass product which has been thoroughly standardized in terms of species, strain, composition, method of cultivation and raw material used, and which has been recognized as an acceptable material from governmental health organizations, cannot be changed easily, at least not without further tests on toxicity and feed or food quality. With new products, however, as with lactose-grown yeast, it is worthwhile to keep ecological factors in mind in order to be able to maintain a culture under production conditions as self-selective as possible.

A. Influence of Temperature and pH Value

In Table 7 the above reasonings have not always been applied. Comments are included, when available, showing that it was possible to maintain an adequately selected yeast even under non-aseptic operation provided a pH below 4.5 was established.

Specific mixed cultures of bacteria and yeast were maintainable only at pH 5 (Bechtle and Claydon, 1971), or pH 5.5 (Lembke *et al.*, 1975). It is of interest to note that some organisms, like *C. intermedia*, are extremely resistant to pH value. While there is usually no need to grow the organism at as low a pH as 2.8 (Mikota, 1973), the fact that high growth rate and yield can be maintained at this acidity could be of considerable advantage. In continuous steady-state culture over long periods, enrichment of acid-tolerant organisms, be it unwanted yeasts or bacteria, could happen. If then, the pH can be dropped to very low values, it is likely that the contaminating organisms will disappear. In extreme cases, the procedure of acid-washing yeasts (at pH values below 2), which is common practice in yeast technology in continental Europe, could of course be applied without risk in acid-resistant strains.

In whey containing about 4.4% lactose, or with a small part of it as lactic acid, and ammonium sulphate as a nitrogen source, the pH if uncorrected will drop to 2.5 to 2.8 even if no significant amounts of non-utilizable organic acids are formed (Mikota, 1973). In continuous culture, long-term adaptation at this pH can be observed. At a dilution rate of $0.2 \ h^{-1}$, productivity of biomass (*Candida intermedia*) rose gradually from about $3.2 \ g \ l^{-1} h^{-1}$ to about $4.5 \ g \ l^{-1} h^{-1}$ after some 200 h of cultivation. Under these circumstances, a semicontinuous process would not seem to be very favourable as changes in pH value according to the periodically exchanged culture volume would act disadvantageously. The situation is illustrated in Fig. 2. (Mikota, 1973) where different culture volumes have been exchanged after performance under continuous cultivation at low pH values. Removal of one half of the culture and replacement with sour whey increased the pH from about 2.3 to about 3.5. Average growth rate of this culture was smaller $(1.9 \ g \ l^{-1} h^{-1})$ than with an exchange of one third $(2.1 \ g \ l^{-1} h^{-1})$ and, in turn, smaller than with an exchange of one ninth of the culture volume $(2.4 \ g \ l^{-1} h^{-1})$. The maximal yields were 22.7, 23.4, and 25.2 g/l, respectively. If substrate was then fed continuously at a dilution rate $0.2 \ h^{-1}$, all three cultures adjusted themselves to the accustomed pH 2.5–2.8, and yeast contents of 20–22.5 g/l corresponding to a productivity of $4–4.5 \ g \ l^{-1} h^{-1}$.

Of the examples listed in Table 7, it is noticeable that two of the successful producers of lactose-grown yeast on a commercial scale (Bel Fromageries in France and Vienna-Unterkärtner-Molkerei in Austria) use quite low pH values during propagation, i.e. pH 3.2–3.8. While successful operation would be possible at pH 2.8 in the Vienna process,

Table 7

Temperature and pH values used for cultivating lactose yeasts

Author	Organism	pH Value	Temperature (°C)	Remarks
Amundson (1967)	*Saccharomyces fragilis*	4.5	30	Shake flasks
Atkin *et al.* (1967)	*Torulopsis cutaneum*	4.5	25	
Bayer (1977)	*Candida intermedia*	3.5–4.2	32–36	No decontamination required in practice
Bechtle and Claydon (1971)	Mixed: yeasts and bacteria	< 5	32	Pure yeast growth
		5	32	Mixed cultures constantly poorer than yeast alone
		5.3	32	Suitable for mixed cultures
Bel Fromageries Process (Vrignaud, 1976)	*Saccharomyces fragilis*	3.5	38	Production scale
Bernstein and Everson (1973)	*Saccharomyces fragilis*	4.5	30	Large pilot-scale
Klein (1975)	*Candida intermedia*	3.0–5.0	30	Non-aseptic cultivation
Lembke *et al.* (1975)	Lactose-utilizing *Candida utilis*	4.5	25	
	Candida krusei	4.2–5.5	33	
	Candida krusei with lactic-acid bacteria (2 fermenters)	6.2	44	First stage: lactic-acid bacteria; large pilot scale

Table 7 (*cont*)

Temperature and pH values used for cultivating lactose yeasts

Author	Organism	pH Value	Temperature (°C)	Remarks
Mikota (1973)	*Candida krusei* with lactic-acid bacteria	5.5	44	Mixed culture; large pilot-scale
Sav Process (Naiditch and Dikansky, 1960)	*Candida intermedia*	2.8	32	Non-aseptic cultivation
	Saccharomyces fragilis	—	26	Production scale
Simek *et al.* (1964)	*Saccharomyces fragilis*	5.2–5.5	30–32	
Surazynski *et al.* (1968)	*Saccharomyces fragilis*			
	Candida pseudotropicalis	4.5–5.5	30	
	Saccharomyces lactis			
Tomíšek and Grégr (1961)	*Torula utilis*	5.4	30	Production scale
	Torula casei			
	Torula cremoris			
Vienna–U.K.M. Process (Bayer and Meyrath, 1976)	*Candida tropicalis*			
	Candida intermedia	3.4–3.6	32–33	Production scale
Vananuvat and Kinsella (1975)	*Saccharomyces fragilis*	5.0	30	
Wagner (1973)	*Candida intermedia*	2.4–3.2	30	
Wasserman (1960b)	*Saccharomyces fragilis*	5.0–5.7	32	
Weilbuchner (1977)	*Candida intermedia*	4.0	32	Skim milk
Wheast Process (Robe, 1964)	*Saccharomyces fragilis*	5.0	30	Production scale
Zerlauth (1973)	*Candida intermedia*	4.0	30	

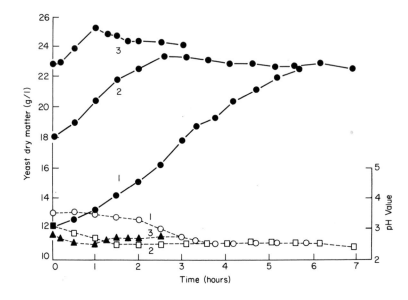

Fig. 2. Time-course of production of cell dry matter (solid circles) and of changes in pH value (open circles) during incubation of batch cultures of *Candida intermedia* with different volumes of culture exchanged for fresh medium (sour whey). Culture volumes exchanged were 0.50 (curves 1), 0.33 (curves 2) and 0.11 (curves 3) of the total volume. From Mikota (1973).

permanent production at this high acidity is not recommended. A pH of 2.8 is considered too low for feeding purposes, if the whole culture liquid is evaporated. Readjustment of pH value after fermentation would, however, increase the salt content, which is not desirable. Hence, adjustment of pH value during cultivation (to between 3.2 and 3.8) is best done with a mixture of ammonium sulphate and ammonia or urea.

With regard to fermentation temperature, most authors used about 30°C. This is likely not to be the result of a determination of temperature optimum, but it is the conventional level for most strains of *Saccharomyces* and *Candida*. *Kluyveromyces* strains have the common habit of growing well at higher temperatures; thus Bel Fromageries operates at 38°C. With *C. intermedia*, cultivation at about 35–36°C is possible.

B. Nutrient Requirements

1. Nitrogen

In Section I it has already been mentioned that whey proteins are hardly utilizable by yeast. Small concentrations of low-molecular weight inorganic and organic nitrogenous compounds cannot satisfy the nitrogen requirement if a crude protein content of some 50% dry matter is expected. Also, it is unlikely that dinitrogen-fixing yeasts have ever been isolated. Nitrogen will therefore have to be the major nutrient added. Cheap nitrogenous compounds, i.e. ammonium ions, mostly as ammonium sulphate, ammonia and urea, will satisfy the nitrogen requirement of yeasts. Rarely, if ever, has addition of amino acids or peptides been found necessary to satisfy their needs. *Candida intermedia* strains isolated by Wagner (1973) and tested by Bayer (1977) were shown to grow perfectly well on fully synthetic media with ammonia as the only nitrogen source. It is common practice to add these nitrogenous compounds slightly in excess of the requirement of the organism. To evaluate the requirements, it is necessary first to add, in trial runs, an excess of nitrogenous compounds and then lower it to the minimum without impairing either growth rate, yield of dry matter or crude protein. Yeasts grown in steady-state culture at an elevated specific growth rate have a higher crude protein (N $\times$ 6.25) content. This increase is partly due to a higher content of ribonucleic acids. In lactose-utilizing yeasts, the same effects can be noted (Bayer, 1977).

2. *Phosphorus*

A number of authors have examined whether the phosphorus content of whey (see Table 3) satisfies the requirements of yeast. Harrison (1967) and Wasserman (1960b) found that *Kluyveromyces* sp. grown in whey requires additional phosphorus as phosphate. For each 50 kg of drymass produced, Wasserman (1960b) added 9 kg of KH_2PO_4. In the Polyvit process (Waeser, 1944), nitric acid was added 'to mobilize' phosphate in whey. Most authors working on whey utilization by yeast on a laboratory scale added phosphate salts, probably from habit or for safety. The question is of some importance in the industrial process since phosphate salts, after all, constitute additional expense. According to Lembkte *et al.* (1975), phosphorus in whey is available at a sufficient concentration (*Candida krusei* together with lactic-acid bacteria). Bayer (1977) showed that *Candida intermedia* could be grown at specific growth rates of up to 0.28 h^{-1} where a higher nucleic acid (and therefore phosphorus) content was noted than at 0.2 h^{-1} without phosphate having been added. There must be *Kluyveromyces* strains which can be grown perfectly well in whey without phosphate addition, since Bel Fromageries using this species, do not list it as one of the nutrient requirements (Vrignaud, 1976; Blanchet and Biju-Duval, 1969).

3. Other Minerals

According to Lembke *et al.* (1975) and Butschek and Neumann (1962) all other minerals, such as Mg^{2+} and other heavy metals in addition to phosphate, are present in whey in sufficient concentrations. Sulphur, if there is an additional need, is normally added as ammonium sulphate. The presence of minerals 'in sufficient concentrations' may not be a sufficiently accurate statement, since interactions between various media components in other systems are well known. The chances are that further optimization will bring about the possibility of somewhat increasing either yield or growth rate.

4. Vitamins

Few yeast species, such as *Saccharomyces kloeckerianus*, are recognized to be fully independent of vitamins for their optimal growth. Milk and whey do have a fair complement of vitamins (see Table 3). Nevertheless, in the media formulations of most authors, some additional source of vitamins is added, e.g. yeast extract or corn-steep liquor. Atkin *et al.* (1967) found that addition of 0.4% yeast extract, or 0.4% corn-steep liquor lowers the doubling time of *Trichosporon cutaneum* from 3.2 to 2 h; exponential growth, up to 80–90% of maximal population density, took place in both cases. Harju *et al.* (1976), working with *Kluyv. fragilis* A.T.C.C. 8608, were also able to increase specific growth rate and yield with either yeast extract or a vitamin complex. Biotin, which is at times present only in low concentrations in whey, added alone, produced the same maximal yield (0.3 kg dry mass/kg lactose) as yeast extract or the vitamin mixture, but the specific growth rate was increased only from $0.15 \, h^{-1}$ (no vitamin supplementation) to $0.27 \, h^{-1}$ as compared to $0.4–0.43 \, h^{-1}$ with a full complement of vitamins.

Some vitamins, such as nicotinic acid, folic acid and thiamin, are enriched in *Kluyv. fragilis* as compared to its content in whey (Wasserman, 1961). The vitamin composition of *Kluyv. fragilis* grown on whey, supplemented with yeast extract, is near enough the same as that of other yeasts (see Section VIII. A). Lembtke *et al.* (1975) added autolysed yeast (*C. krusei*) obtained from their own process. The main purpose, however, was to obtain proteolytic enzymes designed to make whey proteins available to the organism. No mention is made about the vitamin requirements of the yeast used. In the next section, some substrate formulations for culturing yeast on whey indicate that vitamins are often added for growing *Kluyveromyces* sp. on whey. The vitamin requirement certainly varies from strain to strain, and these formulations are not necessarily valid for all strains of the genus *Kluyveromyces*.

Using a strain of *C. intermedia*, Bayer (1977), Klein (1975), Mikota (1973), Wagner (1973), Weilbuchner (1977) and Zerlauth (1973) were able to obtain high yields (50% or higher based on lactose, 50% average protein content) in continuous single-stage culture at high dilution rates without addition of vitamins to sweet or sour whey of various origins, or to skim milk; in fact, specific growth rates as high as 0.28 to 0.3 h^{-1} (Bayer, 1977; Weilbuchner, 1977) were feasible at the above yield.

C. Media Formulations

The final formulation of medium ingredients includes sources of nitrogen and phosphorus and other minerals, vitamins, proteases, acids or alkalies to adjust pH value, and antifoam agents. No universal recommendation with regard to antifoam agents can be made, however, as the demand is dictated very much by the cultivation equipment used. Table 8 lists the various ingredients used by several authors.

D. Oxygen Requirement

The supply of oxygen in biomass production represents a significant part of the production costs. Apart from the influence of the aeration system used in practice, it is of importance to know the specific oxygen requirement of the organism under the particular conditions of cultivation. From the work of Atkin *et al.* (1967), a specific oxygen requirement of 0.39 g oxygen per gram dry mass can be calculated for the lactose-utilizing *Trichosporon cutaneum*. This very low specific oxygen requirement was the result of cultivating the organism in a 1:4 diluted whey supplemented with 0.4% yeast extract or 0.4% corn-steep liquor. Apart from the discrepancy between the quoted oxygen absorption rate

Table 8

Medium formulations for yeasts grown on whey

Reference	Nitrogen (source)	(kg/m³)	Phosphorus (source)	(kg/m³)	Vitamins (source)	(kg/m³)	Acid or alkali (1/m³)		Antifoam
Amundson (1967)	$(NH_4)_2SO_4$ (only inoculum)	1.7	H_3PO_4 2.38 75%	0.92	CSL	6.6	H_3PO_4	2.4	
Bayer (1977)	$(NH_4)_2SO_4$	1.59	—	—	—		KOH		Glanapon 2000
Bernstein and Everson (1973)	NH_3	1.2	H_3PO_4	0.16	yeast extract	3.0	H_3PO_4	0.5	
Waldhof Process: Demmler (1950)	$(NH_4)_2SO_4$ and NH_4OH	1.2	—		—				
Harju et al. (1976)	$(NH_4)_2HPO_4$ $(NH_4)_2SO_4$ NH_4OH	1.24	$(NH_4)_2HPO_4$ or H_3PO_4	0.12	yeast extract	1			
Lembke et al. (1975)	$(NH_4)_2SO_4$ or NH_3	0.93	—		yeast autolysate (dry matter)	0.7	H_2SO_4 H_3PO_4	2.5 0.3	

Table 8 (cont)

Medium formulations for yeasts grown on whey

Reference	Nitrogen (source)	(kg/m³)	Phosphorus (source)	(kg/m³)	Vitamins (source)	(kg/m³)	Acid or alkali (l/m³)	Antifoam
Porges et al. (1951)	$(NH_4)_2HPO_4$ $(NH_4)_2SO_4$ NH_3	0.25 0.68 0.36	$(NH_4)_2HPO_4$	0.21	—	—	NH_4OH	Corn oil and lard
Sav Process (Naiditch and Dikansky, 1960)	Urea $(NH_4)_2SO_4$ (1:1)	1.5	H_3PO_4	0.5	—	—	—	
Surazynski et al. (1968)	$(NH_4)_2SO_4$	2.12	KH_2PO_4	1.14	yeast extract	1	—	
Vananuvat and Kinsella (1975)[a]	$(NH_4)_2SO_4$ Peptone Urea	2.38 1.8 0.32	KH_2PO_4	1.14	yeast extract	2.0		
Vienna-U.K.M. Process (Bayer and Meyrath, 1976)	$(NH_4)_2SO_4$ $(NH_4)OH$	0.8 0.8					NH_4OH	Glanapon 2000 (concentrate)
Wasserman (1960a)	$(NH_4)_2SO_4$	1.06	K_2HPO_4	1.14				

[a] corrected for 4.5% lactose concentration (instead of 2%)

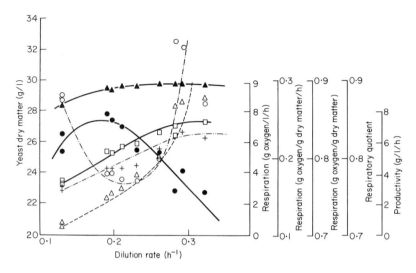

Fig. 3. Changes in yeast dry matter content (●), productivity (□), and gas exchange expressed as g oxygen/l/h (▲), g oxygen/g yeast dry matter/h (△), and g oxygen/g dry matter (○) in continuous cultures of *Candida intermedia* on skim milk. From Weilbuchner (1977).

of the fermenter and the oxygen demand of the culture, it is not known whether *T. cutaneum* grown under these particular conditions does have a protein content of about 50%. Bronn (1966) determined a value of 0.79 g oxygen/g dry mass in *C. utilis*; Cooney (1975) observed a value of 0.68. The specific oxygen requirement of the lactose-utilizing *C. intermedia* was examined in more detail by Weilbuchner (1977) (Fig. 3). Grown in steady-state culture on defatted milk under the same conditions as by Bayer (1977) on whey, with very similar results, Weilbuchner found the minimal specific oxygen requirement to be 0.76 to 0.78 g oxygen/g dry mass at dilution rates of 0.19 to 0.23 h^{-1}. At dilution rates higher or lower than this, the specific oxygen demand was considerably higher. Maximal yield was reached at a dilution rate of 0.19 h^{-1} with 27.8 g/l. The small differences in specific oxygen requirement when the dilution rate is varied between 0.19 and 0.23 1^{-1} could lead to a certain degree of optimization in industrial biomass production, i.e. a smaller yield at a higher biomass productivity might be obtained at a somewhat higher dilution rate with a higher oxygen demand per litre fermenter volume an hour. Furthermore, it is to be taken into account that at a dilution rate of 0.19 h^{-1}, the crude protein content was 55.3% and that of nucleic acids 9.3% and, at 0.23 h^{-1}, the values were 54.4% and 10.2%, respectively. While yield and protein content are the prime factors influencing profitability, and productivity does to a lesser extent, the biological and technological advantages of a higher dilution rate, i.e. less contamination by foreign organisms, might still throw the balance towards the use of the higher dilution rate in practice. It is clear that differences of this magnitude will have to be settled in the plant under the particular conditions of operation over longer test periods. Techniques for supplying oxygen to the culture will be dealt with in Section VI (p. 254).

E. Type of Whey

Sweet whey contains considerably less lactic acid than sour whey (see Table 2, p. 209). The former, if untreated and stored for several hours, as is necessary for maintaining a steady flow rate to the fermenter, will rapidly show an increase in lactic acid concentration. Thus, utilization of both lactose and lactic acid for biomass production is of significance. Butschek and Neumann (1967) reported that lactic acid can be utilized with equal yield as lactose by yeast. Bernstein and Everson (1973) found sour whey as good a substrate for yeast production as sweet whey. Bechtle and Claydon (1971) obtained highest yeast cell counts with acid whey (quarg and cottage cheese manufacture), with sweet whey (cheddar cheese) a little less, and with chemical wheys (mineral acids added for casein manufacture) they obtained the lowest cell counts. In mixtures of cottage- and cheddar-type whey, the slight proportional improvements were due to additions of cottage-type whey. The use of chemical wheys for biomass production has also the disadvantage of increasing the ash content of the product if the whole culture is evaporated. Bayer (1977) and others in the Vienna group found no significant differences in yield between sweet and sour whey when yield was determined as weight biomass using *C. intermedia*.

For experimental purposes and in order to have a reproducible medium composition for a given run of tests, it is common practice to have a stock of either whey powder or partly evaporated whey. Bechtle and Claydon (1971) showed that concentrating the whey has no deleterious effect on composition when processed at a low temperature. They also point out, however, that concentrating by a factor of more than three results in inhomogeneity due to lactose crystallization and sedimentation.

F. Foam Control

Biomass production on whey in submerged aerated cultures is always accompanied by foaming. This is to be attributed to lactalbumin and lactoglobulin in the whey. In deproteinized whey, foam production is certainly less. In practice, foam control is achieved by chemical agents as well as specific mechanical foam breakers or particular fermenter design. The latter are described in a subsequent section. Chemical agents for foam control are given in the following examples: Atkin *et al.* (1967) added 0.1 ml/min to 240 ml of culture of a 1:5 dilution of Dow Antifoam C, while Bechtle and Claydon (1971) found that Hodag K-58 und K-60 among the food-acceptable antifoam agents to be most appropriate. Vanauvat and Kinsella (1975) controlled foam automatically by a 10% solution of FG-10, Dow Corning Corp.; Bayer (1977) used Glanapon 2000 (Busetti and Co., Wien, Austria) in a 10% solution for automatic foam control.

IV. PRETREATMENT OF WHEY

Depending upon a number of circumstances (deleterious effect of whey proteins on fermentation, type of product to be obtained, contamination problems, waste disposal), whey is to be subjected to various kinds of treatments prior to its use as a substrate.

A. Decontamination and Sterilization

Usually it has been considered necessary to subject whey to some form of decontamination process prior to fermentation. Sterilization for yeast production at a pH value below 5 is not practised. Hanson *et al.* (1947) steamed whey from 5 to 40 min at a pH value between 1 and 4 (preferably 1.5 to 3.5). Various cultures of yeast produced somewhat higher yields by using this form of pretreatment. The fermentations were carried out at an initial pH of 5 or higher. Amundson, (1967) adjusted the pH to 4.5 and steamed at 93°C for 5 minutes. Additional phosphoric acid was added to adjust the pH to 3.5 for fermentation. Bernstein and Everson (1973) adjusted the pH value to 4.0 with hydrochloric acid and heated for 45 min at 80°C. Surazynski *et al.* (1968) heated 20 min at 65°C. Bechtle and

Claydon (1971) used 'cool' sterilization by adding hydrogen peroxide at 1% concentration and heating at 65°C for one hour. Beef catalase free of bacterial contamination was used to destroy residual peroxide. Atkin *et al.* (1967) first centrifuged whey to remove large cheese particles followed by sterile Seitz filtration. Some of the above-mentioned decontamination processes, i.e. peroxide treatment or sterile filtration, would not be economically practicable on an industrial scale. While decontamination, in principle, is apt to render the fermentation process more reliable, it was not found necessary in the Vienna-Unterkärtner Molkerei process (Bayer and Meyrath, 1976) using the fast-growing *C. intermedia* at pH 4. The counts of lactic-acid bacteria in native whey did not significantly increase during yeast production and, over two years operation, did not interfere with either yeast productivity or yield.

Bernstein and Everson (1973), fermenting at pH 4.5 with prepasteurization, did not find any sign of contamination by lactic-acid bacteria. Amundson (1967) operating batchwise carried out 14 consecutive runs under non-aseptic conditions (except the first) without significant bacterial contamination. Wasserman (1960b) found that, in semicontinuous culture with *Kluyv. fragilis* at an averaged 4-hour residence time, no contamination problem occurred under non-aseptic conditions. Equally, Müller (1948) had confirmed that neither moulds nor bacteria were able to develop significantly at pH 5 if efficient yeasts were used under non-aseptic conditions.

For quick detection of contamination, Bechtle and Claydon (1971) recommended Gram stains and reactions of culture fluid on litmus milk. If the whole culture liquid is to be evaporated, care has to be taken not to increase the salt content too much. Thus, adjustment to a pH lower than 3.5 for heat treatment will require alkali for fermentation and/or a pH adjustment of the final product in amounts larger than is necessary to supply nitrogen in the form of ammonia or urea.

B. Deproteinization

In some processes (e.g. Bel Fromageries; Vrignaud, 1976) deproteinization is considered necessary for successful fermentation with *Kluyv. fragilis*. In some form or other (probably excessive foaming), whey proteins interfere in the fermentation process. Wasserman (1960a) found that *Kluyv. fragilis* grew as well in raw whey as in heat-treated deproteinized whey; so did Bayer (1977) using *C. intermedia*. If deproteinization was desired, Wasserman (1960a) clarified the medium (cottage cheese whey) first by filtration through sintered glass at a pH of 4.6 adjusted to pH 6.0 with sodium hydroxide, autoclaved at 121°C for 10 minutes, cooled, the pH adjusted to 4.5 and centrifuged. Deproteinization is usually carried out by heating to boiling at the isoelectric point (pH 4.65); decontamination is then carried out at the same time. According to Vrignaud (1976), the temperature of sweet whey (pH 6.4–6.8) is raised by a heat exchanger to 80°C, the pH adjusted quickly to 4.4 and the temperature raised within a few minutes to 87–90°C. The Wheast Process operates similarly.

Precipitated protein is recovered by sedimentation at gravity at a concentration of 25–30% dry matter, roller- or spray-dried and usually sold separately at a price higher than that of yeast in view of the very high protein content. The yield of protein is in the order of 5 g dry mass per litre. According to Bayer (1977), the essential cationic composition of whey is not changed by deproteinization (heat at iso-electric point).

Lembke *et al.* (1975) investigated the feasibility of deproteinization in their dual-culture process. After lactic fermentation, when a pH 4.6–4.7 (near the iso-electric point) was reached, the mash was heated to 72°C by heat exchanger and then to 92–95°C by steam injection for 20 minutes, when protein precipitation took place. Without separating the proteins, the mash was cooled to 25°C and transferred into a fermenter, similar to the Waldhof process, where yeast production took place. The authors claim that the expected advantages, i.e. ease of recovery of whey proteins, together with yeast in the centrifuges, were not fully realized. There were deposits in the unavoidable storage tanks prior to and after yeast production, as well as redispersion of proteins into a pseudocolloidal form, preventing complete recovery by centrifugation. Deproteinization (hot-acid) after yeast production, and prior to separation, resulted also in losses in the centrifuges due to yeast plasmolysis and partial solubilization of the cell contents.

In the French Bel process, carried out industrially, care is taken to pasteurize whey immediately, as it is available prior to any other processing. Normally the whey is then subjected to batch deproteinization, requiring equally large storage vessels as in the primary storage, then stored again in order to be able to feed the fermenters continuously. In the Vienna-Unterkärtner Molkerei process, experience has shown that the primary storage need not be preceded by a pasteurization process. The obvious lactic-acid bacteria development, during an approximate 24-hour storage, does not harm the yeast production process. The loss of energy from the substrate by this partial fermentation is insignificant.

C. Particular Additives

Apart from the additives (mineral acids, nitrogen and phosphorus sources, vitamins, antifoam agents) mentioned earlier, other materials may be added to achieve particular effects. Thus, deproteinization may not always be desirable, but mobilization of these proteins for biomass production may well be. Since no yeast with sufficient extracellular proteolytic activity has been discovered, Lembke *et al.* (1975) added yeast autolysate to their mixed bacteria-yeast culture. The autolysate was prepared by mixing 10 kg of recovered yeast (*C. krusei*, 20% dry mass) with 10 kg of ammonium sulphate, allowing it to autolyse for 24 hours at 45°C, and adding it to the fermenter at a rate of 5 kg yeast autolysate per tonne of whey.

Simek *et al.* (1964) added molasses at a rate of 10 g/l. This addition should serve a dual purpose, namely increasing the productivity of the fermenter and supplying vitamins for growth of *Kluyv. fragilis*. Bühner *et al.* (1975) considered it would be beneficial to raise the total sugar content to 8% by addition of molasses. In a plant using 150 tonnes of whey per day, production costs of biomass (dried) could then be decreased from the calculated 50–60 Deutsche marks/100 kg to about 30 Deutsche marks/100 kg. It has to be realized, however, that this procedure is applicable only if very efficacious and efficient aeration systems are available.

V. BIOMASS PRODUCTION KINETICS AND GENERAL PROCESSES USED

In practice, a continuous fermentation (biomass production) process followed by continuous biomass recovery is to be aimed at. While whey is produced discontinuously, it is nevertheless advantageous to carry out the fermentation and all other necessary processes continuously. If not, all the usual additional expenses in terms of fermenter size, labour requirement, pumps, storage and buffering tanks will occur.

A. Problems of Continuous Culture

Steady-state biomass production process on whey has met with greater difficulties than were anticipated. From a successful batch operation, it cannot be concluded that steady-state fermentation will be equally successful. In the Bel process, it is stressed that it is important to possess strains (*Kluyv. fragilis*) regularly producing high yields, in continuous operation, implying that not all strains are suitable for continuous operation, at least, not under the particular conditions of operation used.

The results obtained in batch cultures with *C. intermedia* could not be transferred into continuous cultivation (Bayer, 1977). Using deproteinized sweet whey at 4% lactose concentration with 0.75% ammonium sulphate and 0.1% yeast extract added, 18 g dry mass/litre were obtained in 9 h with a maximum yield of 21 g/l after 9.75 hours. Starting from a pH of 4.65, the pH value rises first to about 6 (4 h) then decreases strongly, and is normally kept constant at pH 4 by addition of

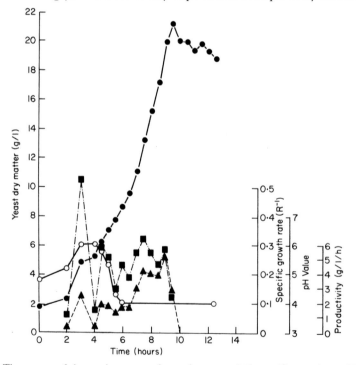

Fig. 4. Time-course of changes in content of yeast dry matter (●), specific growth rate (■), pH value (○), and productivity (▲) in a batch culture of *Candida intermedia* growing on sweet whey. From Bayer (1977).

alkali. Maximal specific growth rate (μ) is about 0.3 h^{-1} (up to 18 g/l). Maximal productivity is 5.4 g l^{-1} h^{-1} (Fig. 4). From these reproducible observations, one would infer that continuous steady-state culture would be possible at dilution rates up to 0.3 h^{-1}. Using the same substrate, and starting continuous culture after about 18 h of batch growth, the above yields could not be obtained at any dilution rate. The optimum obtained was at a dilution rate of 0.22 h^{-1} when the steady-state biomass concentration was about 10 g/l (Fig. 5). The same results were obtained in synthetic lactose media.

Butschek (1962) and Vrignaud (1976) reported that a one-stage continuous culture of lactose-utilizing yeasts (*Kluyveromyces* spp.) could be carried out only at lactose concentrations lower than that of native whey. In fact, *C. intermedia* reacts similarly in a 2% lactose whey medium supplemented with ammonium sulphate and yeast extract. At a dilution rate of 0.18 h^{-1}, steady-state concentrations of more than 10 g/l were obtained. A characteristic feature of the culture grown at lower sugar concentration is that less neutralizing agent is required.

If the culture is allowed to acidify (with ammonium sulphate as the nitrogen-source) to a pH value of 2.5 to 2.8, in native whey at 4% lactose, and operated continuously at this acidity without pH regulation, Mikota (1973) found that *C. intermedia* produces a high-level steady-state biomass concentration (higher than 20 g/l). In further tests, Bayer (1977) was able to attribute the poor yields in continuous culture at a pH value of 4 to mineral ions, in particular sodium. Semicontinuous cultures (exchange of

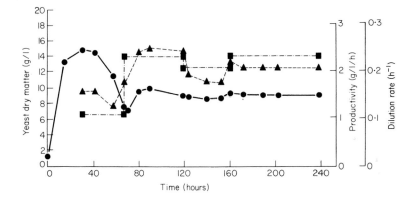

Fig. 5. Changes in the content of yeast dry matter (●), and productivity (▲), in continuous cultures of *Candida intermedia* grown at different dilution rates (■). Cultures were grown on sweet whey in the presence of sodium ions, with the pH value controlled at 4.0 with sodium hydroxide.

either 0.75 or 0.50 of culture volume by fresh substrate in 5 and 3 h intervals respectively) performed well if potassium hydroxide, instead of the previously used sodium hydroxide, was used to regulate pH value. Yields higher than 20 g/l were obtained (Fig. 6). The use of sodium hydroxide as neutralizing agent in semicontinuous culture, while producing higher yields (15 g/l) than in continuous (10 g/l) at the same effective dilution rate, was not fully satisfactory.

Increasing the sodium concentration by adding disodium hydrogen phosphate and maintaining pH value at 4 by sodium hydroxide, resulted in steadily decreasing yields in semicontinuous operation until virtual wash out was reached (Fig. 7). In continuous culture, the use of potassium hydroxide as a neutralizing agent and the omission of sodium phosphate did not produce yields as high or as reliable as the semicontinuous version. When, however, a salt mixture of 0.4% K_2HPO_4, 0.36% $MgSO_4$. $7H_2O$ and 0.1% $CaCl_2.2H_2O$ was added as well as the usual ammonium sulphate and yeast extract, a steady-state concentration of biomass of about 22 g/l at a dilution rate of 0.2 h^{-1} could be obtained. Later tests showed that both dipotassium hydrogen phosphate and yeast extract could be omitted provided either potassium

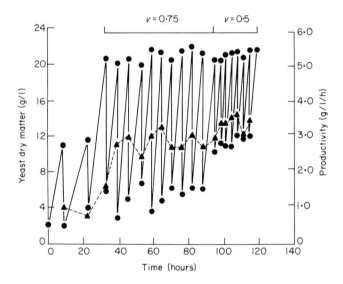

Fig. 6. Changes in the content of yeast dry matter (●), and productivity (▲), in semicontinuous cultures of *Candida intermedia*, grown on sweet whey, with 0.75 of the culture volume (*v*) exchanged for fresh medium after 5-hour intervals, and 0.50 of the culture volume for fresh medium after 3-hour intervals. From Bayer (1977).

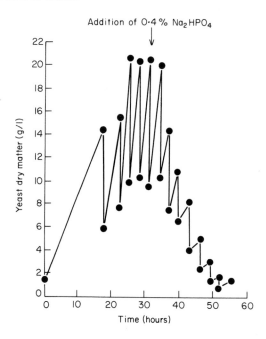

Fig. 7. Time-course of production of dry matter in a semicontinuous culture of *Candida intermedia* as affected by addition of 0.4% (w/v) of disodium hydrogen phosphate. The pH value of the culture was maintained at 4.0 with sodium hydroxide. After Bayer (1977).

or ammonium hydroxide was used as a neutralizing agent and magnesium sulphite or calcium chloride was added. The performance of these continuous cultures was in agreement with the predictions from batch-culture results, whereas omission of these salts from continuous cultures gave lower yields than predictable from the corresponding batch cultures with identical salt composition: a remarkable divergence of the usually accepted feasibility of transfer from batch-culture tests to continuous culture (see also Meyrath *et al.*, 1973), despite the observations by Butschek (1962), Lembke *et al.* (1975) and Bayer (1977) that, according to the composition of yeast with regards to mineral elements, whey composition should be adequate.

The effect of dilution rate in chemostat culture with *C. intermedia* was described in Section III D. If the oxygen supply is adequate, it can be concluded that in practice a dilution rate of 0.28 h^{-1}, a productivity of 6.5 g l^{-1} h^{-1} and a yield of about 22 g/l at 40 g/l of lactose can be maintained. Temperature and pH value are then the only factors to be regulated. The Bel process with *Kluyv. fragilis* functions continuously also,

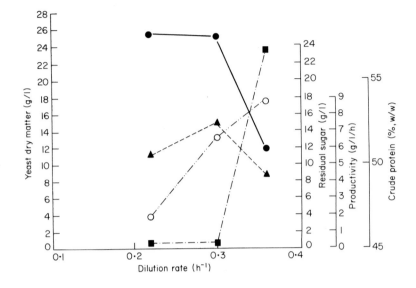

Fig. 8. Effect of dilution rate on the content of dry matter (●), residual sugar (■), crude protein content of cells (O), and productivity (▲), in chemostat cultures of *Candida intermedia* growing on sweet whey. From Bayer (1977).

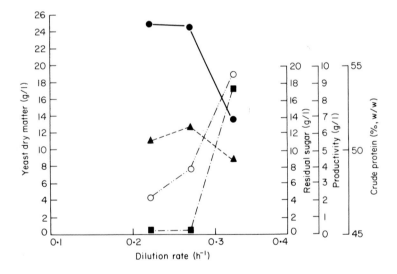

Fig. 9. Effect of dilution rate on the content of dry matter (●), residual sugar (O), crude protein content of cells (■), and productivity (▲), in chemostat cultures of *Candida intermedia* growing on sour whey. From Bayer (1977).

with these two parameters kept under control. With lactose concentration being adjusted to 3.4% (Vrignaud, 1976), the productivity at a dilution rate of 0.25 h^{-1} is about 4.25 g l^{-1} h^{-1}. The effect of dilution rate on yield and productivity with *C. intermedia* in sweet and sour whey is shown in Figs. 8 and 9 (Bayer, 1977).

B. Modern Industrial Processes

1.   The Bel Fromageries Process

Blanchet and Biju-Duval (1969) and Vrignaud (1976) characterized the Bel process as follows in order to fulfil the aims of set standards of the product (see Fig. 10): (1) immediate treatment of whey as it comes out of the cheese plant; (2) separation of the non-assimilable whey proteins discontinuously; 75% of the whey proteins are recovered, corresponding to 5 g protein per litre; (3) intermediate storage of the deproteinized whey, adjustment of the lactose concentration to 3.4% and addition of mineral salts followed by storage at 25°C; (4) continuous transfer of medium into the fermenter where the steady-state culture of *Kluyv. fragilis* is kept at 38°C, pH 3.5, and under aeration. In one plant, a net fermentation volume of 22 to 23 m³ is maintained in each fermenter equipped with a Le François–Mariller-type aeration system; 1700 to 1800 m³ air/hour is supplied by compressors; 5600 to 6000 l of substrate per hour are added to the fermenter, corresponding to a residence time of about 4 hours; residual sugar concentration is less than 1 g/litre; (5) the effluent of the fermenter is passed through centrifuges to recover the yeast, resuspended in water and centrifuged again; further concentration of yeast is achieved by continuous rotary filters; (6) the yeast cake is plasmolysed (thermolysed) at 83–85°C resulting in liquefaction of the solid mass; the destruction of lipases occurring at this stage is desirable to prevent the final product from becoming rancid. This solubilization is achieved partly by thermal plasmolysis, partly by heat-stable proteases and ribonucleases breaking down the polymers; (7) the thermolysed material is roller-dried to about 95% solids followed by bagging.

2.   The Kiel Processes

Lembke *et al.* (1975) and Moebus and Lembke (1971) have proposed the following variants of their mixed-culture processes.

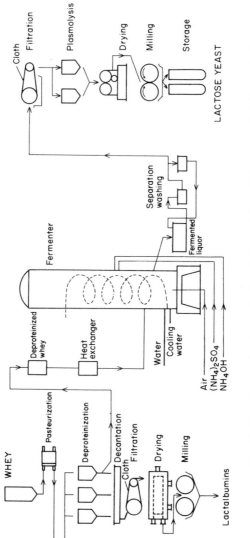

Fig. 10. Flow diagram showing production of whey protein and *Kluyveromyces lactis* biomass in the Bel Fromageries process. From Blanchet and Biju-Duval (1969) and Vrignaud (1976).

(a). Semicontinuous yeast production preceded by semicontinuous lactic fermentation.

1. The initial pH of whey is adjusted if necessary to 5.2 with ammonia, inoculated heavily with *L. bulgaricus*, the temperature is maintained at about 44°C, the pH kept constant at 4–4.5 by addition of ammonia until all the lactose is fermented (anaerobically) to lactic acid, a process requiring from 7–11 hours. About 125 litres of concentrated ammonia solution are required per tonne of whey, about one half of the fermenter volume is transferred into a second fermenter for yeast production, and fresh whey is added to the first.

2. Yeast production (*C. krusei*) with lactate as carbon source in the second fermenter under aerobic conditions is carried out semicontinuously by lowering the fermenter content from 6.5 to 2.5 tonnes and replacing the volume by effluent from the first fermenter, the fermentation time being 7 hours; addition of fresh substrate to the yeast fermenter is regulated by pH value, the lack of increase in pH value signals total utilization of substrate (lactate); the temperature is kept at 40 to 45°C.

3. Further treatment of the yeast produced can be carried out as described for the Bel process; lactic-acid bacteria, however, cannot be recovered by the commercial centrifuges available.

(b). Continuous lactic fermentation followed by continuous yeast production.

1. Lactic fermentation (*L. bulgaricus*) in the first fermenter is grown under gentle aeration with addition of ammonia (pH 4.8); partial yeast development takes place.

2. Yeast production (*C. krusei*) takes place with strong aeration in the second fermenter; further addition of ammonia is not necessary. The following data characterize this two-stage process:

	1st stage (*Lactobacillus bulgaricus*)	2nd stage (*Candida krusei*)
Temperature (°C)	44	44
pH value	4.8	6.2
Dilution rate (h^{-1})	0.07	0.14
Air (vol./vol./min)	0.12	0.47

(c) Single-stage continuous process of mixed lactic fermentation (*L. bulgaricus*) and yeast production (*C. krusei*) under the following conditions: temperature 44°C; pH = 5.5; dilution rate = 0.12 h^{-1}; air

0.35 vol./vol./min; yield, 22 g/l (at 41 g lactose/l. The productivity can be calculated to be 2.64 g $l^{-1}h^{-1}$.

3. The Vienna Process

This process is characterized by the use of a fast-growing acid-resistant strain of *C. intermedia*. In a dairy in Klagenfurt, Austria (Unterkärtner-Molkerei), no pretreatment of the sweet whey is carried out. A flow-diagram of the process is shown in Fig. 11. Ammonia and ammonium sulphate are added for satisfying the nitrogen requirement and for adjusting the pH value. The productivity is limited to about 4.5 g l^{-1} h^{-1} at a yield of about 24 g/l; higher productivities, as shown in previous sections, would be possible from the biological point of view. No separation of the yeast is carried out at present; instead the whole culture is concentrated to a dryness content of 12%. The costs of a secondary effluent treatment after recovery of yeast and protein would be higher than the evaporation costs.

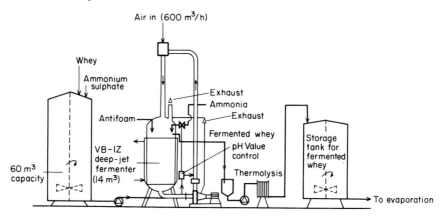

Fig. 11. Flow diagram showing production of fermented whey concentrate using *Candida intermedia* in the Vienna-Unterkärtner-Molkerei process. From Bayer and Meyrath (1976).

C. Inoculum Development

Even if continuous fermentation is practised, the problem of inoculum development arises periodically for process restarting, be it as a consequence of a general clean-up of the plant, for the sake of repairs, or otherwise. Since a biomass-production unit on whey is normally attached to a dairy or a cheese plant, there will be trained personnel available to maintain the cultures.

1. Stock Cultures

These can be kept in the usual way, either on slants at about 4°C, freeze dried, or frozen in liquid nitrogen. Yeasts can be subcultured on yeast-malt extract-agar or whey-agar supplemented with ammonium sulphate, with or without addition of yeast extract. Lactic-acid bacteria cultures can be subcultured on whey-agar supplemented with calcium carbonate or on any of the currently recommended media. Most lactic-acid bacteria are fairly air-tolerant and will normally grow on slants from which oxygen has not been removed. Cultures of both yeasts and lactic-acid bacteria are grown at their optimal temperature for 24 to 48 hours and can then be kept near freezing point for some four weeks before subculturing.

2. Inoculum Build-up

In order to carry out an industrial-scale fermentation, the inoculum is gradually built up, care being taken to use a large inoculum, i.e. 5 to 10% culture volume, at each stage.

Starting from a slant culture of *Kluyv. fragilis*, the following example can be quoted as a typical process of inoculum build-up (Amundson, 1967).

Stage	Shake culture	Laboratory fermenter	Seed tank
slants	6 × 200 ml cultures in 2000 ml conical flasks	3 × 14 litres (net)	500 gal
	12 h, 30°C	8–10 h, 30°C	8–10 h, 30°C
	2% lactose,		
	0.3% yeast extract	whey medium	whey medium
	0.4% ammonium sulphate,		
	0.05 M phosphate,		
	trace elements		
	(Mg, Mn, Co, Zn, Fe, Na, K)		

While Amundson (1967) carried out pilot tests in a 500-gallon fermenter,

Table 9

Yields and productivities obtained by growing microbes on whey

Author	Organism	Cultivation process	Lactose concentration used (%)	Average productivity (g l^{-1} h^{-1})	Biomass (per 100 g lactose)	Crude protein in dry mass (%)	Whey protein (g/l)	Whole fermented liquor concentrated (% solids)	% crude protein in dry mass	% ash in dry mass	Remarks
Amundson (1967)	Saccharomyces fragilis	batch	lactose with lactic acid 5.0		32	~ 50%	1.4				0.6% corn-steep liquor added
Atkin et al. (1967)	Trichosporon cutaneum	continuous	1.2	3.0–10.5	75	21.86					Diluted whey supplemented with 0.4% corn-steep liquor or 0.4% yeast extract
	Saccharomyces fragilis N.R.R.L.-Y1109	continuous	1.2	1.5–4.6	37	49.81					
Bayer (1977)	Candida intermedia	continuous	4.8	7.5	53	51.8					
Bechtle and Claydon (1971)	Mixed culture: lactic-acid bacteria										
	with yeast	batch	5.1	0.69	43.3	50.9					
	yeast only	batch	5.1	0.64	49.8	41.4					
Bel process (Blanchet and Biju-Duval, 1969; Vrignaud, 1976)	Saccharomyces fragilis	continuous	2.8	4.0	50	min. 47%	4.5				
Chapman (1966)	Saccharomyces fragilis	batch	4.4 lactose with 0.6% lactic acid		19	45–60					

Table 9 cont.
Yields and productivities obtained by growing microbes on whey

Author	Organism	Cultivation process	Lactose concentration used (%)	Average productivity (g l⁻¹h⁻¹)	Biomass (per 100 g lactose)	Crude protein in dry mass (%)	Whey protein (g/l)	Whole fermented liquor concentrated			Remarks
								(% solids)	% crude protein in dry mass	% ash in dry mass	
Fabel (1949)	*Oidium lactis*	batch	4.5		50	50					
Kiel process; (Lembke *et al.*, 1975)	Mixed culture; *Candida krusei* *Lactobacillus bulgaricus*	continuous	4.1	2.7	52	50.7 58.4					44°C
Lefrançois (1966)	*Klyveromyces fragilis* *Klyveromyces lactis*	continuous	3.6	2	50			95	5.1	16	
Sav process; (Naiditch and Dikansky, 1960)	*Saccharomyces fragilis* *Candida pseudo-tropicalis*	continuous	4.2	2	36–42			95	5.1	16	
Simek *et al.* (1964)	*Saccharomyces fragilis*	continuous	3.8–4.2	2–3				99	8.6	22	1% molasses added
Tomíšek and Grégr (1961)	Mixed culture; *Candida utilis* *Candida pseudo-tropicalis* *Candida tropicalis*	semi-continuous	2.99–4.6	2	40–54		2.5–4				
Vienna-U.K.M. Process (Bayer and Meyrath, 1976)	*Candida intermedia*	continuous	4.2	4.5	54		7.8	92	47	16	
Waldhof process; Demmler, 1950	*Candida utilis*	continuous	3.8	1.5	32	52.5	3–4				

this fermenter can be used to inoculate a 10-fold larger fermenter, which in many instances is already small production scale. The seed-tank volume can be kept small if a continuous yeast production is carried out at this stage, the biomass recovered by a centrifuge, kept cool and accumulated until an amount of yeast has accumulated corresponding to 5–10% of the expected yield in the industrial-scale fermenter. This concentrated yeast cream (10–12% dry mass) is transferred to the fermenter. Seeding the fermenter takes place at the beginning of whey transfer, and the yeast develops already while filling up.

D. Yields and Productivities

In Table 9, a compilation of both laboratory- and industrial-scale procedures, under various methods of cultivation, various substrate compositions and various organisms used, is presented. The figures contained therein are self-evident and notable features have been discussed in previous sections.

VI. EQUIPMENT AND INSTRUMENTATION

On the laboratory scale, any of the commercially available fermenters are satisfactory as long as there is sufficient turbulence and oxygen transfer. For industrial-scale economy in terms of capital and running costs, simplicity and sturdiness of all pieces of equipment are factors to be considered.

A. Material

Stainless steel, resistant to acid solutions, is available at reasonable prices. It is the common material used nowadays for storage vessels, piping, fermenters and sometimes for pumps. Autoclaving is not required, hence wall thickness can be kept as low as required merely for static reasons. Mild-steel lines with various acid-resistant material would be cheaper in capital costs, but lower heat transfer and necessity of maintenance

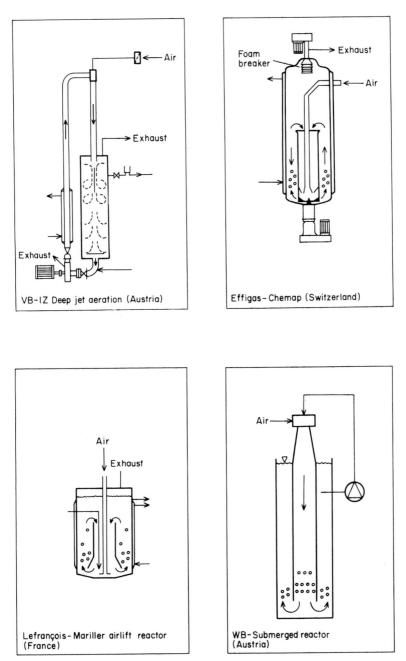

Fig. 12. Schematic representation of four fermenter types used for biomass production from whey.

prevent it from being favoured in practice. Fibre-glass tanks and pipes would be perfectly feasible from the biological point of view (no necessity for high-temperature treatment) but, again, heat transfer in the fermenter, being poor, involves the use of additional cooling coils which are less and less favoured.

B. Fermenters

Some of the types of fermenter tried out with lactose yeasts are shown in Fig. 12. Apart from the problem of foaming, production of yeast from whey presents no greater problem in terms of aeration than from other raw materials. The need for sufficient oxygen transfer has to be met, otherwise development of lactic-acid bacteria will present a problem. From the results here quoted, an oxygen transfer rate of about $4 \text{ g l}^{-1}\text{h}^{-1}$ is to be considered as minimum to avoid a deficiency of dissolved oxygen. As shown in the section on oxygen requirement, a fast-growing yeast in the steady state needs a transfer of about $6.5 \text{ g l}^{-1} \text{ h}^{-1}$. The energy requirement to achieve this oxygen transfer is critical. The lowest specific energy requirement quoted by some firms is 0.5 kWh per kg of dissolved oxygen.

In the Bel process, François–Mariller-type fermenters are used. It is not known whether the need to dilute whey to about 34 g lactose per litre is due to the inability of the fermenter to achieve a higher oxygen-transfer rate or because of better stability of the culture in steady state, as discussed previously. This fermenter, operating with air-lift, has a low efficiency of oxygen utilization but nevertheless is fairly economic in terms of energy requirement.

Newer developments of air-lift type fermenters, such as the I.C.I. and Uhde-Hoechst closed-loop fermenters, are considerably more effective in terms of oxygen transfer. They are designed for large-scale operation and, by a strongly increased air-lift effect and heights of 30 m and more, achieve very high circulation rates, with increased oxygen-transfer rates. Foaming problems are said to be less troublesome with this fermenter than with many other types.

The Chemap system (Effigas) has been used on pilot scale in the Kiel process. It is not known whether the low productivity (about $2.6 \text{ g l}^{-1} \text{ h}^{-1}$) of the Kiel system in mixed cultures is due to the particular process or to the inability to achieve higher oxygen transfer in the fermenter.

Effigas aeration is usually combined with the mechanical foam breaker Fundafom. The total energy requirement of the system (fermenter and foam breaker) is comparatively high.

The deep-jet aeration system of I. Z. Böhlen (G.D.R.), marketed by Vogelbusch (Austria), is characterized by the use of an external pump for rapid circulation of the mash. The mixture air-liquid is pumped to a level considerably higher than the holding vessel. The velocity of the falling mash through a shaft, accelerated by its slightly conical shape, entrains air from the mixing chamber downwards. Turbulence is achieved in the fermenter by the high-velocity central stream. The specific gravity of the mix is about 0.5. The pump, designed specially for transfer of air–liquid mixtures, carries out a partial degasification. No mechanical foam breaker is used, and the requirement for chemical antifoam agents is minimized. Oxygen-transfer rates of 4.5 g $l^{-1}h^{-1}$ at an energy consumption rate of 1 kWh per kg have been achieved on industrial-scale yeast production from whey. According to the manufacturer, both oxygen transfer and energy consumption can be improved considerably by modifications to the pump.

The Wagner-Biro (Austria) fermenter has been tried for yeast production from whey on a pilot-plant scale. Its characteristic is that, by passing the mash at high velocity through a comparatively narrow neck, air is drawn in by the under pressure created. The size of the air bubbles in the mash is regulated both by the velocity of the mash and the size of air-inlet jets. The uniform size of gas bubbles obtained is advantageous both for oxygen transfer and degasification. In the fermenter, of slim design, a countercurrent effect is created by the different velocities of the liquid and gas phase. The liquid phase in its fast downward move carries with it the air bubbles (at lowered velocity) and must be prevented from rising by liquid-flow velocity. Fresh mash thus continuously extracts the oxygen from the bubbles during their stay in the fermenter, and a high degree of oxygen utilization of the air put in is achieved. Oxygen-transfer rates of 7 g $l^{-1} h^{-1}$ have been measured in yeast production from whey. Degasification can be carried out to more than 90% by a cyclone installation before the mash reaches the pump. This is an advantage as conventional pumps can be used for circulation of the mash. Energy requirement in large-scale systems has been quoted to be not higher than 0.5 kWh per kg of oxygen transferred. For a more detailed discussion of various fermenter types, see also Katinger (1977).

C. Biomass Recovery

The recovery of filamentous organisms is comparatively cheap as rotary filters, low in capital and running costs, are adequate. Single-cell organisms are normally recovered by centrifuges. Bacteria, small in size, incur exorbitant costs in centrifugal energy. Yeasts can be recovered by centrifuges at energy requirements of 50–100 Wh per kg dry mass. Capital costs for centrifugal separators are the order of 8 to 10% of the plant-equipment costs. The yeast concentration in the mash has to be in the order of 10 g/l as a minimum to ensure economy. The slurry obtained after centrifugation has a concentration of 120 to 150 g/l. The use of strongly flocculent yeasts is advantageous in the recovery process, provided a density of about 120 g/l is obtained in reasonable time (see Meyrath, 1975). If they are not to be recovered separately from the mash, they present a problem in concentrated substrates like whey, where they are not required in a recirculation system to increase fermenter productivity. For non-flocculent yeast, Gasner and Wang (1970) have examined edible flocculating agents for their recovery.

Yeasts, at a concentration of 120 g/l, can be distributed directly to users (usually piggeries) provided the slurry is acidified to a pH ≤ 4.5. Formic, citric or lactic acids can be used for this purpose. The sale of wet yeast is particularly feasible for yeasts from whey, as most plants are comparatively small, and high drying costs can be avoided. If the biomass is to be dried, further concentration of the yeast slurry, up to 280 to 300 g/l, is desirable for economic reasons, prior to drying as a spray or on steam-heated rollers. This additional concentration is carried out in some cases by filtration, mostly by evaporation. Amundson (1967) proposed a self-cleaning clarifier for recovery of *Kluyv. fragilis* from whey, which in one operation yields a concentration of 30% solids. According to Robe (1964), a similar 'continuous desludging centrifuge' is used in the Wheast process whereby 14 to 15% solids concentration is obtained.

An alternative method to yeast separation is to use the whole fermented liquor as a feeding material. The solids content is too low for direct use. On the other hand, it is not always necessary to evaporate to dryness (which is expensive) if users are available in a geographical radius of some 150 km. As the material is then often destined for piggeries, it can be used as liquid-feeds. The advantage is obvious; there is no Biochemical Oxygen Demand in the effluent. For economic considerations (see Section IX). Drying the yeast-containing liquor on rollers presents corrosion problems in view of its acidity. Neutralization is not favoured, as ash content would undesirably increase, apart from the costs of chemicals.

VII. PRODUCT COMPOSITION AND USAGES

A. Composition

Various qualities of biomass can be derived from whey. Not only will the composition change with the organism used, but the method of cultivation, the nature of added compounds and recovery procedures will also have an influence. Nucleic acid content tends to increase with the specific growth rate of cultures. The amount of mineral components, such as ammonium sulphate as a source of nitrogen, combined with a required neutralization, will increase the mineral content of the mash and, therefore, also increase markedly the ash content of the biomass, if it is not washed after centrifugation, and particularly if the whole mash is evaporated. Table 10 shows several examples of gross biomass composition and the effect of washing after the first recovery stage (usually centrifuging) as well as the influence of the method of recovery. Biomass (yeast) recovered by centrifuges will normally contain 12–15% dry mass. Therefore a considerable amount of residual mash components will be enclosed. The mineral content of the dried product will rise and, nearly in proportion, the protein content will decrease. If the whole product is evaporated, the ash content will increase drastically. If this material is used for feeding purposes, care has to be taken to be very economic with the addition of minerals during fermentation (see Section IV.C).

A comparison of the detailed composition (amino acids and vitamins) of yeasts and fermented whey products determined in various laboratories for a number of organisms used is shown in Tables 11 and 12.

B. Feeding Trials

Numerous feeding tests have been carried out with yeast produced from whey. Surazynski et al. (1967b), using a Wistar strain of rat, found that replacement of isoelectric casein from milk protein (in the presence of soybean oil, wheat starch and vitamins) with 8.5, 17.0, 25.7 and 34.3%,

Table 10

Gross composition of biomass obtained by growing cultures on whey. Effect of washing and of recovery method

Author (process)	Treatment	Moisture	Crude protein	Percent (w/w)				
				Lipids	Minerals	Carbohydrate	Lactose	Lactic acid
Amundson (1967)	washed	2.6	54.35	10.34	5.68		2.43	
	unwashed	2.0	25.89	8.58	18.61		5.75	0.8
Bel Process Blanchet and Biju Duval (1969) Vrignaud (1976)	washed	6	47	9	7	31		
	unwashed				9.3–11			
Bernstein and Everson (1973)	washed	3–4	45–55		5–7			
	unwashed	3–4	42–50	2	8–11			
	and evaporated	3–4	32–40	2–3	15–20			
Moebus and Lembke (1975)	washed	8.9	45.3	—	9.4			
Lembke et al. (1975)	washed with whey protein	6.2	55.9					
	unwashed and evaporated				11			
					25			
Sav Process (Naiditch and Dikansky, 1960)	evaporated		25–36	4–5	16–20		30–17	3.0–4.5
Simek et al. (1964)	evaporated: 1% molasses added	1	54	4	22			
Vienna-U.K.M. Process (Bayer and Meyrath, 1976)	evaporated	9.34	41.9	5.9	16.7		0.63	
	roller-dried	88	50	—	10		1	—
Wheast Process Robe (1964)	washed with whey protein	5–7	58–60	1–2	8–10	22–26		

Table 11

Amino-acid composition of biomass produced from whey

Amino acid (g/16 g nitrogen)	F.A.O. pattern	Amundson (1967) Yeast	Amundson (1967) Yeast whey protein	Aitken et al. (1967) Trichosporon cutaneum	Aitken et al. (1967) Saccharomyces fragilis	Bernstein and Everson (1973) Saccharomyces fragilis	Blanchet and Biju-Duval (1969) Saccharomyces fragilis	Delaney et al. (1975) Saccharomyces fragilis	Schulz & Oslage (1975) Candida krusei Lactobacillus bulgaricus	Skupin et al. (1974) Propionibacterium shermanii	Surazynski et al. (1967) Yeast protein preparation	Surazynski et al. (1968) Candida pseudotropicalis	Surazynski et al. (1968) Saccharomyces lactis	Vienna-UKM process Bayer & Meyrath (1976) Candida intermedia	Vienna-UKM process Bayer & Meyrath (1976) Saccharomyces fragilis	Wasserman (1961) Streptococcus faecalis	Wasserman (1961) Escherichia coli
Lysine	4.2	11.1	10.4	8.41	11.14	6.9	6.65–7.3	8.0	7.6	7.98	8.86	7.4	7.7	6.7	11.14	7.2	3.8
Histidine	—	4.0	3.0	2.47	3.98	2.1	1.2–2.25	2.0	2.1	—	1.82	1.7	2.0	—	3.98	1.3	1.2
Threonine	2.8	6.5	7.5	4.69	5.57	5.8	5.2–5.9	5.3	4.8	5.46	5.28	3.6	4.9	5.7	5.57	3.3	3.1
Cystine/Cysteine	—	—	3.8	—	—	—	0.95–1.2	1.7	1.6	—	1.01	1.4	1.0	1.7	—	—	—
Valine	4.2	7.8	7.4	5.69	5.72	5.4	5.3–6.95	5.6	5.0	5.63	6.50	6.5	6.3	5.3	5.72	4.5	6.0
Methionine	2.2	1.6	2.0	1.97	1.57	1.9	1.2–1.4	1.5	1.7	1.44	2.29	1.8	1.4	1.3	1.57	1.9	2.0
Isoleucine	4.2	6.0	6.1	4.44	5.05	4.0	4.45–6.45	4.8	4.6	5.04				5.1	5.05	4.1	4.3
Leucine	4.8	9.6	11.0	7.15		6.1	7.45–7.65	8.1	7.5	10.36	14.48			7.8		4.7	8.3
Tyrosine	2.8	3.4	4.1	4.21	4.57	2.4		3.9	—	2.51	4.44	4.8	4.0	—	4.57	2.2	—
Phenylalanine	2.8	5.1	4.4	7.42	5.05	2.8	3.75–4.3	4.2	4.8	3.65	5.65	4.5	4.5	4.0	5.05	2.8	2.8
Tryptophan	1.4	—	2.3	—	—		1.3–1.35	1.7	—	—	1.69	1.5	1.5	—	—	—	—
Proline	—	—	—	4.21	—			4.2	—	—	—	3.5	4.0	3.7	—	2.6	—
Glycine	—	—	—	5.94	4.24			3.7	3.6	—	—	5.1	4.9	3.1	4.24	4.6	—
Arginine	—	7.4	5.7		7.37		4.8–5.0	4.9	—	4.77	—	—	—	3.6	7.37	3.4	4.2
Aspartic acid	—	—	—	8.9	10.4			9.4	4.2	—	—	—	—	9.8	10.4	7.6	—
Serine	—	7.0	6.8	5.45	5.21			4.7	—	—	—	4.8	4.3	5.1	5.21	1.9	—
Glutamic acid	—	—	—	12.37	15.34			13.8	—	16.24	—	11.5	11.0	14.7	15.24	10.1	—
Alanine	—	—	—	7.91	7.21			5.8	—	—	—	7.3	6.6	5.2	7.21	5.7	—

Table 12

Vitamins in two types of commercially produced lactose yeast

	Bel Process	Vienna-U.K.M. Process
	(mg/100 g dry wt)	
Thiamin (B_1)	1.77	0.26
Riboflavin (B_2)	5.6	7.6
Pyridoxin (B_6)	1.02	0.68
Nicotinic acid	54	22.2
Pantothenic acid	9.15	—
Biotin	0.022	0.035
Folic acid	0.80	—
Folinic acid	0.38	—
Cobalamine (B_{12})	0.0004	0.0031

respectively, of lactose–yeast–protein concentrate (35% protein and probably about 20% minerals) gave weight increases after 40 days of 92, 79.2, 59.2 and 42.8% of the control, respectively. The sum of protein administered was constant. In these tests it is suspected that the sulphate content of the product tested was too high. In similar experiments, Booth *et al.* (1962) came to the conclusion that inclusion of 20% of the protein diet from whey-yeasts (separately recovered and washed) gave satisfactory body gain. Bernstein and Everson (1973), who tested casein, centrifuged 'fermented whey mass' and 'evaporated fermented whey mass', and found that the last showed slower growth rates in rats, and that casein and centrifuged biomass were equally good. Frahm and Lembke (1975), using the Wistar strain, concluded that 10% of yeast protein grown on whey (washed) can be included in the protein diet. Lembke and Baader (1975) carried out feeding trials on farm animals and on egg production. Replacement of fishmeal and soya in terms of protein by biomass from whey (*L. bulgaricus* and *C. krusei*, washed) resulted in equal or better carcass weights. For broiler production, 20% lactose biomass was recommended. In egg production, more than 20% can be used and whey biomass can serve as sole source of nitrogen. In beef production, lactose biomass was equal to conventional sources of protein.

Schultz and Oslage (1975) found that supplementation of lactose yeast with methionine increased net protein utilization in feeding trials with rats. Both the biological value and net protein utilization of lactose biomass were increased in combination with cereals. In pig-feeding trials with a diet in which soybean meal had been replaced by whey biomass

(*C. intermedia*, whole mash evaporation), produced after the Vienna–Unterkärtner-Molkerei process, the slaughter value of the animals was increased (personal communications from users).

C. Users

For dietetic purposes, government regulations in many countries permit use of *Saccharomyces* species only. Whether *Kluyveromyces* spp., the newly defined genus for the previously described species *Sacch. fragilis*, is to be regarded as a 'saccharomyces yeast' will be a matter for debate. This clearly shows how much (or how little) value is to be attributed to defining a genus or a species fit for human consumption. Anyway *Kluyv. fragilis* produced by Bel Fromageries has been used for more than 10 years for dietetic purposes (Vrignaud, 1976). This product is also renowned for its particularly low content of nucleic acids. Vrignaud (1976) showed that the ratio of nucleic acids to protein in meat is 20:100 whereas, in the lactose-yeast, this ratio is only 12:100. The preparation can be used in the form of powder or tablets as additives to vegetables, white sauce, soups and mince. It is recommended for convalescent patients and babies, in slimming and in physical sports. In addition to flavouring qualities, lactose-yeast powder provides texture in the preparation of liver and meat paste and in biscuit making.

No doubt the particular raw material, i.e. a milk derivative, has done much to popularize the use of lactose yeast for dietetic purposes. In sulphite waste-liquor yeast, for example, there may be justified skepsis in terms of the possible presence of lead and arsenic in yeast. Fabel (1949) reported large-scale trials of *Oidium lactis* mycelium as additive to sausages and fillings. In animal feeding the Bel product is recommended, particularly for calves, piglets, carnivores (pets and fur production) and race horses. For slaughter calves, 6–8% lactose yeast is included in the diet, and 12–15% of skim milk can thus be replaced. For breeding animals, the calves are administered 15% of lactose yeast.

Amundson (1967) carried out tests with minks and dogs, the former requiring a particularly well balanced amino-acid composition for high-quality fur production. Dried whey (large lactose content) produced a laxative effect with 2–3% in the ration. Minks fed washed lactose-yeast outgrew the controls after a short adaptation period and reached full weight one month ahead when the average weight was 40 g higher than that of the controls fed on conventional protein sources. With dogs, palatability was increased significantly when lactose yeast replaced brewer's yeast (3% added). Lactose yeast, as a sole source of protein, could in fact replace other proteins.

Muller (1969) refers to results of several authors who used lactose yeast. The Sav product is said to be suitable for pigs and calves and, in particular, for poultry; the Wheast product has been used successfully for turkey, fish, mink, calves and convalescent animals.

Summarizing, it can be concluded that lactose yeast can replace other conventional sources of protein beneficially and that in many instances, such as its use for carnivores or in well-balanced rations for pigs, calves and poultry, excellent results are obtained.

VIII. ECONOMY

There are many different aspects to the feasibility of biomass production from whey. Considering the fact that virtually one half of the world whey production is wasted, with disastrous effects on surface waters, one wonders why nothing had been done about this a long time ago. This is all the more striking as many of the 'milk nations' have to rely on very large imports of proteinaceous feedstock. While the European Economic Community produces 75% of its requirements in grains (wheat, barley), some 80% of the proteinaceous feeds have to be imported (Bühner et al., 1975); more than 15 million tonnes (1 of fish meal, 14 of plant origin) are imported by the Community.

The 1973 crisis of soybean supply to Europe, which brought with it extreme price rises, fostered a strong desire by the appropriate countries for a higher degree of independence in terms of protein feedstock. Single-cell protein, always on the verge of feasibility, became a more realistic proposal than ever before. It is clear that the price of single-cell protein cannot be defined on the basis of occasional crises, but it is also considered that the potential of alternatives and the production of even small amounts of protein feeds will lessen the risk of being subjected to excessive price speculations and racketeering. According to Bühner et al. (1975), lactose yeast is particularly desirable for young pigs in the 20–50 kg range. Hence, a higher price will be paid. Furthermore, a general addition of 2–3% yeast to all proteinaceous feedstock is fully acceptable to the feed mills and to the users, attributable to the extra vitamin content. This proportion in the feeds has very little effect on the final price of mixed feeds even if a higher price (e.g. 10%) has to be paid for the microbial biomass.

Adequate price is the result of good marketing. The French product, well standardized and extensively tested, with excellent results in calf-raising, is paid about double the amount of 'conventional' feed yeast in France and other European countries. The fact that prime quality meat is obtained in a shorter time is good enough to justify a higher price. In other fields, this example is being imitated, e.g. by the Danish producer of enzymes and pharmaceuticals Novo (Windahl, 1977). The procedure implies that extensive and costly trials have to be carried out, an expense which only large companies can afford. Small firms whose waste problem may often be no less costly (in proportion) obviously will have to rely on state support, in terms of research both for the effluent treatment trials and for the feeding tests. In the dairy industry, an alternative is to act as a community or a federation if the buying of existing know-how is considered too expensive or otherwise is not feasible. The whole problem of price structure is of course fluctuating permanently. Time will come when valuable and well-paid products derived from effluents will have the effect of putting a price onto the raw material (the effluent), which has then turned valuable instead of creating disposal costs.

Even if a good price is obtained for the product in the end, efforts will have to be made to keep production costs as low as possible. In this respect, much can be done by keeping the process simple, avoiding expensive labour costs and energy requirements as far as is possible. In biomass production from whey, a big handicap is often the smallness of the plant required, involving relatively higher capital and labour costs. The drying of the material is particularly expensive on a small scale (some 24% of production costs; Bühner et al., 1975). Depending upon the localization, a cheese plant producing 30 000 litres of whey per day may not be able to distribute the native whey directly to the farmers; hence methods will have to be sought which justify an upgrading process of the whey on this smallish scale.

According to Muller (1969), production of lactose yeast in Australia is commercially sound at a volume of 1000 tonnes per annum if 200–220 Australian dollars per tonne is paid for the product (a price which was close to the current proteinaceous feeds). The Bel process, with a good price for the product, working on a 300 000 litres of whey per day basis (about 2300 tonnes of yeast per annum) is reported to be economically sound (Vrignaud, 1976). Bernstein and Everson (1973) considered that

Table 13

Investment costs (in Deutsch Marks) for biomass production in the Kiel process. From
Bühner *et al.* (1975)

Items	20 000 litres whey/day (DM)	150 000 litres whey/day (DM)
Substrate preparation	132 060	168 000
Fermentation	424 000	917 840
Separation	154 000	378 980
Drying	322 000	1 170 820
Building	31 000	150 000
Total	963 960	2 785 640

biomass production from whey would require a minimal production of
4000–8000 tonnes per annum in order to be economical. Based on large-
scale pilot tests by Lembke *et al.* (1975) and Moebus and Lembke (1975),
Bühner *et al.* (1975) examined the costs according to the Kiel process. The
fixed costs for two small plants were calculated as shown in Table 13; the
production costs are shown in Table 14. The variable costs of production
include mainly nutrients, additives and energy (electromotive power and
fuel). For the 20 000 litres per day plant, the fermentation represents 40%
and drying 24% of the production costs. The costs for whey have been
assumed to be zero.

According to Drews (1975), the now disused Waldhof plant near

Table 14

Production costs (in Deutsch Marks per tonne) for biomass production by the Kiel
process. From Bühner *et al.* (1975)

Items	20 000 litres whey/day (DM)	(%)	150 000 litres whey/day (DM)	(%)
Variable costs	673	42.5	405	55.2
Depreciation	426	27.0	164	22.4
Interest	213	13.0	82	11.2
Personnel	190	12.0	51	6.8
Repair	85	5.5	32	4.4
Total	1587	100	734	100

Mannheim producing 1080 tonnes of biomass per annum had production costs of DM 787 per tonne (92% dry) of biomass. The production costs of a material with 20% dry mass, corresponding to an approximate production of 5130 tonnes per annum, were DM 104 per tonne, i.e. DM 500 based on 100% dry mass. High investment costs and high energy demands were stated to be the cause for the failure of the venture. From this it would result that, under German conditions, a biomass project from whey could not be viable at production costs between DM 730 and 780 per tonne and the minimal plant size would have to be on the basis of 150 000 litres of whey/day capacity. There are, however, at least three aspects to consider: (1) are there better alternatives of using or upgrading whey? If not, the costs for biological purification of the effluent would certainly be of an order surpassing the biomass production costs by a factor of at least 10; (2) to what extent could the production costs be lowered? (3) to what extent could appropriate marketing guarantee a better sales price? These problems cannot be examined in detail here, but the facts in other countries certainly show that profitable projects can be realized.

As an example, the calculation derived from a small Austrian plant (30 000 litres per day) is shown (Table 15).

The product, a 12% concentrate of the evaporated fermented mash, has been used for liquid feeding in large-scale piggeries for two years. This procedure is feasible under Austrian conditions for plant sizes up to at least 120 000 litres whey per day. Depending upon the circumstances, several alternatives are possible, i.e. separate recovery of whey proteins (hot acid) and of biomass, followed by biological treatment of the residual waste; ultrafiltration to recover functional proteins and separate recovery of biomass; drying of biomass if large-scale installations (from 150 000 litres of whey per day upwards) are possible. In this example, the simplification of the process has done much to lower production costs. Further simplification is still possible in small and very small plants, where evaporation costs up to about 12% dry mass, or recovery of the biomass would involve comparatively large costs on this production scale. The use of strongly flocculent yeasts, which can produce a dry mass content of about 12% in the sediment after 1–2 hours residence time, would save some 17–20% on investment and about 20% on operation costs. Agreements can also be reached with the equipment manufacturers for small-size plants to use local personnel for part-manufacture of equipment and for the erection of the plant.

Table 15
Production costs of fermented whey concentrate (12% dry mass, 50% crude protein)

	Requirement per 1000 litre concentrate	Price per unit (Austrian Schillings)	Price per 1000 litre concentrate (Austrian Schillings)
Whey	3000 litre	0.02/litre	60.00
Nutrients			
Ammonium sulphate	10.5 kg	3.2/kg	33.60
Ammonia (25%)	9.6 litre	3.0/litre	28.80
Cleaning agents (NaOH)			3.6
Antifoam	120 g	30.0/kg	3.6
Cooling water			
Fermenter	19 m³	1.2/m³	22.8
Evaporation	25 m³	1.2/m³	30.00
Electromotive power			
Fermentation (0.5 kWh/kg oxygen)	35 kWh	0.79/kWh	27.65
Evaporation	48.4 kWh	0.79/kWh	37.92
Steam	0.462 tonnes	256/tonne	118.3
Personnel	0.3 Cat.A	106.68/h	32.00
	0.3 Cat. B	67.69/h	20.31
Laboratory			15.80
Depreciation (10% of machinery and equipment = 126 000 Austrian Schillings/a)	—	—	42.00
Interest (7% of one half of investment costs = 63 000 Austrian Schillings)	—	—	21.00
Repair (1.5% of total investment costs = 27 000 Austrian Schillings/a)	—	—	9.00
		Total	506.38

Calculation bases:
Plant costs: 1.8 Million Austrian Schillings
Yield of biomass, dry mass (50% crude protein): 23–24 g/l
Productivity based on liquid content of fermenter: 4.6–4.8 g $l^{-1}h^{-1}$
Gross fermenter volume (equal to twice the liquid content): 14m³
Energy requirement for aeration and agitation: 0.5 kWh/kg oxygen
Continuous one-step operation
No pretreatment of whey
Product sold in liquid form after concentration of the fermented liquid (12% dry mass) within a radius of 200 km
Personnel from the existing dairy were integrated.

Table 16

Production costs of fermented whey concentrate (12% dry mass, 50% protein in dry mass)

	Austrian Schillings per 1000 l of concentrate
Whey	60.00
Nutrients together with antifoam and cleaning agents	78.60
Cooling water	31.00
Electromotive power	66.00
Steam	118.30
Personnel	53.00
Laboratory	15.80
Depreciation, capital costs and repairs	72.60
Total	495.30

Calculation bases:

Yield of biomass, dry mass (50% crude protein)	23–24 g/l
Productivity based on liquid content of fermenter	4.6–4.8 g l^{-1} h^{-1}
Gross fermenter volume (equal to twice the liquid content)	14 m^3
Energy requirement for aeration and agitation	0.5 kWh/kg oxygen
Steam requirement for evaporation	23 kg per 1000 l of evaporated water

Personnel from the existing dairy were integrated.

In conclusion it can be stated that biomass production from whey is economically sound also on small-scale operation, furnishing high-grade feeds and reducing or abolishing completely the problem of effluent disposal.

REFERENCES

Amundson, C. H. (1967). *American Dairy Review* **29**, 94.
Atkin, C., Witter, L. D. and Ordal, Z. J. (1967). *Applied Microbiology* **15**, 1339.
Bayer, K. (1977). *Doctoral Thesis*: University of Agriculture, Vienna.
Bayer, K. and Meyrath, J. (1976). *Zentralblatt für Bakteriologie und Hygiene* **252**, 37.
Bechtle, R. M. and Claydon, T. J. (1971). *Journal of Dairy Science* **54**, 1595.
Bernstein, S. and Everson, T. C. (1973). *Food Processing Waste Management* Cornell Agricultural Waste Management Conference.
Berry, R. A. (1923). *Journal of Agricultural Science* **13**, 218.
Blanc, B. (1969). International Dairy Federation Seminar on Whey Processing and Utilization at Weihenstephan, G.F.R.
Blanc, B. (1974). *Laitier Romand 87/89*, **3**.

Blanchet, M. and Biju-Duval, F. (1969). International Dairy Federation Seminar on Whey Processing and Utilization at Weihenstephan, G.F.R.

Booth, A. N., Robbins, D. J. and Wasserman, A. E. (1962). *Journal of Dairy Science* **45**, 1106.

Bronn, W. K. (1966). Symposium Arbeitsmethoden und Aktuelle Ergebnisse der Technischen Mikrobiologie, Berlin.

Bühner, T., Hollman, P. and Meinhold, K. (1975). *Sonderheft Berichte der Landwirtschaft* **192**, 720.

Butschek, G. (1962). *In* Die Hefen II, (F. Reiff, ed.). Verlag Hans Carl, Nürnberg.

Butschek, G. and Neumann, F. (1967). *In* Die Hefen II, (F. Reiff, ed.). Verlag Hans Carl, Nürnberg.

Chapman, L. J. P. (1966). *New Zealand Journal of Dairy Technology* **1**(3), 78.

Cooney, C. L. (1975). *In* Symposium on Continuous Culture of Microorganisms, (D. W. Tempest, ed.). p. 146. Oxford.

Davidov, R. B., Gul'ko, L. E. and Faingar, B. I. (1963). *Izyvestia Timiryazev sel'skokhoz Akademia* **5**, 166.

Delaney, A. M., Kennedy, R. and Walley, B. D. (1975). *Journal of the Science of Food and Agriculture* **26**, 1177.

Demmler, G. (1950). *Die Milchwissenschaft* **4**, 11.

Drews, S. M. (1975). *Sonderheft Berichte der Landwirtschaft* **192**, 755.

Edelsten, D. (1974). *Die Milchwissenschaft* **29**, 348.

Fabel, K. (1949). *Die Milchwissenschaft* **4**, 203.

Frahm, H. and Lembke, A. (1975). *Sonderheft Berichte der Landwirtschaft* **192**, 599.

Gasner, L. L. and Wang, D. I. C. (1970). *Biotechnology and Bioengineering* **12**, 873.

Gordon, W. G. and Whittier, E. O. (1965) *In* 'Byproducts from Milk' (B. H. Webb and E. O. Whittier, eds.). The AVI Publishing Company Inc., 1970.

Hanson, M. A., Rodgers, N. E. and Meade, R. E. (1947). United States Patent 2 465 870.

Harju, M., Heikonen, M. and Kreula, M. (1976). *Die Milchwissenschaft* **31**, 530.

Harrison, J. S. (1967). *Process Biochemistry* **2**, 41.

Janicka, T., Maliszewska, M. and Pedziwilk, F. (1976). *Acta Microbiologica Polonica* **25**, 205.

Johnstone, D. B. and Pfeffer, M. (1959). *Nature, London* **183**, 992.

Katinger, H. W. –D. (1977). *Proceedings of the Fourth Symposium of the Federation of European Microbiology Societies*, Vienna, B28.

Klein, W. (1975). Diploma Thesis: University of Agriculture, Vienna.

Lefrançois, L. (1966). *Zentralblatt für Bakteriologie und Hygiene* **196**, 99.

Lehninger, A. L. (1975). 'Biochemistry', p. 643. Worth Publishers Inc., New York.

Lembke, A. and Baader, W. (1975). *Sonderheft Berichte der Landwirtschaft* **192**, 903.

Lembke, A., Moebus, O., Grasshoff, O. and Reuter, H. (1975). *Sonderheft Berichte der Landwirtschaft* **192**, 571.

Lodder, J. (1970). 'The Yeasts, A Taxonomic Study'. North Holland Publishing Co., Amsterdam.

Majchrzak, R. (1976). *In* Rynek Bialka ze Źródel Nietradycyjnych, (B. Limbs, R. Majchrzak and W. Bednarski, eds.). Polska Akademia Nauk, Instytut Rozwoju wsi i Rolnictwa.

Meyrath, J. (1975). *Process Biochemistry* **10**(3), 20.

Meyrath, J., Bayer, K., Braun, R., Mikota, P. and Han. H. E. (1973). Symposium Technische Mikrobiologie, Berlin.

Mietke, M. and Dubrow, H. (1944). *Deutsche Molkerei-und Fettwirtschaft* **37**, 290.

Mikota, P. (1973). Diploma Thesis: University of Agriculture, Vienna.

Moebus, O. and Lembke, A. (1975). *Kieler Milchwirtschaftliche Forschungsberichte* **27**, 3.
Müller, W. R. (1948). *Die Milchwissenschaft* **4**, 147.
Muller, L. L. (1969). *Process Biochemistry* **4**, 21.
Naiditch, V. and Dikansky, S. (1960). French Patent, 1235 978.
Negaard, J. (1975). Doctoral Thesis: Eidgenössische Technische Hochschule, Zürich.
Olling, Ch. C. J. (1963). *Netherlands Milk Dairy Journal* **17**, 176.
Pedziwilk, F., Janicki, J. and Nowakowska, K. (1970). *Acta Microbiologica Polonica* **2** (19), 299.
Porges, N., Pepinski, J. B. and Jasewicz, L. Z. (1951). *Dairy Science* **34**, 615.
Robe, K. (1964). *Food Processing, Chicago* **25**, 95.
Scheurer, F. (1968). *Deutsche Molkerei-Zeitung* **36**, 1572.
Schulz, M. E. (1969). International Dairy Federation Seminar on Whey Processing and Utilization. Subject 1, Addendum p. 3, Weihenstephan, G.F.R.
Schulz, E. and Oslage, H. J. (1975). *Sonderheft Berichte der Landwirtschaft* **192**, 607.
Šiman, J. and Mergl, M. (1961a). *Prumysl Potravin* **12**, 155.
Šiman, J. and Mergl, M. (1961b). *Zpravy Vyzkumneho Ustavu Mlekarenskeho Praha* **9**, 10.
Simek, F., Kovács, J. and Sárkány, J. (1964). *Tejipar* **13**, 75.
Skupin, J., Pedziwilk, F., Giez, A., Jaszewski, B., Trojanowska, K. and Nowakowska, K. (1974). *Roczniki Technologii i Chemii Zywnosci* **24**(1), 17.
Surazynski, A., Chudy, J. and Aleksiejczyk, Z. (1967a). *Journal of Nutrition and Dietetics* **4**, 118.
Surazynski, A., Chudy, S., Pozmanski, J. and Mieczkowski, M. (1967b). *Journal of Nutrition and Dietetics* **4**, 306.
Surazynski, A., Poznanski, S. and Wojciech, A. (1968). *Die Milchwissenschaft* **23**, 484.
Tilbury, R. H., Davis, J. G., French, Sh., Imrie, F. K. E., Campbell-Lent, K. and Orbel, C. (1974). *Proceedings of the Fourth International Symposium on Yeasts*, p. 265. Vienna.
Tomišek, J. and Grégr, V. (1961). *Kvasny prumysl* **7**, 130.
Ullman, A. and Schwartz, M. (1973). Offenlegungsschrift, G.F.R. 2 328 016.
Vananuvat, P. and Kinsella, J. E. (1975). *Journal of Food Science* **40**, 336.
Vringnaud, Y. (1976). *Die Österreichische Milchwirtschaft* **31**, 405.
Waeser, B. (1944). *Chemiker Zeitung* **7**, 120.
Wagner, K. (1973). Diploma Thesis: University of Agriculture, Vienna.
Wasserman, A. E. (1960a). *Journal of Dairy Science* **43**, 1231.
Wasserman, A. E. (1960b). *Dairy Engineering* **77**, 374.
Wasserman, A. E. (1961). *Journal of Dairy Science* **44**, 379.
Weilbuchner, J. (1977). Diploma Thesis: University of Agriculture, Vienna.
Windahl, B. (1977). *Proceedings of the Conference on Effluent Treatment in the Biochemical Industries*, London.
Wix, P. and Woodbine, M. (1958a). *Dairy Science Abstracts* **20**, 537.
Wix, P. and Woodbine, M. (1958b). *Dairy Science Abstracts* **20**, 621.
Wix, P. and Woodbine, M. (1959a). *Journal of Applied Bacteriology* **22**, 14.
Wix, P. and Woodbine, M. (1959b). *Journal of Applied Bacteriology* **22**, 175.
Zerlauth, G. (1973). Diploma Thesis: University of Agriculture, Vienna.

9. Biomass from Cellulosic Materials

CLAYTON D. CALLIHAN AND JAMES E. CLEMMER*

Department of Chemical Engineering, Louisiana State University, Baton Rouge, Louisiana, U.S.A.

I. INTRODUCTION

Cellulose is produced over most of the land surface of the earth by photosynthesis. The amount of biomass produced by conversion of carbon dioxide and water depends among other factors on the intensity of the sun, the annual rainfall and available nutrients. The biomass per acre generally increases as the distance from the equator decreases. Maximum cellulose production per unit of land area generally occurs in the

*Present address: Department of Chemical Engineering, University of Mississippi, Oxford, Mississippi, U.S.A.

271

K

semitropical rain forests near the equator. Sugar-cane bagasse grows very rapidly in latitudes from $0°$ to $30°$ where high rainfall occurs, yielding biomass productions as high as 60 short tonnes per acre per year on specially planted lands.

Typical biomass products of photosynthesis contain about 50% cellulose, about 25% lignin and approximately 25% hemicelluloses, xyloses and pentoses. If micro-organisms are to consume these materials rapidly when plants die, they prefer certain types of sugars but, of course, some micro-organisms consume all plant materials. If this were not so, we should currently be buried in compounds such as C_5 sugars and lignins.

A. Substrate Preference of Micro-Organisms

Cellulose is a more easily digestible portion of the biomass, produced by photosynthesis, than lignin and hemicellulose. It is not, however, nearly as easily consumed as is sucrose, fructose and other water-soluble components that are biosynthetic products of living plants. The solubility of a polysaccharide is largely determined by the length of the chains which bind the residues together, the number of branches, and the degree of orientation or alignment of the individual residues.

Sucrose is readily metabolized by invading organisms and converted to products such as carbon dioxide, alcohol and methane, depending on whether it is aerobically or anaerobically consumed. This phenomenon is easily seen in freshly harvested sugar-cane bagasse. Most of the sucrose is removed by crushing and extraction, but usually 5–6% is left in the stalks. If the bagasse is baled and stacked into piles immediately upon discharge from the sugar mill, large quantities of acetone, alcohol and acetic acid are released by fermentation in the piles with subsequent pungent odours. The centres of the bales become very hot where exothermic decay is occurring rapidly. Once the soluble carbohydrates have been consumed, decay slows down and may stop completely, depending upon the moisture content. Higher molecular-weight carbohydrates remain behind, undergoing little if any decay. Apparently those organisms that consumed the sugar did not have the ability to consume the anhydroglucose or xylose residues that make up the repeating units of the polymeric chains. The amount of starch present in sugar-cane bagasse is slight; however, other green plants produce sizeable quantities of starch as nutritional reserves. Typical of these would be cassava and potato.

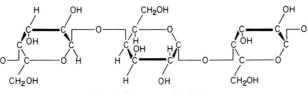

Fig. 1. Structure of cellulose.

The principal difference between starch and cellulose is in the manner of connecting the basic glucose units together. In starch, the glucose residues are connected predominantly by the α-1, 4 linkage, while for cellulose the linkages are β-1, 4.

In order for micro-organisms to consume rapidly any given substrate, that material must be of low molecular weight and water-soluble. Native cellulose and native starch usually do not meet either of these criteria, so that their consumption is either postponed indefinitely, or proceeds very slowly if all of the necessary conditions are satisfied.

A high molecular-weight polymer, such as starch or cellulose, is usually insoluble because of the strong intermolecular attractive forces between adjacent hydroxyl groups along the chains. Even small simple molecules like water show the effects of these Van der Waals attractive forces, and this effect is reflected in their elevated boiling temperatures and high latent heats of vaporization. Figure 1 shows the positions of the hydroxyl groups in cellulose molecules. These groups cause large forces which are primarily responsible for the delay in microbial consumption of both starch and cellulose.

B. Hydrolysis of Biopolymers

When glucose residues are hooked together in linear chains, either through the α or β linkage, they become water-insoluble when more than three or four units are fastened together (branched amylopectin may be an exception). Therefore, in order to consume either substrate, the first step that must be initiated by the potential organisms is to depolymerize the chains and break them down to glucose. This depolymerization, or hydrolysis, is performed on starch by an enzyme referred to as amylase and on cellulose by an enzyme called cellulase. In nature, where symbiotic cultures abound, amylase or cellulase may be produced by one

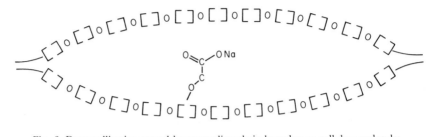

Fig. 2. Decrystallization caused by protruding chain branches on cellulose molecules.

organism which consumes only a small portion of the hydrolysate, while another organism (or organisms) consumes the remainder.

Advantage is taken of symbiotic growth for production of yeasts from starch in the Symba process (Jarl, 1969). Here *Endomycopsis fibuliger*, which has the ability to produce large amounts of amylase, is grown symbiotically with *Candida utilis*, a fast-growing yeast with a good amino-acid composition in its protein. In the Louisiana State University process for growing bacteria on cellulose (Callihan and Dunlap, 1969), *Cellulomonas flavigena* is grown because of its ability to produce cellulose-degrading enzymes, while *Alcaligines faecalis* is grown symbiotically because of its rapid growth and good protein value.

The molecular weights of both cellulase and amylase enzymes are usually rather large depending, of course, on which particular fraction one is talking about. But, generally, they range in size from about 25 000 to about 200 000 in molecular weight. This means that the enzymes have bulk or size, and that they must in some manner make contact with those molecules available to them. If the carbohydrate molecules are tightly held together by secondary bonds, then the relatively large enzymes cannot penetrate the matrix to make the required contact for carrying out the hydrolysis step.

C. Pretreatment Requirements

It follows as a consequence of the special requirements of the enzymes that, in order rapidly to carry out the hydrolysis required for microbial consumption, decrystallization of some kind should be carried out in a pretreatment step. Figure 2 shows that the attractive forces between chains can be cancelled by substituting a relatively large pendant group for one of the 3-hydroxyl groups on the glucose mesomer units. Limited substitution of the hydroxyl groups causes solubilization of cellulose chains, and water-soluble cellulose and starch derivatives have been made with molecular weights of over a million. However, the reason they are soluble is because of the prevention of most of the hydrogen bonding by substitution of the hydroxyl groups, and because of the potential energy requirement of a precise distance of separation between adjacent donor and acceptor of hydrogen bonds. Close approach is prevented by the bulk of the substituted carboxymethyl groups.

A further complication delaying enzymic attack occurs in native cellulose due to the presence of the high molecular-weight polymer lignin. This highly coloured aromatic material generally wraps itself around the tiny fibrils that hold cellulose molecules in native cellulose, and protects them from attack by invading enzymes. It would appear that, before rapid microbial consumption of cellulose or amylose (the linear chains of starch) can be carried out, a pretreatment step is necessary that effectively both decrystallizes the glucose polymers and depolymerizes lignin.

II. HYDROLYSING TECHNIQUES

A. Enzymic Hydrolysis

Many different pretreatments have been attempted to accomplish these required physical changes in native cellulose, and a great many have been successful. A good example of a mechanical technique that gives the required molecular alterations is done at the U.S. Army Natick Laboratories by Mandels and Weber (1969). These researchers subject native cellulose to several hours of ball milling. The product from this energy-intensive step can then be hydrolysed quite rapidly by cellulase enzymes produced from a fungus, *Trichoderma viride*, as discussed on page 277. By using a cell-free extract of extracellular enzymes, they are able to produce reasonably concentrated solutions of glucose. A highly automated pilot plant has been built by these workers to carry out enzymic hydrolysis of cellulose to produce glucose.

Investigators at the University of California at Berkeley (Cysewski and Wilke, 1970; Wilke *et al.*, 1976) have used the concept developed by the Natick researchers to produce yeast and alcohol from cellulose. They do this by using a cell-free enzyme extract to convert cellulose to glucose and then, in separate fermentation vessels, produce yeast aerobically for single-cell protein or anaerobically for alcohol. Ethanol has a large demand now as a fuel extender. It is interesting to note that the yeast they were growing for alcohol production was *Saccharomyces cerevisiae* which will only ferment about 70% of the sugars in the hydrolysate. They proposed to produce S.C.P. by growing torula yeasts on the remaining xylans because most micro-organisms will not metabolize xylan hydrolysates. Since these organisms are facultative aerobes and do not produce alcohol, they were to be harvested for their protein value.

James R. Frederick and his coworkers at Iowa State University (Frederick *et al.*, 1977), recognizing the inability of cellulolytic organisms to hydrolyse hemicelluloses present in native cellulosic wastes, particularly in corn and larchwoods, isolated the enzymes responsible for the hydrolysis. They found the enzyme system surprisingly complex with more than ten separate enzymes taking part in the breakdown. *Beta*-Xylosidase, the enzyme that catalyses the last step, namely the production of xylose from short xylo-oligosaccharides, was extensively purified and characterized. It was found to be quite stable at temperatures up to 70°C at pH 4. Cellobiase, on the other hand, was reported by Han (1969) to have been rapidly de-activated above 60°C, although it had very good hyrdolytic activity at 55°C at a pH value of 5.5.

B. Chemical Hydrolysis

Other investigators (Han and Callihan, 1974) have attempted to carry out an acid or base hydrolysis of cellulose as a pretreatment to microbial digestion. The results have been rather disappointing. Acid tends to attack the amorphous regions first, leaving the less accessible crystalline regions behind. Since enzymes have difficulty penetrating crystalline regions, the net results are that either the hydrolysis must be carried to completion, which is expensive, or a large portion of the initial substrate must be discarded.

C. Cellulase Enzymes

Although a pretreatment of some sort would generally be desirable, its use is of less importance in some cases than in others, depending upon the completeness of the available system of enzymes. Certain cellulase enzyme systems have a C_1 enzyme, postulated by Reese and his coworkers (1950) as being the swelling enzyme whose function is to swell or decrystallize the cellulose. Figure 3 shows the steps involved in enzymic hydrolysis of cellulose as proposed by these investigators. Their conclusions were based on results obtained using enzymes from the fungus *Trichoderma viride* attacking untreated cellulose. Subsequent research has

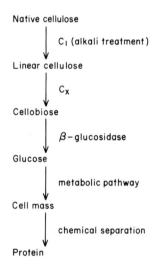

Fig. 3. Microbiological degradation of cellulose.

shown that not all cellulase systems contain the C_1 or swelling enzyme. As a matter of fact, the bacterial system for cellulose digestion investigated at Louisiana State University does not contain a C_1 enzyme and, in order to obtain reasonable digestion rates, the cellulose must be swollen by other techniques.

It should be pointed out that, even if the particular system of cellulase enzymes being contemplated for use contained C_1 enzymes, this does not negate the need for a pretreatment. It only says that the effect of any given pretreatment is of less importance where a C_1 enzyme is part of the system. It is also possible to obtain improvement in the rate of microbial consumption with almost any mechanical or chemical treatment.

III. CELLULOSE AVAILABILITY

Typical farm crops such as corn, beans, maize, peanuts and cereal grains are grown for their fruiting product and the stalks are invariably discarded. Yet the stalks and roots represent a good source of carbohydrate for microbial production of single-cell protein. As an example, sucrose in sugar cane represents about 13% by weight of the cane, and world production of sugar from sugar cane is about 50 million tonnes per year. That means that approximately 335 million tonnes of bagasse alone are available for S.C.P. production. Vinasse from sugar beets also represents a sizeable amount of available substrate. Currently, a large portion of bagasse is burned to generate power to run sugar mills, while most of the vinasse is incorporated into animal feeds, but both materials could be made available if microbial protein were needed badly enough.

In addition to bagasse and vinasse, there are many other agricultural residues which are simply left in the fields where they do very little good in terms of utility. Stalks from soybeans, corn, rice, barley, oats, wheat and cotton represent tremendous quantities of cellulose that could be converted to S.C.P. One can conclude then that the only thing holding up S.C.P. production from cellulosic wastes must be the technological problems associated with production.

IV. SINGLE-CELL PROTEIN

A. Indirectly from Cellulose

Numerous methods for converting cellulose into useful protein have been reported. These may be grouped into four general types of processes: (1) chemical hydrolysis of cellulose to glucose and growth of micro-organisms on the hydrolysate; (2) enzymic hydrolysis of cellulose to glucose and growth of S.C.P. on glucose; (3) symbiotic cellulose hydrolysis by one intact organism and conversion of hydrolysis products to S.C.P. by another intact organism; (4) hydrolysis of cellulose and biomass production by a single organism.

Acid hydrolysis and growth of the food yeast, *Candida utilis*, were developed and utilized by the Germans in World Wars I and II. The German Scholler or weak-acid process for hydrolysing cellulose was upgraded at the U.S. Forest Products Laboratories, and a few wood-sugar plants were built in the United States. Meller (1969) did an in-depth economic analysis on acid hydrolysis and growth of yeast on the resultant sugar. He determined a production cost of the order of 30 U.S. cents per pound for the yeast produced.

A recent study by Bobletton *et al.* (1976) of continuous hydrothermal degradation of straw revealed that treatment can be selectively carried out, so that first hemicellulose and then cellulose are converted. This is accomplished at two temperatures, namely 218°C for hemicellulose, then 260°C for cellulose degradation. This yielded xylose and arabinose at the lower temperature and glucose together with cellobiose at the higher. Though *C. utilis* was shown efficiently to utilize the four sugar products, only 15% of the straw hydrolysate was recovered as sugars.

Enzymic hydrolysis of cellulose and cellulase production by several fungi have been studied extensively at the U.S. Army Natick Laboratories by Mandels and Weke (1969); Ghose and Kostick (1970); Ghose (1969); Mandels *et al.* (1974); Reese and Mandels (1971); Reese (1969); and Huang (1975). Ghose and Das (1973) described a continuous process based on work by Ghose and Kostick (1970). This process consists of pulverizing the cellulosic material to a size of less than 25 microns and heating to 800°C in an oxidizing atmosphere followed by enzymic saccharification at pH 4.8 and 50°C. Enzymes and cellulose are recycled and the 7.3% syrup is converted to yeast protein by a fermentation step. The plant is designed to hydrolyse 100 metric tonnes of cellulose per day, and to produce 60 metric tonnes of sugar per day, with a volumetric production efficiency of 1.16 grams per litre per hour. Yeast for S.C.P. could be grown on the sugars, or the sugars could be separated and marketed.

Ghose and Das (1973) projected a total production cost of 16, 8 and 6.5 U.S. cents per pound of glucose produced from enzymic hydrolysis for plants of 10, 100 and 250 metric tonnes, respectively, per day. These costs were based on cost for cellulose of two U.S. cents per pound. Other economic evaluations of enzymic hydrolysis are presented in the literature. The processes are similar but, generally, each is unique in its method of preparation or recycle of the enzyme.

Nystrom and Allen (1976) proposed a series of batch fermenters to produce cellulase from *Trichoderma viride*. Each fermenter would be ready for harvest sequentially, and the one being harvested would provide inoculum for another to re-initiate the cycle. The enzyme would be harvested by ultrafiltration, and would contribute 1.7 to 2.2 U.S. cents per pound of sugar to the manufacturing cost based on one-time use of the enzyme. Recycling of 50% of the enzyme could decrease this cost by 25%.

Wilke et al. (1976) projected a cost of the order of 6 U.S. cents per pound for a plant hydrolysing 885 tonnes per day of newsprint. The newspaper was hammermilled to 20 mesh size prior to processing. The digestors were a series of five back-mixed reactors to which enzyme was recycled. Wilke et al. (1976) found that the manufacturing cost for the sugar was very sensitive to the cost of substrate; 3.7 U.S. cents per pound of sugar was added to the cost for each $20 per tonne in substrate cost. A substantial decrease in manufacturing costs is possible with increased mass conversion of the substrate.

Brandt (1975) points out that any economic analysis of enzymic hydrolysis is questionable, since no satisfactorily complete data are available for a continuous process. He feels that a cost of 20 U.S. cents per pound of glucose is more realistic.

B. Direct Production of Single-Cell Protein from Cellulose

Peitersen (1975) developed a method for producing S.C.P. from C. utilis and T. viride when grown symbiotically on unwashed alkali-treated rice straw. He found no difference in the effect of the micro-organisms but, in both cases, he found that the rate of enzyme production increased. The symbiotic cultures consumed 2 to 5 grams of cellulose per litre in seven to ten days of incubation.

There are several proposed processes for growing cellulolytic micro-organisms on cellulose; these are by Crawford et al. (1973); Bellamy (1974); Updegraff (1971) and Callihan et al. (1974). Crawford et al. (1973) grew Thermosperma fusca on tissue-mill waste. Eighty-nine percent of the waste material, initially 5 grams per litre, was converted to a product at a 24-hour generation time and a yield on solid of 35–40%.

A great deal of work has been done at the General Electric Company and at the University of Pennsylvania on growth of thermophilic actinomycetes on waste cellulose (Updegraff, 1971; Zabriskie et al., 1975). The organism is grown at 55°C and pH 7.5–7.8. Seventy to eighty percent of alkali- or ammonia-treated cellulose is utilized in the fermentation. A separation problem occurred in that 50% of the protein is extracellular.

Table 1

Alkali treatment parameters and cell-growth parameters for sugar cane bagasse.
From Callihan and Dunlap (1971)

ALKALI TREATMENT

$Y_s = 0.95 - 1.80$ N $- 0.02t$
$88 < T < 121°C$
$0.5 < t < 0.6$ h
$0.0 < N < 0.2$ grams of sodium hydroxide per gram of dry bagasse
$2 < L/S < 20$ grams of liquid per gram of dry bagasse
$0.65 < Y_s < 0.95$

CELL GROWTH

1st Order kinetics to cell concentration of 5 g/l

$$t_{md} = 5\text{--}7 \text{ h}$$

Zero-order kinetics for cell concentration of 5 to 20 g/l

$$\frac{dC}{dt} = 1.0 \text{ g/l--h}$$

$$\text{Yield} = 0.48 \, \frac{\text{g cell mass}}{\text{g cellulose consumed}}$$

Cellulose consumption $\left(\dfrac{S_0 - S}{S_0}\right) = 0.83$ (batch cultures)

$$\text{Base requirement} = 0.34 \, \frac{\text{g sodium hydroxide}}{\text{g cell mass produced}}$$

$$\text{Substrate loading} = 10.0 < S_0 < 40.0 \text{ g/l}$$

Y_s = substrate yield after alkai treatment

t_{md} = mass doubling time

S_0 = initial substrate concentration

$88 < T < 121°C$ = the temperature range for which this equation is useful

Updegraff (1971) grew *Myrothecium verrucaria* on ball-milled newspaper. Fifty to seventy percent of the pulp was consumed, and the unconsumed pulp and fungus were recovered, together yielding a product of about 30% protein. The productivity of the fermentation was given as 0.11 grams per litre-hour. The advantages include easier heat removal and exclusion of many contaminants by utilizing higher temperatures.

C. The Louisiana State University Process

The Louisiana State University process involves aerobic growth of a cellulose-degrading bacterium, which was isolated by Han and

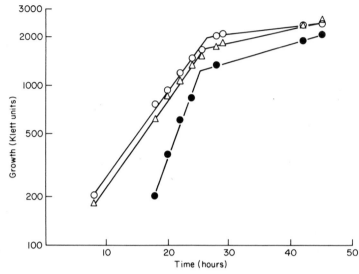

Fig. 4. Time-course of growth of *Cellulomonas* sp. on alkali-treated bagasse. △ indicates growth, and ○ indicates sodium hydroxide consumed (mmol. 10); ● indicates solids consumed (g/l. 10^2).

Srinivasan (1968) and identified as a member of the genus *Cellulomas*. After further characterization, the organism was called *Cellulomonas uda* (A.T.C.C. 21399). However, *Cellulomonas uda* is not at present recognized as a separate species of the genus. Therefore, the organism used in this research will be referred to as *Cellulomonas* sp.

Dunlap (1969) and Callihan and Dunlap (1971) developed a hot-alkali treatment for growth of *Cellulomonas* sp. on sugar cane bagasse. Irwin (1974) and Callihan and Irwin (1974) followed with a study of the parameters involved in alkali treatment and its effect on growth of *Cellulomonas* sp. The results of these studies are summarized in Table 1. It is postulated that the hot-alkali treatment replaces the C_1 function of the cellulase enzyme system which is missing from *Cellulomonas* sp. (Han and Srinivasan, 1968; Srinivasan and Fleenor, 1972).

Evidence indicates that sodium hydroxide treatment hydrolyses the ester bonds which hold lignin and hemicelluloses around the cellulose fibres in a netlike fashion. This creates surface in the pores as well as forming ionic bonds between the sodium ions and the hydroxyl groups of cellulose, which swells and hydrates the fibres (Bellamy, 1974; Kirk, 1971).

Irwin (1974) found that growth of *Cellulomonas* sp. on alkali-treated sugar cane bagasse was related to the amount of bagasse remaining

Table 2

Overall material balance and process parameters of the Lousiana State University process for producing single-cell protein

1.00 g Bagasse (dry wt.)

ALKALI TREATMENT

$T = 93°C$

$t = 1$ h

$N = 0.01$ (g NaOH/g bagasse)

$Y_s = 0.70$ (grams bagasse solids recovered after treatment and washing)

$N^*_{RX} = 0.05$ (g NaOH/g bagasse consumed in reaction)

0.70 g Washed Bagasse (dry wt.)

FERMENTATION

$t_d = 6.0$ h (cell mass-doubling time)

$Y = 0.40$ (g cell mass per g bagasse consumed)

$\dfrac{S_0 - S}{S_0} = 0.85$ (g bagasse consumed per g bagasse present initially)

base $= 0.14$ (g base/g cell mass produced) required

0.24 g cell mass (dry wt.)
(0.12 g protein, dry wt.)

insoluble after alkali treatment and washing (Y_s). The yield on alkali treatment was found to be a strong function of the ratio of the weight of alkali to the weight of dry material treated, and a weak function of the treatment time. The relationship and limits of application are given in Table 1 and typical growth curves are shown in Figure 4.

Bagasse is not a very pure product; in fact, it only contains about 50% cellulose, the remainder being lignin and hemicellulose. This means that two pounds of bagasse feed must be used to get one pound of cellulose. Furthermore, the micro-organism has metabolic energy requirements, as do all living systems, so that the yield of cells from a pound of cellulose is only one half a pound, the remainder coming off as carbon dioxide. This means that, from one pound of bagasse feed, only about a quarter of a pound of dry cells would be produced at 100% conversion.

Based on eight years of research funded by $1 500 000, researchers at the Louisiana State Univeristy (L.S.U.) have designed a process that could be used to produce protein commercially on a large scale. The flow sheet for the process is shown in Figure 5. Table 2 shows the process parameters and overall material balance for the L.S.U. process.

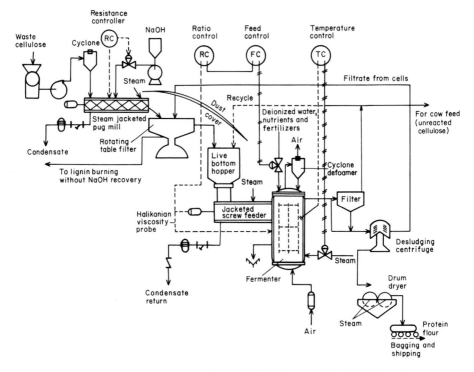

Fig. 5. Flow sheet for producing single-cell protein (*Cellulomonas* sp.) in the Louisiana State University process.

An economic analysis was made for a plant producing 100 tonnes of biomass per year. The results are shown in Table 3. These data would indicate a somewhat higher cost for S.C.P. than for soybean meal. However, if the comparison is made on a protein basis, S.C.P. produced by the L.S.U. process looks somewhat better because the cell mass is about 55% protein while soybean is only about 43%. The product is a pale yellow powder. The mineral content of the product is listed in Table 4 and the amino-acid composition in Table 5 (p. 287). The amino-acid

Table 3

Economics of the Louisiana State University process
for production of single-cell protein

Basis: 90.10⁶ kg/year (50% protein)

Capital Investment	Total cost in U.S. dollars
Fermentation	$35 000 000
Recovery	10 000 000
Drying	3 000 000
Cooling	1 500 000
Others	6 500 000
Working capital	4 000 000
Total capital investment	$60 000 000

Product cost based on:	Cost/lb of product in U.S. cents
Raw Materials	
Bagasse ($10/tonne @ 50% moisture)	4.2
Caustic soda (50% liquid)	1.4
Phosphoric acid	0.4
Hydrochloric acid (36.5%)	0.3
Anhydrous ammonia	1.4
Nutrients	1.4
Total raw materials	9.1
Operating Costs:	
Utilities	
Fuel @ $1.75/MM Btu	2.2
Electricity @ 1.7¢/Kwh	1.0
Water	0.1
Total utilities	3.3
Controllable fixed costs	
Non-controllable fixed costs	
Depreciation @ 6.67% of investment per year	1.9
Taxes and insurance @ 2% of investment per year	0.1
Total non-controllable fixed costs	2.0
Total manufacturing cost	16.0
Marketing and distribution cost	1.5
Return on total capital investment @ 10% after tax	6.0
Total product cost (including return on investment)	23.5

Table 4

Composition of *Cellulomonas* sp.

Component	Composition (% dry mass)
Protein	
by Kjeldahl (total nitrogen × 6.25)	55.6
by ion exchange	48.2
by Beckman amino-acid analysis	47.0
Fat	0.7
Ash	4.8
Digestible portion of the lysine present	96.0
Elemental	
Carbon	42.10
Nitrogen	8.40
Phosphorus	1.13
Potassium	0.13
Iron	0.02
Sulphur	0.11
Magnesium	0.04
Calcium	0.07
Sodium	0.23
	(parts per million)
Zinc	29
Copper	0
Cobalt	1.3
Manganese	2.5

composition is on a par with soy meal in the critical amino-acids (methionine, threonine, valine, leucine, isoleucine, arginine, histidine, phenylalanine and lysine) which are not synthesized by higher animals. The other essential amino acid, tryptophan, was not assayed.

Note added in Proof

In the last few years, the developed nations have gone from having an excess of energy to a gross deficiency. Many investigators are looking at living plants as solar collectors. Essentially all of the products of photosynthesis, that is, sugars, starch, cellulose, hemicellulose and part of the lignin, can be converted microbially to either alcohol or volatile fatty acids which in turn can be used as fuels. The only basic difference between S.C.P. production, as shown in this chapter, and fuel production is that the latter is done anaerobically. The conditions required for pre-treatment, hydrolysis and fermentation are all essentially the same for production of either product. Only, less nutrients are required in anaerobic fermentation because very little biomass is produced.

Table 5

Amino-acid composition of single-cell protein, *Cellulomonas* sp., produced in the Louisiana State University process, compared with those from other sources

Organism: Substrate: Dried by:	*Alcaligenes* sp. glucose vacuum oven	Yeast (Y$_c$ 13) glucose vacuum oven	*Cellulomonas* sp. glucose vacuum oven	*Cellulomonas* sp. cellulose lyophilized	*Cellulomonas* sp. cellulose lyophilized	*Cellulomonas* sp. cellulose lyophilized	Soy meal (44% protein)
Amino acid							
Lysine	5.7	6.0	4.0	4.0	4.1	5.8	5.0
Histidine	2.8	3.3	3.8	4.2	4.4	2.2	2.7
Arginine	6.9	3.6	7.5	7.1	7.1	7.5	6.9
Aspartic acid	10.1	0.9	8.6	8.2	8.6	9.1	11.7
Threonine	5.0	4.9	5.3	5.3	5.5	5.6	4.4
Serine	4.3	4.6	4.5	4.8	5.4	3.7	6.4
Glutamic acid	13.7	19.1	16.7	14.1	14.5	17.2	20.4
Proline	4.2	4.6	4.6	4.9	4.8	4.8	5.6
Glycine	5.3	4.8	5.5	5.4	5.5	5.9	4.3
Alanine	8.2	16.8	10.3	10.7	10.0	10.5	4.5
Valine	6.8	4.6	6.7	6.4	6.4	8.7	4.5
Methionine	2.2	1.0	1.3	1.2	1.2	1.2	1.2
Isoleucine	5.2	6.7	3.7	3.5	3.5	3.3	4.8
Leucine	8.9	6.8	8.7	10.1	8.6	8.4	8.5
Tyrosine	5.9	7.3	6.5	6.7	6.9	2.8	4.1
Phenylalanine	4.8	4.9	2.4	3.5	3.6	3.4	5.1
	100.0	99.9	100.1	100.1	100.1	100.1	100.1
Grams of amino acid per 100 grams of sample dry weight	49.7	26.9	44.1	45.4	48.5	48.4	45.6

REFERENCES

Bellamy, W. D. (1974). *Biotechnology and Bioengineering*, **16**, 869.
Bobletter, O., Niesner, R. and Rohr, M. (1976). *Journal of Polymer Sciences*, **20**, 2083.
Brandt, D. (1975). *In* 'Cellulose as a Chemical and Energy Resource', (C. E. Wilke, ed.), p. 275. John Wiley and Sons, Inc. New York.
Callihan, C. D. and Dunlap, C. E. (1969). *Compost Science* **10**, 1 and 2, 6.
Callihan, C. D. and Dunlap, C. E. (1971). 'Single-Cell Protein from Cellulosic Wastes'. Document of United States Environmental Protection Agency.
Callihan, C. D. and Irwin, G. H. (1974). 'Growth of Bacteria on Alkali Treated Celluloses for Food Production', American Chemical Society Meeting, Atlantic City, New Jersey.
Callihan, C. D., Dunlap, C. E. and Irwin, G. H. (1974). 'Recycle of Cellulosic Waste by Conversion to Protein'. *Global Impacts of Applied Microbiology*, IV; p. 94. Sao Paulo, Brazil.
Crawford, D. L., McCoy, E., Harkin, J. M. and Jones, P. (1973). *Biotechnology and Bioengineering* **15**, 833.
Cysewski, G. R. and Wilke, C. R. (1976). *Biotechnology and Bioengineering* **28**, 1297.
Das, K. and Ghose, T. K. (1973). *Journal of Applied Chemistry and Biotechnology*, **23**, 829.
Dunlap, C. E. (1969). Ph.D. Dissertation: Louisiana State University.
Frederick, J. R., Fratzke, A. R., Oguntemein, G. B., Frederick M. R. and Reilly, P. J. 1977. B. F. Ruth Symposium, p. 83. Iowa State University Chemical Engineering Department, Ames, Iowa.
Ghose, T. K. (1969). *Biotechnology and Bioengineering* **11**, 239.
Ghose, T. K. and Kostick, J. A. (1970). *Biotechnology and Bioengineering* **11**, 239.
Han, Y. W. and Srinivasan, V. R. (1968). *Applied Microbiology* **16**, 1140.
Han, Y. W. (1969). Ph.D. Dissertation: Louisiana State University.
Han, Y. W. and Callihan, C. D. (1969). *Applied Microbiology* **27**, 159.
Huang, A. A. (1975). *Biotechnology and Bioengineering* **18**, 1421.
Irwin, G. H. (1974). Ph.D. Dissertation: Louisiana State University.
Jarl, K. (1969). *Food Technology* **23**, 1009.
Kirk, T. K. (1971). *Annual Review of Phytopathology* **9**, 185.
Mandels, M. and Weber, J. A. (1969). *In* 'Advances in Chemistry', (G. J. Hajry and E. T. Reese, eds.), American Chemical Society Publications, No. 95.
Mandels, M., Hontz, L. and Nystrom, J. (1974). *Biotechnology and Bioengineering* **16**, 1471.
Meller, F. H. (1969. 'Conversion of Organic Solid Waste into Yeast: An Economic Evaluation', United States Department of Health, Education, and Welfare, Rockville, Maryland.
Nystrom, J. M. and Allen, R. K. (1975). *Process Biochemistry*, December, p. 27.
Peitersen, N. (1975). *Biotechnology and Bioengineering* **17**, 361 and 1291.
Reese, E. T., Sui, R. G. H. and Levinson, H. S. (1950). *Journal of Bacteriology* **59**, 485.
Reese, E. T. (1969). *In* 'Advances in Chemistry', Series No. 95. American Chemical Society, Washington, 26.
Reese, E. T. and Mandels, M. (1971). *High Polymer*, **5**(5), 1079.
Srinivasan, V. R. and Fleenor, M. B. (1972). *Developments in Industrial Microbiology* **13**, 47.
Updegraff, D. M. (1971). *Biotechnology and Bioengineering* **13**, 77.
Wilke, C. R., Cysewski, G. R., Yang, R. D. and Von Stockar, U. (1976). *Biotechnology and and Bioengineering* **18**, 1315.
Zabriskie, D. W., Armigir, W. B. and Humphrey, A. E. (1975). *Abstracts 95th Meeting of the American Society of Microbiology, New York*.

10. Biomass from Carbohydrates

A. J. FORAGE AND R. C. RIGHELATO

Philip Lyle Memorial Research Laboratory,
P.O. Box 68, Reading, England.

I. INTRODUCTION

An estimated two million tonnes of single-cell protein, mostly yeast, are produced annually in the world. Production is mainly from cane and beet molasses with about 500 000 tonnes from hydrolysed wood wastes and corn trash, and from sulphite-waste liquor. The largest single producer is Russia, manufacturing nearly one million tonnes (National Science Foundation, 1975), much of which is produced from cellulosic materials. Other centrally planned economies are also major producers, though relatively small amounts are manufactured in Europe, United States, Taiwan and South Africa.

The main market for microbial protein is animal feed. In free-market economies, its price is set primarily against soybean meal which has fluctuated markedly in both supply and demand and hence selling price. This, coupled with the relatively high cost of most of the carbohydrates which can be used as substrates, makes S.C.P. production in the major free-market economies a commercially unattractive venture except under special circumstances. Factors of critical importance are a high price for protein and use of low-cost carbohydrates, usually wastes, as raw materials.

Over the last decade, the technology to produce S.C.P. has been extended to the purpose of pollution control. Rather than production of protein, the objective has been to remove the biological oxygen demand from industrial, food processing, and agricultural wastes. In these circumstances, the main product of the manufacturing process is, of course, clean water, S.C.P. being the by-product which can be sold and thus offset at least part of the effluent treatment costs. Although the principles of microbial growth and the technologies employed are common to both objectives, the economic constraints in deciding whether S.C.P. technology is appropriate to effluent treatment are different.

Table 1

Approximate costs for single-cell protein production from molasses and sulphite waste liquor[a]

Process	Yeast /Molasses	Mould/Sulphite-waste liquor
Variable costs (% of total)		
Carbohydrate source	35	15
Other materials	15	30
Aeration		15
Cooling		2
Recovery and drying	16	4
Miscellaneous		2
Labour	9	9
Fixed costs	23	23
Organism	*Candida utilis*	*Paecilomyces varioti*
Y carbohydrate (kg/kg)[b]	0.5	0.47
Y oxygen (kg/kg)	1.4	—
Productivity (kg/m³−h)	4	4

[a] Data based on Moo Young (1977) and Forage and Righelato (1978).
[b] Y values indicate the efficiency of conversion of substrate into product.

Where protein production is the aim, the economic feasibility of an S.C.P. plant depends largely on the cost of alternative protein sources. Where pollution control is the objective, the economic viability is measured against other effluent-treatment schemes, although the financial return from the sale of the product and thus prices of alternative protein sources will influence the assessment.

In this review, we consider the aspects of growth of microbes on sugars and starches that are pertinent to S.C.P. production, and then discuss separately the raw materials and technology used in production of protein and pollution control.

II. PHYSIOLOGY OF BIOMASS PRODUCTION FROM CARBOHYDRATES

A breakdown of the costs of S.C.P. production from sulphite-waste liquor can be used to demonstrate the economically critical steps in the process (Table 1). The physiological criteria to which the economics can be related are usually considered to be: (i) product concentration; (ii) rate of product formation, productivity; (iii) efficiency of conversion of raw materials to product, yield; and (iv) quality and hence the value of the product itself. In the case of S.C.P., product concentration *per se* has relatively little effect on the process economics, the costs of recovery of the microbial biomass, filtration, centrifugation or flocculation are, within limits, independent of biomass concentration and form a small part of the total process cost (Table 1). On the other hand productivity, yield and quality of the product are critical and are discussed individually. Where the raw material is a waste stream, the ability of the chosen microbe(s) to cope with the complex mixture of substrates that most waste or by-product streams comprise is an additional and important criterion.

A. Yield

Except for brewer's or distiller's yeast, in which the biomass is a by-product, all commercial S.C.P. processes are aerobic, anaerobic yields of cells from carbohydrates being very much lower than aerobic yields. Since carbon and oxygen together represent the major cost of S.C.P. production, the efficiency of utilization of these substrates in synthesis of biomass is of the utmost importance. The precise requirements for substrates depend on the type of biomass that is to be produced. In general, the biomass is maximized for protein content, a condition which is achieved by using high growth rates and carbohydrate-limited growth conditions. Biomass with high lipid content has been produced under nitrogen-limited growth conditions but is not now undertaken commercially. The following discussion therefore refers to biomass containing a high content of protein.

The elemental composition of *Candida utilis* (Harrison, 1973) can be taken as typical of yeasts and moulds grown for protein feed: C, 0.47; H, 0.06; O, 0.31; N, 0.09; P, 0.02 plus trace quantities of metals. In addition to the requirement for substrates to be assimilated into cell matter, there is a substantial demand for energy to synthesize the highly ordered cellular macromolecules and maintain the metabolic integrity of the cell. In aerobic cultures, most of the energy is gained by oxidation of the carbohydrate source to carbon dioxide. The energy so gained is dissipated as heat, which must be removed from the fermentation. Provision of oxygen and removal of heat are both expensive operations, and so the energetics of growth are important determinants of the efficiency of an S.C.P. process. An approximate mass balance based on numerous empirical data shows that one gram of biomass requires approximately 2 g carbohydrate and 0.7 g oxygen (equation 1). Oxygen uptake and heat evolution have been shown to be stoicheiometrically related at approximately 3 kcal/g cell mass or 4 kcal/g oxygen (Cooney *et al.*, 1968).

$$\underset{2.0}{C_6H_{12}O_6} + \underset{0.7}{O_2} + \underset{0.1}{N,P,K,Mg,S} \rightarrow \underset{1}{Biomass} + \underset{1.1}{CO_2} + \underset{0.7}{H_2O} \qquad (1)$$

Although the maximum yield constants for many yeasts and moulds are similar to the maxima observed with bacteria, as with bacteria they vary considerably with substrate, growth rate and perhaps other conditions such as pH value and temperature (Pirt, 1965; Hempfling and Mainzer, 1975; von Meyenburg, 1969, Mason and Righelato, 1976). One of the most important sources of variation is specific growth rate. Assimilation of energy-yielding substrates can be described in terms of a growth-dissociated, maintenance component, m, and a growth-associated component, Y_G (equation 2):

$$-\frac{ds}{dt} = \times \left(\frac{\mu}{Y_G} + m\right) = \frac{1}{Y}\frac{dx}{dt} \qquad (2)$$

Where ds/dt is the rate of use of the limiting substrate,
  x the cell concentration (gl^{-1}),
  Y_G the yield (cells produced/substrate consumed), and
  μ the specific growth rate (h^{-1}).

Data from a range of yeast and mould cultures have been found to conform to equation 2, which predicts a decreasing yield with decreasing growth rate (Mason and Righelato, 1976). Whilst the physiological significance of the term m is generally considered to be cell maintenance or turnover of components, it is artifactual in that it is generated algebraically. It could also represent, for example, substrate consumption for extracellular product formation or a continuous change in Y_G value with growth rate. Nonetheless, the analysis given by equation 2 is valuable in that it can be used to predict substrate demands, yields and heat production as a function of growth rate. Some values of Y_G and m for carbohydrate and oxygen in glucose-limited chemostats are given in Table 2.

Table 2

Growth yields and maintenance rates for glucose and oxygen in glucose-limited chemostat cultures. From Mason and Righelato (1976)

| | Glucose | | Oxygen | |
Organism	Y_G (g/mol)	m (mmol/g/h)	Y_G (g/mol)	m (mmol/g/h)
Saccharomyces cerevisiae	100	0.03	36	0.62
Torula utilis	—	—	45	1.3
Aspergillus nidulans	110	0.11	68	0.55
Aspergillus awamori	95	0.09	40	0.60
Penicillium chrysogenum WIS 54–1255	81	0.11	50	0.74
Penicillium chrysogenum JWI	87	0.13	40	0.85

The dry weight of the yeast was assumed to be twice the protein content.

It would appear from equation 2 that maximum yields would be obtained at growth rates close to the maximum; in practice, however, incomplete substrate assimilation and large changes in the value for Y at high growth rates are common. It is well established that growth rate is a function of substrate concentration; hence high growth rates result in a relatively high residual substrate concentration in the fermenter. The growth rate at which virtually complete substrate utilization is obtained depends on the form of the saturation curve, and the operating growth rate is determined by a balance between the requirements for high conversion efficiency (Y) and complete assimilation of substrate. Particularly in waste-treatment processes, the latter, i.e. removal of biological oxygen demand, may be of overriding importance.

With carbohydrate substrates, changes in metabolism towards a less efficient utilization of the substrate at high growth rates are frequently observed, even when the carbohydrate is the growth-limiting substrate. The yeast *Saccharomyces cerevisiae* exhibits oxidative metabolism of glucose at growth rates below 0.25 h^{-1} in glucose-limited cultures, but at higher growth rates shifts progressively to the anaerobic dissimilatory pathway, with ethanol and carbon dioxide as the major products and very low growth yields (von Meyenburg, 1969). Mason and Righelato (1976) found a similar shift from oxidative to glycolytic energy generation at high growth rates in a strain of the mould *Penicillium chrysogenum*. The same authors observed the highest yields on glucose and oxygen under oxygen-limited conditions, which supported the suggestion by Ryder and Sinclair (1972) that maximum yields might be obtained under dual oxygen- and carbon-limitation.

The approximate quantities of inorganic nutrients for biomass production are given in Table 3. Like the requirements for carbohydrate and oxygen, these vary with growth conditions, particularly growth rate. The nitrogen content, a critical feature in selling biomass, is at a

Table 3

Inorganic nutrient requirements and yields of micro-organisms

Element	Supplied as	Yield (g cell/g element)	Organism	Reference
N	NH$_3$	10–26	*Candida utilis*	Brown and Rose (1969)
		8–9	*Penicillium chrysogenum*	Mason and Righelato (1976)
P	PO$_4^{3-}$	84	*Penicillium chrysogenum*	Mason and Righelato (1976)
K	K$^+$	60–120	*Klebsiella aerogenes*	Tempest (1969)
Mg	Mg^{2+}	600–900	*Candida utilis*	Tempest (1969)
		2100	*Penicillium chrysogenum*	Mason and Righelato (1976)
S	SO$_4^{2-}$	440	*Penicillium chrysogenum*	Mason and Righelato (1976)

maximum at high growth rates, as is the protein content in some instances (Herbert, 1958; Khmel and Andreeva, 1969). The nucleic acid content of cells, which is related to nitrogen, potassium and magnesium, also increases with growth rate (Neidhardt, 1963) and is the main reason for higher nitrogen contents at higher growth rates. Nucleic acid levels in yeast and moulds are generally lower than in bacteria and, except in the manufacture of functional protein preparations, no attempt is made to lower them. The nitrogen content of feed yeast and high quality mould S.C.P. is usually 8–9%; the true protein content is 40–45% and nucleic acid 7–10%. Nitrogen is supplied as ammonia, ammonium salts or urea. For the remaining inorganic nutrients, the yields given in Table 3 may be used as a guide. However, competitive effects on the ion uptake mechanisms and precipitation in the medium can affect the quantity that has to be added to the medium, which must be assessed empirically.

Table 4

Reported maximum specific growth rates of some fungi on simple media

Species	Growth rate $(\mu_{max}; h^{-1})$	Temperature	Reference
Neurospora crassa	0.35	30	Zalokar (1959)
Fusarium graminareum	0.28	30	Anderson *et al.* (1975)
Aspergillus nidulans	0.36	37	Trinci (1969)
Aspergillus niger	0.30	30	Righelato *et al.* (1976)
Geotrichum lactis	0.35	25	Trinci (1971)
Penicillium chrysogenum	0.28	25	Morrison and Righelato (1974)
Achlya bisexualis	0.80	24	Griffin *et al.* (1974)
Saccharomyces cerevisiae	0.45	30	Von Meyenburg (1969)
Candida utilis	0.40	30	Forage (1978)

B. Productivity

Productivity, the quantity of biomass produced per unit size of plant, determines the capital investment required for a given output of product. Since fermenters and ancillary equipment comprise the major cost of biomass plants, fermenter productivity is a critical component in process design. It is for this reason that continuous rather than batch cultures are invariably used in modern processes. In continuous culture, productivity

$\left(\dfrac{\mathrm{d}x}{\mathrm{d}t}\right)$ is given by Dx where x is organism concentration in the effluent stream, and D the dilution rate. In a simple continuous culture without feedback of cells, at steady state $D = \mu$ (specific growth rate). The physiological limit to productivity is thus the maximum growth rate of the organism used (Table 4), and the maximum concentration of organism that can be obtained. Theoretically, the maximum cell concentration would be that at which the cells occupy the whole volume of the culture, i.e. at about 200 g dry weight/litre. However, in practice, productivities are an order of magnitude less than a theoretical maximum of about $100 \, \mathrm{g \, l \, h^{-1}}$. The reasons for this lie in the difficulties of transfer of substrates to the cells, and toxic products from them.

The substrates or products presenting the greatest transfer problems are those whose concentrations in the culture fluid, determined by their solubility or toxicity, are low compared with the quantities that must be transferred during cultivation. Sugars and inorganic salts generally present no difficulties as they are readily soluble at the necessary concentrations and, given adequate bulk-mixing, can be dispersed homogeneously. In contrast, the mass transfer of oxygen is frequently the process that limits productivity. About 0.7 g oxygen are required for each gram of biomass, but the solubility of oxygen from air, about $0.0075 \, \mathrm{g \, l^{-1}}$ at S.T.P., permits only a small proportion of the requirement to be in solution at any one time. Heat presents a similar problem. At 3 kcal per g biomass produced, a cell concentration of $20 \, \mathrm{g \, l^{-1}}$, which is common to many S.C.P. processes, results in production of 60 kcal per litre. Since the maximum that can be tolerated is usually less than $5 \, \mathrm{kcal \, l^{-1}}$, equivalent to a temperature rise of $5°C$, cooling must be provided to the fermenter to remove surplus heat. For the chemical engineering aspects of mass and heat transfer, the reader is referred to reviews by Taguchi (1971) and Banks (1975). The microbiological parameter of critical importance to productivity is the oxygen demand, determined by the oxygen-yield constant of the micro-organisms.

The oxygen-yield constant is a measure of the oxygen, and hence heat, that must be transferred to obtain a given quantity of biomass. Conventional fermenters achieve oxygen-transfer rates in the order of $4-8 \, \mathrm{g \, l^{-1}}$ with yeast cultures; hence, with 0.7 g oxygen per g cell, yeast productivity is in the order of $5-10 \, \mathrm{g \, l \, h^{-1}}$. Rates in this range are claimed for several processes (Moo Young, 1977; Vogelbusch, 1971). The oxygen and heat yields of cells grown on sugars are higher than those of cells grown on hydrocarbons, the only other substrates in industrial use for biomass production; thus carbohydrates have an important processing advantage (Moo Young, 1977). Cooling may nonetheless be a problem, particularly in hot climates, in which case fermentation at temperatures considerably above ambient would be an advantage. Yeasts are mesophiles, none growing at temperatures much above 40°C. Thermophilic organisms could present a considerable advantage where cooling is a problem, but none is grown commercially at present.

The morphology of the micro-organism is of similar importance as it is a determinant of the mass-transfer and heat-transfer coefficients. Yeasts, whose shapes vary between spheres and short rods, produce culture broths with low viscosities and Newtonian flow behaviour; they are, therefore, relatively easy to mix. Filamentous fungi, on the other hand, present several problems. In the pellet form, the culture viscosity can be quite low, but the pellets present long diffusion paths for substrates and quickly become oxygen-limited in the centre (Pirt, 1967; Taguchi, 1971). The open filamentous growth form produces broths with high viscosities and pseudoplastic flow behaviour; this gives low mass-transfer coefficients and requires high power inputs to achieve good bulk mixing (Roels et al., 1974). For optimum mass transfer, short hyphal filaments are preferred (Carilli et al., 1961).

C. Organisms

In this section, processes using carbohydrates as substrates are described and the choice of organism is discussed. In most cases, yeasts are used; they exhibit relatively high carbohydrate and oxygen-yield constants, high specific growth rates and a unicellular morphology. They are moreover an acceptable high-value animal feed used as a source of protein and vitamins. In recent years, processes using moulds have been developed and come into commercial operation. In some cases, the filamentous growth form is an important process advantage, as is the ability of many moulds to utilize a wide range of substrates. The costs of nutritional and toxicological evaluation of new feed sources are high and will probably restrict introduction of new microbial species as S.C.P. However, in cases where they have been tested, the nutritional properties of mould S.C.P.s were comparable to those of fodder yeast (Table 5).

Table 5

Composition of some micro-organisms used as single-cell protein

Analysis	Paecilomyces varioti	Fusarium graminareum	Candida utilis
Dry matter (%)	96	94.2	91
Crude protein (% N × 6.25)	55	54.1	48
Fat (%)	1	1.0	1.35
Ash (%)	5	6.1	11.2
Lysine (g/16 g N)	6.5	3.5	7.2
Methionine (g/16 g N)	1.9	1.23	1.0
Cystine and cysteine (g/16 g N)	1.0	0.75	1.0
References	Forss and Passinen (1976)	Duthie (1975)	Forage (1978)

III. CARBOHYDRATE SUBSTRATES USED IN SINGLE-CELL PROTEIN PROCESSES

The chemical nature and physical form in which the substrate is available influences the choice of organism and process design. The organisms chosen must not only be capable of utilizing the substrate, but must do so at a rate which would minimize the size of plant and thereby keep capital costs as low as possible. The form in which the substrate raw materials are available bears on the type of technology used (solid or deep-culture fermentation) as well as the need for treatment stages to make the substrate available for assimilation.

Although the action of bacteria, moulds and yeasts has been the basis of foods such as yoghurt, cheese, wine and bread, only yeasts have been used to any extent for S.C.P. production. Yeast production from molasses, sulphite-waste liquor and whey has been commercial practice throughout the world for many years, but the yeasts commonly used, *Candida* spp. and *Saccharomyces* spp., cannot utilize polymeric carbohydrates such as starch and cellulose unless these are hydrolysed to their component sugars. Moulds, on the other hand, are generally able to use a wider range of carbohydrates (Table 6) and have been used increasingly where there are relatively more polymeric than simple sugars or when there is a heterogenous mixture, as is often found in food-processing wastes. Even though a micro-organism may be able to metabolize the different sugar residues which may comprise the substrate, it may assimilate them at different rates. It has been shown, for example, that an increase in the glucose-chain length can lower the maximum specific growth rate of *Fusarium graminareum* (Table 7). It is therefore necessary to establish the cause of rate limitation with substrates containing heterogeneous mixtures of carbohydrates, so that the S.C.P. process can be designed accordingly. This is particularly true when the aim of the process is effluent treatment, and the product is clean water with S.C.P. as the by-product. In this case, the efficiency of the fermentation is measured by the percentage decrease in Biochemical Oxygen Demand of the waste stream over five days (BOD_5). The conditions resulting in maximum lowering of BOD_5 may not be the same as those resulting in maximum S.C.P. yields or productivity. In this dicussion, the use of submerged fermentation technology for utilization of simple sugars and for polysaccharides will be discussed. In a third section, the use of solid-culture fermentation which is dealt with in Chapter 5 in this volume will be considered in the context of S.C.P. production.

Table 6

Ability of fungi to grow on single carbon sources

Substrate	Candida utilis	Kluyveromyces fragilis	Aspergillus niger M1	Fusarium moniliforme M4
Glucose	+	+	+	+
Sucrose	+	+	+	+
Maltose	+	−	+	+
Cellobiose	+	+	+	+
Lactose	−	+	+	+
Starch	−	−	+	+
Cellulose	−	−	−	−

Table 7

Effect of glucose-chain length on the maximum specific growth rate of *Fusarium graminareum* at 30°C. From Anderson *et al.* (1975)

Compound	Number of glucose units	Value for μ max (h^{-1})	Doubling time (h)
Glucose	1	0.28	2.48
Maltose	2	0.22	3.15
Maltotriose	3	0.18	3.85

A. Simple Sugars

1. Molasses

Molasses is the traditional source of simple sugars for production of the yeasts *Candida utilis* and *Saccharomyces cerevisiae*. It is a source not only of carbohydrate but also of a range of inorganic and organic nutrients and vitamins. The composition and properties of molasses vary according to source, country of origin, factory and season. Fortunately, this variation does not markedly affect S.C.P. production as it does the production of, for example, citric acid (Steel *et al.*, 1955) or baker's yeast, a high-quality product with special properties required in baking.

In developed countries, the main yeast product from molasses is baker's yeast. In addition to the constraints on the quality of molasses that can be used, the technology employed in production of baker's yeast is different from that for production of feed S.C.P. Baker's yeast is cultured such that it has good functionality in the making of bread, and therefore should not be considered as a microbial product intended as a protein source for food and feed. For information on the production of baker's yeast, the reader is referred to the reviews by White (1954), Burrows (1970), Sobkowicz (1976), and to Chapter 2 in this volume (p. 31).

The main sources of molasses are as a by-product of cane- and beet-sugar refining with a smaller amount from citrus fruit processing. Typically, cane molasses has a higher concentration of invert sugar than beet although the overall sugar contents are similar (Table 8). The nitrogen contents range from 0.5% to 2.5% being generally higher in beet molasses because of appreciable quantities of asparagine and betaine.

Table 8

Analysis of beet- and cane-molasses. From Rhodes and Fletcher (1966)

Content (%)	Beet molasses	Cane molasses
Sucrose	48.5	33.4
Raffinose	1.0	—
Invert	1.0	21.2
Ash	10.8	9.8
Organic non-sugars	20.7	19.6
Nitrogen	1.5–2.0	
Water	18.0	16.0

Molasses requires little or no pretreatment before it is used for S.C.P. production. In some processes, the mineral content is lowered by clarification, which consists of acidification with sulphuric acid to pH 4, heating to 90°C for 30 min and centrifugation to remove the inorganic precipitate and suspended organic matter. Additionally, beet molasses is heated and aerated to remove sulphur dioxide which would otherwise inhibit yeast growth. The molasses liquor is diluted to a sugar concentration 4–6% and supplemented with sufficient nitrogenous nutrient and phosphate such that the carbohydrate is the limiting nutrient in the fermentation. No other nutrient requirements and trace elements are added, as the concentrations necessary for yeast growth are usually satisfied by those present in the molasses liquor. The liquor may require pasteurization or sterilization before it is fed to the fermenter, usually at dilution rate 0.2–0.3 h^{-1}. Depending on the strain, fermentation is controlled between pH 3.5 and 4.5 and temperature 25–35°C. The yeast is harvested by centrifugation, washed and dried to give a product containing 45–50% crude protein. Commercial operations are widespread with major developments in Russia, Taiwan, Cuba and South Africa.

2. Whey

Whey is the liquid effluent of cheese and casein manufacture, most of which is normally run to drain. The BOD_5 value is commonly 60 000–70 000 mg l^{-1} and results mainly from the lactose present. Some whey is fed to pigs and calves and small quantities are used in food beverages. With the growth in the cheese industry, the volumes of whey produced, which was 74 million tons in 1973 (Food and Agricultural Organisation, 1974a), have exacerbated the pollution problem. In order to control the situation, new legislation has been implemented in several countries which, with high sewage-treatment charges, is forcing industries to solve the pollution problems they cause. One solution with whey is the use of the lactose to produce S.C.P., and several processes are operated commercially in the United States, Europe and Australia. In contrast to molasses utilization, these processes are designed to be methods of pollution control, removal of Biochemical Oxygen Demand being a critical parameter.

A typical analysis of liquid whey is shown in Table 9. About 70% of the solids is lactose and 15% protein. In some instances, it is worthwhile separating the protein since it can be sold as a high-value supplement for food and thereby help to make the overall treatment of whey more economical. Protein separation can be achieved by ultrafiltration (Pace and Goldstein, 1975), ion exchange (Jones, 1975), or coagulation by acidification to the isoelectric point (pH 5) and heating to 95°C (Moulin et al., 1974). The deproteinized whey is then used as substrate for production of yeast S.C.P. The whey liquor is sterilized, supplemented with phosphate and nitrogenous nutrient and fed to the fermenter. The fermentation pH value is controlled at about 3.5 and the temperature between 25–30°C. Three types of yeast S.C.P. are currently produced, *Kluyveromyces fragilis* (Powell and Robe, 1964; Moulin et al., 1974), *Torula cremoris* and *Saccharomyces lactis* (Naditch and Dikanksy, 1960; Société des Alcools du Vexin, 1963).

Table 9

Composition of liquid whey. From Food and Agricultural Organization (1974a)

	Content (%, w/v)
Water	93.1
Protein	0.9
Fat	0.3
Lactose	5.1
Ash	0.6

The main economic constraint on whey utilization is its seasonal production. There are simpler alternatives to the pollution problem, such as drying, provided that there is a market for the product. However, since whey production is increasing and there is stricter legislation governing the discharge of pollutants, and since drying costs are high, S.C.P. production may become more economically attractive. Further information on S.C.P. production from whey appears in Chapter 8 (p. 207).

3. Sulphite-Waste Liquor

As with whey and molasses, sulphite-waste liquor requires very little pretreatment before fermentation. Its production is not seasonal, and the liquor is available in large quantities making it economically a good substrate for S.C.P. production. Approximately 100 million tonnes of spent sulphite liquor are produced as waste from wood pulp mills. The liquor has a BOD of $25\,000-50\,000$ mg l^{-1} and therefore is a serious pollution hazard. About 50% of the organic load is lignosulphonate which is recalcitrant to microbial utilization; only 25% is monosaccharides (Table 10). Therefore, although 90% of the BOD can be removed by fermentation, the decrease in total organic carbon is low. Invariably, further treatment of the fermented liquor is required before it can be discharged to water courses. Normal practice is to evaporate and burn the fermented wash or convert the residue into lignin products.

Table 10

Organic composition of spent spruce sulphite liquor. From Forss and Passinen (1976).

		Content (%, w/v)
Lignosulphuric acids		43
Hemilignin compounds		12
Incompletely hydrolysed hemicellulose and uronic acids		7
Monosaccharides		
D-glucose	2.6	
D-xylose	4.6	
D-mannose	11.0	
D-galactose	2.6	
L-arabinose	0.9	22
Acetic acid		6
Aldonic acids and other substrates		10

Two types of S.C.P. processes are in commercial operation. The majority of processes, for example, in North America, Europe and Russia, produce *Candida utilis* (Kretzschmar, 1962; Peppler, 1970). As with S.C.P. production from molasses, these processes have been established for several years and at least one in Germany was reported to be operating during World War II (Robinson, 1952). More recently, the Pekilo process, using the mould *Paecilomyces varioti*, has been developed in Finland (Forss and Passinen, 1976). As an effluent treatment process this is more efficient than the yeast processes, since the mould consumes the acetic acid which is also present and which may be particularly high in concentration in effluents from hardwood pulping.

In outline, the fermentation processes are similar. Sulphur dioxide is stripped from the waste liquor either by aeration or by steam-stripping at

L

pH 1.5–3.0. The pH value of the liquor is adjusted to the optimum for microbial growth and supplemented with nitrogenous nutrients and phosphates. In the Pekilo process, potassium and magnesium salts are also added. After fermentation, harvesting is by centrifugation for yeasts and rotary vacuum filtration for the mould. Since mechanical dewatering is more efficient with the rotary vacuum filter, less water must be removed at the drying stage. Thus, recovery of the mould product is less costly than the yeast product. The amino-acid profiles of the protein, and vitamin content of the mould and yeast, are similar, although the crude protein content of the mould product (at 55–60%) is somewhat higher than for the yeast product.

Replacement of the sulphite process by the Kraft or sulphate process in wood pulping technology has altered the available volumes and chemical composition of pulp-mill effluents. This change has already resulted in closure of S.C.P. production facilities in some North American mills and will probably restrict the future expansion of S.C.P. production from wood pulp effluents.

4. Food Processing Effluents

Effluents from several food-processing industries, such as confectionery, soft drinks and jam preserve manufacture, contain simple sugars readily assimilable by yeasts. Although the concentrations of carbohydrates in the effluent leaving the factories are usually low (less than 0.2% in some cases), streams from the manufacturing processes can be isolated which have high concentrations and are the cause of most of the BOD load. Utilization of these streams for production of S.C.P. requires less capital investment than treatment of the total factory effluent, since smaller volumes are processed. The processing may also lower water charges if the main use of water was for dilution of the effluent before discharge. Furthermore, processing of specific streams may help in control of effluent quality for the S.C.P. process to ensure a safe and standard product.

A process making *Candidia utilis* from confectionery effluent is being commercialized in the United Kingdom and should be on stream late in 1978 (Forage, 1978). The effluent contains 3–4% solids, mainly glucose and sucrose (Table 11), and has a Chemical Oxygen Demand of 30 000–40 000 mg 1^{-1}. Although appreciable amounts of starch are present, 70–80% is hydrolysed to sugars by the low pH value of the

Table 11

Composition of confectionery waste
solids

	Content (%, w/w)
Sucrose	55
Glucose	16
Starch	22
Gelatine	3.5
Caramel	2
Organic acids	1
Coconut	0.5

effluent and heat of sterilization before fermentation. The other components of the stream remain after fermentation and are discharged with a small amount of starch and dextrins not utilized. The process will treat 140 m³ effluent per day by continuous fermentation at a dilution rate of 0.3 h⁻¹, pH 4 and a temperature of 34°C. Pilot-plant work has demonstrated that the process will remove 75% of the Chemical Oxygen Demand, equivalent to 81% of the Biochemical Oxygen Demand, and produce 1.5 tonnes dry yeast product daily. The product is nutritionally similar to other *Candida* yeast S.C.P., containing 48% crude protein and a high concentration of B-group vitamins.

B. Polysaccharides : Starches

Much work has been done on the use of cellulose and cellulose hydrolysates for S.C.P. production. As this is discussed elsewhere in this volume (Ch. 9, p. 271), only the use of starch will be considered here.

The cost of food-grade starches makes it economically unrealistic at present to use them for S.C.P. production. With intensive cultivation of the nutritionally poorer but high-yielding strains of cassava and sorghum, it is possible that starch prices will fall and that S.C.P. production from them will become economical (MacLennan, 1975). Despite this constraint, Rank, Hovis, MacDougall, initally with the Du Pont Corporation, have been developing a process to produce mould S.C.P., namely *Fusarium graminareum*, from wheat, maize or potato starch

(Anderson *et al.*, 1975; Duthie, 1975). The project is claimed to be economically viable if a texturized protein product is made for food use.

A practical disadvantage of using pure starch is that the polymer gelatinizes on sterilization and results in highly viscous broths when the concentration is greater than 2% (w/v). High viscosities at the fermentation stage result in poor mixing and oxygen transfer, and thus poor microbial growth. The viscosity problem could be overcome by supplying starch as the whole-corn meal but, because of the particulate nature of the substrate, sterilization is more difficult. Little detail is published of the Rank, Hovis, MacDougall process except that it is a continuous-culture fermentation operated at dilution rate $0.1 \ h^{-1}$, and employs the high technology necessary to ensure that the product is of food-grade standard (Anderson *et al.*, 1975; Solomons, 1969).

A simpler process was developed to upgrade the protein content of barley grains used in non-ruminant feeding (Reade and Smith, 1975). Using the fungus *Aspergillus oryzae*, a non-aseptic batch fermentation was developed giving a product containing up to 40% crude protein. Barley grain, either steam-cracked or rough-ground, was supplemented with potassium dihydrogen phosphate and a nitrogen source, either urea or ammonium sulphate, inoculated and fermented at pH 3.5 and 30°C. Maximum protein contents were obtained after 30 h, when the medium contained 2% barley. Greater yields of protein were realized with higher barley concentrations but only if the incubation period was extended to about 72 h. The process was scaled up to a one cubic-metre fermenter but, although the fermentation was successfully operated non-aseptically, protein yields were low.

A similar low-technology process is being developed jointly by the University of Guelph, Canada and Centro International de Agricultural Tropical, Colombia, fermenting, aseptically, cassava starch with an asporogenous thermophilic strain of *Aspergillus fumigatus*. An advantage of using a thermophilic organism is that the process can be operated at 45–50°C, which lowers the costs of cooling water to the fermenter. The mould product contains 47% crude protein and has been found nutritionally satisfactory when replacing soybean meal in rat diets (Gregory *et al.*, 1976).

Processes producing S.C.P. from starch-containing substrates which have been commercialized are those which treat effluents from the processing of potatoes, corn and other starch-containing foods. The Symba process was developed by the Swedish Sugar Company and the

Chemap Company to treat effluent and solid wastes from potato processing. The process is based on the symbiotic culture of the yeasts *Endomycopsis fibuliger* and *Candida utilis*. The effluent is fed, after nutrient addition and sterilization, to a fermenter containing the amylase-producing *E. fibuliger* which degrades starch to sugars. This broth, containing sugars, dextrins and *Endomycopsis* cells, is then fed continuously to a second fermenter containing *C. utilis* which has a faster growth rate and assimilates the sugars more rapidly than *E. fibuliger*. The rate limitation of the system is hydrolysis of starch to sugars (Jarl, 1969). The BOD_5 value of the waste water is lowered by fermentation from 10 000–20 000 mg l^{-1} to 1000–2000 mg l^{-1}. The process treats 20 m^3 h^{-1} effluent h^{-1} producing 250 kg dry yeast h^{-1} and has been operating since 1973 (Skogman, 1976).

A simple process with comparatively low capital investment was developed by the Denver Research Institute to lower the BOD_5 value of corn-waste effluent from a Green Giant Company factory in Minnesota. The waste contained only 0.35–0.4% solids, 40% of which was carbohydrate, which accounted for 80–90% of the BOD_5 value (Table 12). The effluent was supplemented with phosphate and nitrogenous nutrients and fed continuously, without sterilization, into aeration ponds or oxidation ditches inoculated with *Trichoderma viride* or *Gliocladium deliquescens*. The process was operated at dilution rate 0.05 h^{-1}, and the pH value was controlled to 3.7 by addition of sulphuric acid. About 95% of the BOD_5 value was removed by fermentation with the production of

Table 12

Chemical analysis of corn-waste before and after batch culture of *Trichoderma viride* 1–23. From Church *et al.* (1972)

| | Analytical value (mg/l) | |
	before fermentation	after fermentation
Chemical Oxygen Demand	2030	210
Biological Oxygen Demand	1640	100
Nitrogen (Kjeldahl)	48	0.8
Protein	50	2
Carbohydrates	1360	70
Chlorides	784	240
Total phosphate	31	5
Soluble phosphate	29	3
Total solids	3560	405
Ash	660	233

1 g mould l^{-1}. In developing the process, a decision was made to use filamentous moulds as these could be harvested by filtration, this being simpler and less costly than using centrifuges which are required for unicellular microbes.

Several other projects are being developed to treat food-processing effluents. Most are designed as low-technology processes where capital and operating costs are minimal, for example the joint Denver Research Institute and Central American Research Institute project on fermentation of coffee-waste water using *Aspergillus oryzae* (Rolz, 1975). Because of the heterogeneous mixture of substrates encountered, the majority of these S.C.P. processes use moulds which could be effectively used in the treatment of a wide range of effluents.

C. Solid Substrates

For centuries, solid-substrate fermentation has been practised to various degrees of sophistication, from the simple composting of vegetable waste and ensiling of hay to the level of operation in the Oriental fermentation industry for manufacture of, for example, soy sauce, saké and tempe. There has been a realization in the West that the technology may be usefully applied to new situations. There are practical difficulties in preparation of solid raw materials for fermentation. Because of its particulate nature, usually only part of the assimilable carbohydrate is immediately available to the organism. An important step in the process, therefore, is comminution of the solid into a form which increases the availability of the substrate, yet does not render the material difficult to handle as a solid matrix. Care must also be taken to ensure that the inoculum and nutrients required for growth are evenly dispersed in the solid mass to obtain rapid even growth and penetration and maximum substrate utilization.

As discussed earlier, a constraint in the productivities of S.C.P. processes is the transfer of substrates particularly oxygen, to, and products such as heat, from, the cells. Where granular materials are used, these problems are more acute. Transfer of oxygen and nutrients within the particles is poor and results in lower productivities than those obtained with submerged cultures. In some cases, oxygen depletion may be so severe that protein yields are decreased and by-products, such as ethanol and organic acids, are formed. Good bulk mixing of the fermenting mass aids gas exchange and heat removal, but generally does not help in removal of other products of catabolism which, at high concentrations, may inhibit microbial growth. Though temperature can be controlled by periodic addition of cool water, it is difficult to control pH value as addition of acid or alkali can result in widely varying differences in local pH values. Some measure of control can be exercized by the use of buffers or nutrients such as urea which, when utilized, do not result in accumulation of protons which lower the pH value. For example, if ammonium sulphate was used as a nitrogen source, sulphate radicals would accumulate and lower the pH below the optimum for growth and, thus, result in decreased fermentation rates and a high sulphate content in the product.

The invasive growth pattern of moulds makes them the preferred microbes for solid fermentations. They generally produce a wide range of enzymes which degrade carbohydrates, and a filamentous growth form adapted to invasion of solids. They are therefore able not only to grow in the interparticle spaces but also to break down the structure of the particles and thereby utilize more substrate than yeasts or bacteria. A disadvantage is that, if bulk mixing is necessary to dissipate heat and aid gas transfer, the action may shear the mycelial strands. Therefore, mixing is usually carried out periodically, imposing a minimum of shear. The main advantage of solid-substrate fermentation is the decrease in energy, particularly at the drying stage, compared to the requirements for submerged-culture processes. The main disadvantage results from the poor mass-transfer rates experienced in fermenting solid matrices which ultimately leads to lower yields and protein production rates.

Several projects have been described to upgrade the protein content of starchy materials, such as cassava and sorghum (Stanton and Wallbridge, 1972; Raimbault, 1977). In essence, the processes consist of moistening or steam-cracking the dry substrate, which is then supplemented with nitrogenous nutrients and phosphate and inoculated with fungal spores, commonly species of *Rhizopus* or *Aspergillus*. The mass is incubated at controlled temperatures with, if high protein contents are required, forced aeration and agitation. In cassava fermentations with *A. niger*, Raimbault (1977) obtained dried product containing 25–30% true protein, but normally protein contents of only 10–20% are achieved.

The technology has also been applied to citrus peel, the solid residue from the fruit and juice canning industry (Forage and Righelato, 1977). Citrus peel was minced into 2 mm cubes and dried to between 50 and

60% moisture. The material was supplemented with nitrogenous nutrients and phosphate, inoculated with spores of *A. niger*, incubated and harvested by simply drying the mass to give a product containing 15–20% crude protein. Several aspects were shown to be important in determining the efficiency of the process. Moisture contents and temperatures outside the quoted range resulted in very low productivities. The concentration of spores, as would be expected, was significant in determining the rate of fermentation. Concentrations of 10^8–10^9 spores per gram citrus peel resulted in maximum protein contents after 36 hours. Lower concentrations of spores resulted in longer fermentation times to achieve the same content of protein in the product. The time of harvest was critical as prolonged incubation resulted in sporulation which, on a commercial scale, is not desirable as the spores may be inhaled by operators. Because the product of solid fermentation has a comparatively low protein content, it would be useful as a feed for ruminants rather than monogastric animals. Citrus peel itself has been used for many years in cattle feeds in the U.S.A.; potentially, therefore, a protein-enriched citrus meal should be acceptable provided its nutritional safety is demonstrated. A similar process has been developed using the yeast *C. utilis* (Lequerica and Lafuente, 1977). The advantages of these processes are that they are simple, require little capital investment and incur low operating costs. Therefore, they are particularly suitable where relatively small quantities of wastes are available.

IV. FUTURE

The constraints on S.C.P. production are economic rather than technical. In free markets, the price of cultivated carbohydrates is sufficiently high to make S.C.P. produced from them economically uncompetitive, given alternative protein sources. The dietary demand for starches and sugars, and for protein, are forecast to increase at a mean rate of 5.3% per year (Food and Agriculture Organization, 1974b). Production forecasts suggest an increase in supply of only 2.7% per year. With this excess of demand as well as high costs of production, it is likely that prices of carbohydrates will remain high and not change relative to those of proteins. In the immediate future, therefore, S.C.P. production from cultivated carbohydrates will not be economic. The alternative is the use of low-value or waste materials, and it is in this area that we envisage most of the new developments in the industry will occur.

Several of the processes described were developed as methods to treat effluents with production of a saleable product to provide a return on the operation. Because of new legislative pressure, it is possible that new processes of this type will emerge. Compared to conventional effluent-treatment schemes, a microbial protein plant has two advantages; firstly it generally occupies less space—an important factor where plants are sited in built-up areas; secondly, in some cases at least, the overall cost of effluent treatment is lowered as a result of sale of the product. It is socially more acceptable and attractive for a factory to waste less and not burden the environment by discharging so much polluting material.

REFERENCES

Anderson, C., Longton, J., Maddix, C., Scammell, G. W. and Solomons, G. L. (1975). In 'Single Cell Protein, II', (S. R. Tannenbaum and D. I. C. Wang eds.), p. 314, M.I.T. Press, Cambridge, Massachusetts.

Banks, G .T. (1975). In 'Topics in Enzyme and Fermentation Technology', (A. Wiseman, ed.), p. 72. Ellis Horwood, Chichester, England.

Brown, C. M. and Rose, A. H. (1969). Journal of Bacteriology 99, 371.

Burrows, S. (1970). In 'The Yeasts', (A. H. Rose and J. S. Harrison, eds.), Vol. 3, p. 349. Academic Press, London.

Carilli, A., Chain, E. B., Gualandi, G. and Morisi, G. (1961). Scientific Reports of the Istituto Superiore di Sanita 1, 177.

Church, B. D., Nash, H. A. and Brosz, W. (1972). Developments in Industrial Microbiology 13, 30.

Cooney, C. L., Wang, D. I. C. and Mateles, R. I. (1968). Biotechnology and Bioengineering 11, 269.

Duthie, I. F. (1975). In 'Single Cell Protein, II', (S. R. Tannenbaum and D. I. C. Wang, eds.), p. 505. M.I.T. Press, Cambridge, Massachusetts.

Food and Agricultural Organization (1974a). F. A. O. Bibliography List, No. 27–283–74 23, (4) 12.

Food and Agricultural Organization (1974b). Assessment of the World Food Situation, World Food Conference E/CONF. 65/3, F.A.O., Rome.

Forage, A. J. (1978). Process Biochemistry 13, 8.

Forage, A. J. and Righelato, R. C. (1977). Proceedings of the Fifth International Conference on Global Impacts of Applied Microbiology, Bangkok, Thailand.

Forage, A. J. and Righelato, R. C. (1978). Progress in Industrial Microbiology 14, 59.

Forss, K. and Passinen, K. (1976). Paperi ja Puu 9, 608.

Gregory, K. F., Alexander, J. C., Lumsden, J. H., Losos, G. and Reade, A. E. (1976). Food Technology 30, 30.

Griffin, D. H., Timberlake, W. E. and Cheney, J. C. (1974). Journal of General Microbiology 80, 381.

Harrison, D. E. F. (1973). Critical Reviews in Microbiology 2, 185.

Hempfling, W. P. and Mainzer, S. E. (1975). *Journal of Bacteriology* **123,** 1076.
Herbert, D. (1958). *In* 'Continuous Cultivation of Microorganisms', (I. Malek, ed.), p. 45. Czechoslovak Academy of Sciences, Prague.
Jarl, K. (1969). *Food Technology* **23,** 1009.
Jones, D. T. (1975). *Proceedings of the First International Conference on Effluent Treatment in the Biochemical Industries,* London.
Khmel, I. A. and Andreeva, N. B. (1969). *Proceedings of the Fourth Symposium on Continuous Cultivation of Microorganisms,* p. 147. Czechoslovak Academy of Sciences, Prague.
Kretzschmar, G. (1962). *Zellstoff und Papier* **11,** 14.
Lequerica, J. L. and Lafuente, B. (1977). *Revista de Agroquimica y Technologia de Alimentos* **17,** 71.
MacLennan, D. G. (1975). *Food Technology (Australia)* **27** (4), 141.
Mason, H. R. S. and Righelato, R. C. (1976). *Journal of Applied Chemistry and Biotechnology* **26,** 145.
Moo Young, M. (1977). *Process Biochemistry* **12** (4), 6.
Morrison, K. B. and Righelato, R. C. (1974). *Journal of General Microbiology* **81,** 517.
Moulin, G., Galzy, P. and Joux, J. L. (1974). *Proceedings of the Fourth International Congress of Food Science and Technology* **3,** 47.
Naditch, V. and Dikansky, S. (1960). French Patent 1235 978.
National Science Foundation (1975). N.S.F. Number A.E.N. 75-13072.
Neidhardt, F. C. (1963). *Annual Review of Microbiology* **17,** 61.
Pace, G. W. and Goldstein, D. J. (1975). *In* 'Single-Cell Protein, II', (S. R. Tannenbaum and D. I. C. Wang, eds.), p. 330. M.I.T. Press, Cambridge.
Peppler, H. J. (1970). *In* 'The Yeasts', (A. H. Rose and J. S. Harrison, eds.), Vol. 3, p. 421. Academic Press, London.
Pirt, S. J. (1965). *Proceedings of the Royal Society B 163* 224.
Pirst, S. J. (1967). *Journal of General Microbiology* **47,** 181.
Powell, M. E. and Robe, K. (1964). *Food Process* **25,** 80.
Raimbault, M. (1977). *Proceedings of the the Fifth International Conference on Global Impacts of Applied Microbiology,* Bangkok, Thailand.
Reade, A. E. and Smith, R. H. (1975). *Journal of Applied Chemistry and Biotechnology* **25,** 785.
Rhodes, A. and Fletcher, D. L. (1966). 'Principles of Industrial Microbiology', Pergamon Press, Oxford.
Righelato, R. C., Imrie, F. K. E. and Vlitos, A. J. (1976). *Sources, Recovery and Conservation* **1,** 256.
Robinson, R. F. (1952). *Science Monthly* **6,** 149.
Roels, J. A., Van den Berg, J. and Voncken, R. M. (1974). *Biotechnology and Bioengineering* **16,** 181.
Rolz, C. (1975). *In* 'Single-Cell Protein II', (S. R. Tannenbaum and D. I. C. Wang, eds.), p. 273. M.I.T. Press, Cambridge.
Ryder, D. N. and Sinclair, C. G. (1972). *Biotechnology and Bioengineering* **14,** 787.
Skogman, H. (1976). *In* 'Foods from Wastes', (G. G. Birch, K. J. Parker and J. T. Worgan, eds.), p. 167. Applied Science Publishers, London.
Sobkowicz, G. (1976). *In* 'Food from Wastes', (G. G. Birch, K. J. Parker and J. T. Worgan, eds.), p. 42. Applied Science Publishers, London.
Société des Alcools du Vexin (1963). French Patent 80 198.
Solomons, G. L. (1969). 'Materials and Methods in Fermentation'. Academic Press, London.
Stanton, W. R. and Wallbridge, A. J. (1972). British Patent 1277 002.

Steel, R., Lentz, C. P. and Martin, S. M. (1955). *Canadian Journal of Microbiology* **1**, 299.
Taguchi, H. (1971). *Advances in Biochemical Engineering* **1**, 1.
Tempest, D. W. (1969). *Symposium of the Society for General Microbiology* **19**, 87.
Trinci, A. P. J. (1969). *Journal of General Microbiology* **57**, 11.
Trinci, A. P. J. (1971). *Journal of General Microbiology* **67**, 325.
Vogelbusch, Gm.b.H. (1971). 'The Vogelbusch fermenter with the high efficiency aeration device', Vogelbusch, Vienna.
Von Meyenburg, H. K. (1969). *Archiv für Mikrobiologie* **66**, 289.
Whitem, J. (1954). 'Yeast Technology'. Chapman and Hall, London.
Zalokar, M. (1959). *American Journal of Botany* **46**, 555.

11. Biomass from Natural Gas

G. HAMER

Kuwait Institute for Scientific Research, P.O. Box 12009, Kuwait

I. INTRODUCTION

Throughout the history of civilization, micro-organisms have played an important role in the preparation of fermented foods and beverages and, even though they themselves have not found widespread favour as staple foods, they have frequently been consumed, in significant quantities, together with the products that they have been used to prepare. However, as constituents of animal feeds, they have found much more general application. In the nutrition of ruminants, micro-organisms produced in the rumen, which is effectively an *in situ* anaerobic fermenter, make an important nutritional contribution, whilst for non-ruminants a practice has developed whereby micro-organisms are frequently added to mixed and compounded feeds as sources of vitamins and other growth factors, although the contribution of their protein has only infrequently been credited.

Modern techniques for the economic production of eggs, poultry, veal and pork demand the use of strictly controlled conditions, and the use of feeds compounded on an economically sensible basis. Compounded feeds, which usually contain between 10 and 30 per cent protein on a weight basis depending on their proposed use, are designed to satisfy the entire nutritional requirements of either the fowl or the animal to be fed, requirements that vary during the life cycle. Protein, the key ingredient, is traditionally provided by the incorporation of protein-rich materials such as soybean meal, a variety of oil-seed meals and fish meal in the feed. Soybean meal contains some 45 per cent by weight of protein and fishmeals contain about 65 per cent protein. The quality of these proteins depends on their available amino-acid profile and varies according to the processing and storage conditions to which they have been subjected. In Europe, fish meal is frequently considered to be an essential ingredient in the more demanding rations, because of its proportionally higher concentration of the nutritionally essential sulphur-containing amino acids, cysteine and methionine. The total world fish catch has reached a plateau and there is a progressive reduction in the fraction of the catch directed towards meal production because more is being used as human food. Hence, the compounded animal-feed industry depends increasingly on agriculture, primarily soybean production, for its protein. Further examination suggests that there is also a geographical factor in this dependence; the U.S.A., together with a smaller but potentially increasing contribution from Brazil, produces most of the world's soybean crop. However, in order to satisfy the high-quality sector of the compounded feed market, a new bulk protein source of either equal or better quality than fish meal is needed, unless European egg and poultry producers are prepared to accept either supplementation of certain feeds with methionine or lower levels of productivity and, ultimately, profit. The bulk ingredients in all compound feeds are cereals which, of course, contribute some protein to the feed, but are primarily included as energy sources.

Bulk transportation allows large quantities of protein to be moved efficiently and economically from continent to continent, particularly from the U.S.A. to Europe, the U.S.S.R. and Japan. The nature of the soybean makes it ideal for bulk handling. Soybeans and soybean meal are traded as commodities where prices are quoted on a future basis (Teweles *et al.*, 1974). This introduces a highly speculative element with respect to prices, such that relatively small actual, or predicted, fluctuations with respect to supply and demand create considerable fluctuation in prices, such that quoted future prices for a specified month of delivery can vary by as much as 100 per cent over a period of less than three months. The volatility of the soybean futures market has been particularly marked during the past seven years (1972–78), although during the previous decade it was relatively stable. The recent past has not only suggested opportunities for alternative protein sources for inclusion in compounded animal feeds, but also introduced uncertainty with respect to both the prediction of prices during the next decade and the wisdom of basing investment decisions for production of potentially complementary high-protein products by industrial routes on such forecasts. In addition to the volatility of the soybean market, one has also seen massive discontinuities with respect to the price and demand patterns for energy and raw materials during the past six years, a situation that has created problems in both developed and developing countries.

For many years, the international agencies concerned with health, food and agriculture, have done much to publicize the world-wide protein shortage. In the developed countries, an increasing demand for high-quality protein has been seen as overall living standards improved, while in most developing countries the population has grown at a greater rate than the capability for food production, particularly that for high-quality proteins. In fact, there is potentially a direct conflict between the interests of the developed countries and those of the developing countries. In the former, intensive animal husbandry requires continued production of compounded animal feeds using predominantly oil-seed proteins which, although frequently unacceptable for direct human nutrition in the developed countries, are often components in the diet in developing countries. Disparity between supply and demand means that people in the developed countries have either to pay more for meat or eat less, while in the developing countries people starve, sometimes because of the commercial attractiveness of exporting oil-seeds to satisfy the requirements of the compounded animal-feed industry in the developed countries.

Single-cell protein production was proposed as a way to provide an alternative source of protein that could supplement the conventional sources of supply for the manufacture of compounded animals feeds. The incentive for S.C.P. production is to be found particularly in Europe, the

U.S.S.R., Japan, the Middle East and North Africa, where indigenous agriculture is unable to provide the necessary protein-rich raw materials for incorporation into compounded animals feeds, but where a demand exists. It is envisaged that S.C.P. can provide a possible key for economic livestock production in some of these regions. All too frequently, S.C.P. is thought of as a potential substitute for either soybean or fish meals in compounded feeds. Such an approach ignores what S.C.P. really is, and whether or not its incorporation into food for direct human consumption, rather than into compounded animal feeds, would be the more logical way to use it. Unfortunately, S.C.P. frequently meets with disapproval from consumer organizations, even as animal feed, while periodic suggestions that it might be an attractive ingredient for human food are usually rejected on what amount to little more than emotional grounds. The types of industrial processes that have either been developed or are in the process of development are such that any S.C.P. produced by currently proposed technology, both with respect to its chemical and its microbiological safety, will be superior to virtually all of the traditional foods eaten today.

This chapter deals with the science and technology concerned with conversion of natural gas to S.C.P. either directly, by cultivation of methane-utilizing bacteria or, indirectly, by cultivation of methanol-utilizing micro-organisms on methanol produced chemically from natural gas. Other chapters in this volume deal with other feedstocks that are of potential commercial interest for S.C.P. production. Although methane is a gas and methanol a water-miscible liquid, they can, from the microbiological viewpoint, conveniently be considered together, as both are C_1 compounds and are metabolized through the same biochemical pathways.

Natural gas is found in many countries in the world. Although the major constituent of natural gas is methane, it is important to realize that, when considering natural gas as the carbon feedstock for S.C.P. production, one is unlikely to be considering pure methane as the feedstock. Natural gas is found both as associated and as unassociated gas. Associated gas production occurs in oil fields simultaneously with oil production, and the gas invariably comprises both methane and other gaseous alkanes in varying proportions depending on the source of production and on gas-treatment procedures adopted. Unassociated gas production occurs without simultaneous oil production from gas rather than oil fields but, even so, it comprises methane and usually much smaller concentrations of the other gaseous alkanes and is not infrequently diluted with carbon dioxide and nitrogen. Hydrogen sulphide is found both in associated and unassociated gases, but it can be easily removed. In regions where a market exists for liquified petroleum gases, these are frequently separated from associated gas, giving the purified associated gas a composition that is similar to unassociated gas. Methanol can be manufactured by chemical processes for a diverse range of carbonaceous feedstocks including wood, coal, naphtha and natural gas.

II. METHANE-UTILIZING AND METHANOL-UTILIZING MICRO-ORGANISMS

The first reports of microbial methane oxidation appeared more than 70 years ago. Virtually simultaneously, Kaserer (1905) reported the existence of methane-utilizing micro-organisms, and Söhngen (1905) described the isolation of the bacterium *Bacillus methanicus*, subsequently described as *Methanomonas methanica* (Jensen, 1909), its taxonomy, and reported the results of growth experiments where the bacterium utilized methane as its carbon and energy source. Many of the laboratory techniques necessary for cultivation and handling of methane-utilizing bacteria were not developed until relatively recently. In fact, only isolated papers, such as, for example, those by Aiyer (1920) and Hutton and Zobell (1949), referred to microbial methane utilization during the next 50 years, and it was not until Foster and his coworkers started to publish the result of their studies with methane and other gaseous hydrocarbons in 1956 that any marked advances, beyond the observations of Kaserer and Söhngen, were made. Foster (1963) reviewed much of his own and his colleagues work in his A. J. Kluyver Memorial Lecture. However, for comprehensive details of the early studies concerning microbial methane oxidation, the reviews of Silverman (1964) and Coty (1969) should be consulted.

From the viewpoint of S.C.P. production from natural gas, Foster's three major contributions were : (i) the definition, with Dworkin (Dworkin and Foster, 1956), that methane is oxidized by bacteria to form biomass and carbon dioxide through the route:

$$CH_4 \rightarrow CH_3OH \rightarrow HCHO \rightarrow HCOOH \rightarrow CO_2$$

thereby establishing methanol as an intermediate; (ii) the discovery, with Leadbetter (Leadbetter and Foster, 1958) that the methane-oxidizing bacterium, *Pseudomonas methanica*, was capable of co-oxidizing other gaseous alkanes; (iii) the isolation with Davis (Foster and Davis, 1966) of the thermophilic methane-oxidizing bacterium, *Methylococcus capsulatus*.

The significance of these important contributions became evident subsequently. The first indicated that the difficult first step of the oxidation could be circumvented by converting methane into methanol by chemical methods and growing methanol-utilizing micro-organisms rather than the then apparently more fastidious methane-utilizing bacteria for protein production. The second, in demonstrating the co-oxidation capacity of methane-utilizing bacteria, indicated a potential major complication in using natural gas rather than methane as the carbon substrate for pure cultures of methane-utilizing bacteria. The third represented the isolation of a bacterium, strains of which have subsequently proved to be the most suitable methane-utilizing bacteria for use in S.C.P. production. One other important discovery with respect to methane utilization that should not be ignored was the finding, by Davis *et al.* (1964), that a methane-utilizing bacterium, *Pseudomonas methanitrificans*, was able to fix atmospheric nitrogen.

Several papers have suggested that micro-organisms other than bacteria are able to utilize methane for growth. These include the report by Enebo (1967) that a *Chlorella* sp. was capable of consuming methane, and that by Zajic *et al.* (1969) indicating that a *Graphium* sp. was capable of growth on natural gas, although subsequently (Volesky and Zajic, 1971) the *Graphium* sp. was shown to utilize ethane not methane. To date, no independently authenticated reports of methane utilization by micro-organisms other than bacteria exist.

The first reports concerned with methane as a possible carbon and energy substrate for S.C.P. production were published by Wolnak *et al.* (1967) and Hamer *et al.* (1967) and, in the same year, some important data concerning cell yields of bacteria grown on methane were also published by Vary and Johnson (1967). However, the need for a rational basis for identification and nomenclature of the increasing number of methane-oxidizing bacteria that were reported in these and other studies was not satisfied until Whittenbury *et al.* (1970) published their important paper on the classification of obligate methylotrophic bacteria. The primary basis of their classification was the arrangement of the bacteria's membrane structure, Type 1 methylotrophs having the membranes arranged in bundles of vesicular discs, and Type 2 methylotrophs having membranes paired in a layer around the periphery of the bacterium. Lawrence and Quayle (1970) validated this classification by demonstrating that the Type 1 methylotrophs, *Methylococcus* and *Methylomonas* species, possessed the ribulose monophosphate pathway for carbon assimilation, although some also have the serine pathway, whilst the Type 2 methylotrophs, *Methylosinus* and *Methylocystis* species, assimilate carbon by the serine pathway.

Recently, in addition to the obligate methylotrophic bacteria, facultative methylotrophic bacteria have also been reported by Patt *et al.* (1974, 1977). However, such bacteria are unlikely, because of their relatively slow growth rates when oxidizing methane, to have application in a direct route for S.C.P. production from natural gas, although they may find application in an indirect route using methanol as feedstock.

Until relatively recently, methanol has received little more than cursory attention as a growth substrate for micro-organisms, although, during the last 50 years, several bacteria capable of growth on methanol as their sole carbon and energy substrate have been identified. These include *Pseudomonas extorquens* (Bassalik, 1914), *Protaminobacter ruber* (den Dooren de Jong, 1927), *Hyphomicrobium vulgare* (Mevius, 1953), *Pseudomonas methylotropha* (Byrom and Ousby, 1974) and a range of other *Pseudomonas* spp. (Kaneda and Roxburgh, 1959; Peel and Quayle, 1961; Anthony and Zatman, 1964) many of which have been subjected to quite extensive biochemical investigations. These biochemically oriented studies have been comprehensively reviewed by Quayle (1972). Relatively few studies have been directed towards the growth characteristics of methanol-utilizing bacteria.

In the last ten years, largely as a result of the stimulus created by the potential of S.C.P. production from methanol, several yeasts capable of growth on methanol have been isolated. These include yeasts classified in the genera *Kloeckera* (Ogata *et al.*, 1969), *Pichia* (Hazeu *et al.*, 1972), *Hansanula* (Levine and Cooney, 1973), *Candida* (Sahm and Wagner, 1972) and *Torulopsis* (Asthana *et al.*, 1971). Methanol utilization has also been reported as a characteristic of three fungi (Sakaguchi *et al.*, 1975). Growth of yeasts on methanol has been the subject of a recent comprehensive review by Sahm (1977). All methanol-utilizing yeasts exhibit a requirement for biotin and thiamin.

Probably the two most important factors that are encountered when micro-organisms are grown on methanol are the inhibitory nature of methanol to methanol-utilizing micro-organisms and the uncoupling of growth from the oxidation of methanol under conditions where significant methanol concentrations are present in the growth medium (Harrison *et al.*, 1972). An interesting technique for overcoming these problems by supplying methanol to the fermentation as a vapour in the air supply was proposed by Hamer (1968), when cultivation techniques for methanol-utilizing micro-organisms were in their infancy. The problems associated with the drop-wise addition of methanol to chemostat cultures have been discussed by Harrison and Topiwala (1974) and fluctuating conditions have been the subject of a study by MacLennan *et al.* (1971) in which *Pseudomonas* AM1 was the bacterium employed.

III. BIOCHEMISTRY OF METHANE AND METHANOL UTILIZATION

Growth of micro-organisms on methane and/or methanol requires that the micro-organisms are able to generate energy and reduction equivalents from these C_1 compounds. In addition, they must also possess metabolic pathways for synthesis of C_3 skeleton from which a central intermediary metabolite such as pyruvate or phosphopyruvate may be derived. Once synthesis of such central intermediary metabolites is accomplished, there is no evidence to suggest that the main pathways leading from the C_3 skeleton to the main groups of cell constituents will be different from the well established pathways of intermediary metabolism found in other micro-organisms.

The currently accepted view on oxidation of methane by bacteria can be summarized by the following series of reactions:

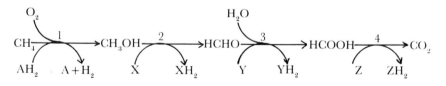

The nature of the electron donor AH_2 and of the acceptors X and Y are still uncertain, and the energy yield of the oxidation of methane to carbon dioxide is unknown, although it is now considered possible to estimate a maximum limit for the energy yield of this sequence of reactions (Quayle,

1972). Microbial methanol oxidation is represented by the latter three steps of the sequence.

According to Quayle (1974), there are three main pathways which accomplish net synthesis of C_3 skeletons from C_1 compounds in bacteria. These are: (i) The ribulose diphosphate or Calvin cycle for carbon dioxide fixation, a pathway typically found in bacteria growing on carbon dioxide; (ii) the ribulose monophosphate cycle of formaldehyde fixation of which three variants have been suggested; (iii) the serine pathway of formaldehyde fixation of which two variants have been proposed. Methane- and methanol-utilizing bacteria employ variants of the ribulose monophosphate cycle and the serine pathway in order to assimilate C_1 substrates.

In the case of the ribulose monophosphate cycle, the three variants proposed differ in the way in which conversion of fructose 6-phosphate to pentose 5-phosphate and either glyceraldehyde 3-phosphate or pyruvate

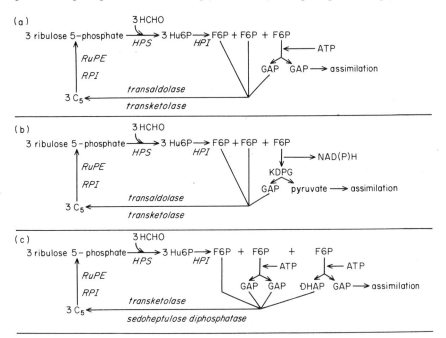

Fig. 1. Schematic representation of the ribulose monophosphate cycle of formaldehyde fixation. (a) FDP variant; (b) KDPG variant; (c) SDP variant. HPS indicates hexulose phosphate synthase, HPI hexulose phosphate isomerase, RuPE ribulose phosphate epimerase, RPI ribose phosphate isomerase, Hu6P D-arabino-3-hexulose 6-phosphate, F6P fructose 6-phosphate, GAP glyceraldehyde 3-phosphate, KDPG phospho-2-oxo-3-deoxygluconate, and DHAP dihydroxy-acetone phosphate. From van Dijken (1976).

is accomplished (van Dijken, 1976). In the case of the serine pathway, glycine, a reduced C_1 unit, and carbon dioxide are converted to malate. In the isocitrate lyase-positive variant of this pathway, malate is activated to malyl-CoA, subsequently cleaved to glyoxylate and acetyl-CoA which are then converted to glycine by a reaction sequence involving citrate synthase, aconitrate hydratase and isocitrate lyase. In the isocitrate lyase-negative variant, synthesis of a C_4 compound is accomplished in a similar way, but neither malate thiokinase nor isocitrate lyase is involved. The ribulose monophosphate cycle and the serine pathways are represented schematically in Figs. 1 and 2.

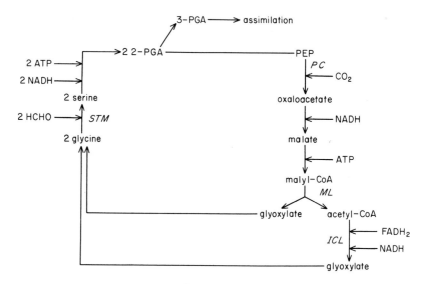

Fig. 2. Schematic representation of the ic 1[+] variant of the serine pathway of formaldehyde fixation. STM indicates serine transhydroxymethylase, ML malyl-CoA lyase, PC phosphoenolpyruvate carboxylase, ICL isocitrate lyase, 2-PGA 2-phosphoglycerate, 3-PGA 3-phosphoglycerate, and PEP phosphoenolpyruvate. From van Dijken (1976).

Methane-utilizing bacteria that use the ribulose monophosphate cycle include *Methylococcus capsulatus* and *Pseudomonas methanica*. The serine pathway is used by *Methylocytis* spp. and *Methylosinus trichosporium*. *Pseudomonas methylotropha* and some non-methane-utilizing obligate methylotrophs employ the former, whilst *Pseudomonas* AM1, *Pseudomonas extorquens* and *Hyphomicrobium* spp. employ the latter. There is also some evidence that suggests that, in *Methylococcus capsulatus*, the serine pathway

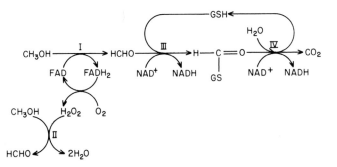

Fig. 3. Pathway of methanol oxidation by methanol-utilizing yeasts. (I) indicates alcohol oxidase, (II) catalase, (III) formaldehyde dehydrogenase and (IV) formate dehydrogenase. From Sahm (1977).

is present alongside the ribulose monophosphate cycle (Whittenbury *et al.*, 1974). On the basis of the energy requirements for various pathways of C_1 assimilation, it has been predicted (van Dijken and Harder, 1975) that a bacterium having the ribulose monophosphate cycle as its main path of carbon assimilation should show a higher growth yield on methane than one with the serine pathway. Experimental proof of this prediction is, of course, extremely difficult, as it is most unlikely that optimal growth with maximum yield coefficients of species exhibiting the two main alternatives will occur under similar growth conditions and at identical growth rates, although a correction can be made for the latter. Significant variations of the yield coefficient can be observed in methane- and methanol-utilizing micro-organisms.

It has been proposed (Sahm, 1977) that methanol-utilizing yeasts oxidize methanol using the pathway represented schematically in Fig. 3. There is also evidence that suggests that some yeasts use the dissimilatory ribulose monophosphate cycle for oxidation of formaldehyde to carbon dioxide. This cycle is shown schematically in Fig. 4. It is suspected that it may also contribute to oxidation of formaldehyde in *Ps. methanica* and *M. capsulatus* (Ström *et al.*, 1974).

IV. EFFICIENCY OF GROWTH ON METHANE AND METHANOL

Within the context of S.C.P. manufacture, Hamer *et al.* (1976) listed the most important physiological factors contributing towards the efficient

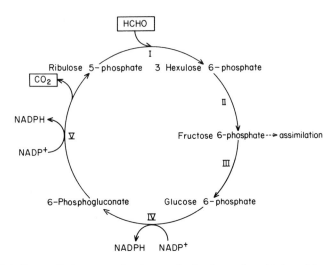

Fig. 4. Dissimilatory ribulose monophosphate cycle for the cyclic oxidation of formaldehyde. (I) indicates hexulose phosphate synthase, (II) 3-hexulose phosphate isomerase, (III) phosphoglucoisomerase, (IV) glucose 6-phosphate dehydrogenase and (V) phosphogluconate dehydrogenase. From Sahm (1977).

production system. These include: (i) yield coefficient; (ii) growth rate; (iii) cell density; (iv) substrate affinity; (v) culture stability; (vi) resistance to contamination; and (vii) thermotolerance. Without doubt the most critical of these physiological factors in any S.C.P. production process is the yield coefficient. The yield coefficient is usually defined as the weight of dry cells produced per unit weight of carbon-energy substrate utilized, but other definitions are encountered. It is a measure of the biological efficiency of the micro-organisms under consideration, and varies with respect to culture conditions. The yield coefficient, as defined above, ignores the protein content of the cells and this must, of course, be taken into account in the evaluation of maximum yield coefficients of potential process micro-organisms for S.C.P. manufacture. For example, when cultures are grown in the presence of excess carbon substrate, storage products in the form of carbohydrates and fats tend to be synthesized within cells and, because of the smaller expenditure of energy for such synthesis, yield coefficients measured on a whole-cell basis must be treated with caution and the cell protein content must be taken into account, particularly if S.C.P. production is the object of the growth process. In the ideal process for S.C.P. manufacture, the only products will be protein-rich biomass, carbon dioxide and water, with conditions adjusted so that protein-rich biomass production will be maximized. The most generally accepted approach concerning the efficiency of carbon substrate oxidation and microbial cell production is the Y_{ATP} concept first proposed for anaerobic growth by Bauchop and Elsden (1960), reviewed extensively for energy yields and growth of heterotrophs by Payne (1970) and applied to the growth of methylotrophs by Harrison *et al.* (1972) and more recently by Anthony (1978). The relative energetic advantages concerning selection of either methane- or methanol-utilizing micro-organisms as the basis of an S.C.P. manufacturing process are still unclear, a point stressed by Skryabin *et al.* (1975). Remarkably little experimental evidence concerning the energetics of the methane → methanol oxidation step for methane-utilizing bacteria is available, although two hypotheses have been proposed with a view to assessing the relative merits of the direct and the indirect routes for S.C.P. production from natural gas. The hypothesis proposed by van Dijken and Harder (1975) postulates an energy-dependent reversed electron-transport system, whilst that proposed by Tonge *et al.* (1974) postulates that a carbon monoxide-binding cytochrome is generated as the reductant for a methane mono-oxygenase system. Both hypotheses have subsequently been further elaborated (Harder and van Dijken, 1977; Tonge *et al.*, 1975, 1977), but the key question of whether or not the first step in methane oxidation is coupled to energy generation remains unanswered. Recently, a general observation on maximum growth yields in relation to the carbon and energy contents of various growth substrates has been reported (Linton and Stephenson, 1978). This introduces the concept of micro-organisms growing on a simple carbon and energy substrate under either carbon or energy limitation, and is pertinent to methane and methanol oxidation.

In the case of micro-organisms that utilize gaseous, but also volatile, liquid carbon and energy substrates, the difficulties of accurately determining yield coefficients must not be underestimated. For methane-utilizing bacteria, details of appropriate techniques for accurate yield coefficient measurements have been published by Barnes *et al.* (1977). It is doubtful if either the yield coefficient data for methane-utilizing bacteria reviewed by Hamer and Norris (1971) or much of the subsequently published data could stand rigorous examination. For methane-utilizing bacteria, Klass *et al.* (1969) proposed a maximum yield coefficient of 1.44, Harwood and Pirt (1972) reported that, for *Methylococcus capsulatus* growing on methane at 37°C, the yield coefficient under methane limitation was 1.01, and under oxygen limitation 0.31, while Wilkinson *et al.* (1974) reported yield coefficients, at 32°C, of 0.99

under oxygen limitation and 0.80 under methane limitation, for a mixed culture in which a Type 2 methylotroph predominated. In contrast with these relatively high values, Vary and Johnson (1967) reported yield coefficients of between 0.50 and 0.62 for methane-limited, and between 0.58 and 0.70 for oxygen-limited mixed cultures growing at 30°C. Further, ammonia was identified as the best nitrogen source for high yield coefficients. In a subsequent study, Sheehan and Johnson (1971) reported yield coefficients of between 0.57 and 0.66 for high cell-density continuous mixed cultures, growing at 45°C with nitrate as a nitrogen source. In other studies, Iwamoto (1969) obtained a yield coefficient of 0.44, and Fukuoka (1972) reported values within the range 0.44 and 0.58.

For methanol-utilizing bacteria, yield coefficient data also vary considerably, particularly between batch and continuous culture. Low yield values for growth on methanol in batch-culture experiments, where methanol is in excess, have frequently been attributed to a failure to account for the evaporation of methanol from sparged cultures. However, a more plausible explanation for low yields in the presence of excess methanol is that uncoupling of either the energy-producing or the energy-utilization process occurs when methanol is in excess, a situation that is usual in batch cultures, but will not occur in methanol-limited continuous cultures (Harrison et al., 1972). There seems to be little feedback regulation of the assimilative and oxidative pathways, so that oxidation proceeds faster than assimilation in the presence of excess methanol, and the energy generated is wasted. As methanol is an inhibitory substrate, such a mechanism could benefit methanol-utilizing bacteria by providing a means for rapidly lowering the methanol concentration in cultures where it is present in significant concentrations. In batch-culture experiments employing a *Pseudomonas* sp., Harrison et al. (1972) found that the yield was decreased from 0.33 to 0.15 as the initial concentration of methanol was increased from 2.5 to 25.0 g l^{-1} whilst, with the same bacterium growing in methanol-limited continuous culture, a yield of 0.40 was achieved. In addition, Harrison et al. (1972) determined yields in continuous culture for several methanol-utilizing micro-organisms. The yields measured were, for *Protaminobacter ruber* 0.25, for *Ps. extorquens* 0.40, for *Torulopsis glabrata* 0.43 and for *Pichia pinus* 0.30. In earlier batch-culture studies using methanol-utilizing mixed cultures, Vary and Johnson (1967) reported a yield of 0.33, and Häggström (1969) a yield of 0.41, whilst for various cultures of pure obligate C_1-utilizing bacteria, Whittenbury *et al.* (1970) claimed yields of 0.40. After an extensive study with *Methylomonas methanolica*, Häggström and Molin (1976) were only able to report a yield of 0.39 under conditions approaching those that might be employed in a production process. This value represented a decrease of 19% compared with yields previously reported for this bacterium (Dostálek *et al.*, 1972; Dostálek and Molin, 1973). A maximum yield of 0.45 was reported for *Pseudomonas* AM1 in continuous culture at 30°C by MacLennen *et al.* (1971). Ignoring results where claims concerning carbon dioxide fixation by methanol-utilizers are made (Babij *et al.*, 1974), the highest yield reported for a pure culture growing on methanol is 0.54. This was achieved in continuous culture with *Pseudomonas* C at 35°C (Battat *et al.*, 1974). The maximum yield for this same bacterium in batch culture was only 0.31 (Chalfan and Mateles, 1972).

In general, yield data for methanol-utilizing yeasts have shown similar patterns to the data for bacteria. For a *Kloeckera* sp. growing at 30°C, Ogata *et al.* (1970) reported a yield of 0.25 for *Torulopsis glabrata* growing at 30°C; Asthana *et al.* (1971) reported a yield of 0.45 compared with 0.43 reported by Harrison *et al.* (1972); for *Candida boidinii* growing at 38°C, Sahm and Wagner (1972) reported a yield of 0.29. For *Hansenula polymorpha* growing in the range 37°C–42°C, Levine and Cooney (1973) reported a yield of 0.36, and van Dijken *et al.* (1976) a yield of 0.38 at 37°C. Yield depression with increasing methanol concentration has been reported for yeasts by Reuss *et al.* (1974) and by Pilat and Prokop (1975).

Cooney (1974) has suggested that the theoretical maximum yield for growth on methanol is 0.60, but very few of the yield values reported here for monocultures approach this value, and the wisdom of using monocultures as the basis for S.C.P. production processes is subject to question, particularly in view of developments that have occurred with use of structured mixed cultures of micro-organisms. There is no valid reason to assume that an industrial fermentation process must be based on a monoculture, particularly when one considers that traditional fermentations for both foods and beverages are frequently based on mixed cultures and septic production methods.

Extensive studies have been carried out on the application of defined structured mixed cultures that grow on methane and on methanol, particularly as a result of evidence that mixed populations were able to grow to greater cell densities with shorter generation times than many pure cultures of methylotrophs. Two interesting preliminary observations were that, in the isolation of obligate methane-utilizing bacteria from either water or water-associated soil samples, methanol-utilizing hyphomicrobia invariably appeared as contaminants (Hamer and Norris, 1971), and that compounds that were potential intermediary metabolites in bacterial growth strongly inhibited growth of *Methylococcus capsulatus* (Eroshin *et al.*, 1968).

In a detailed study of a particularly stable mixed culture that was able to grow readily on methane, Wilkinson *et al.* (1974) demonstrated that two major interactions occurred. In the first of these, *Hyphomicrobium* sp. was shown to be capable of utilizing trace concentrations of methanol produced by, and inhibitory to, the primary methane-utilizing bacterium in the culture whilst, in the second, a *Flavobacterium* sp. and an *Acinetobacter* sp. removed other potentially inhibitory metabolic products. The *Hyphomicrobium* sp. was able effectively to scavange methanol because of its markedly higher affinity, with respect to the primary methane-utilizing bacterium, for methanol (Wilkinson and Harrison, 1973) and also, at lower oxygen tensions, the *Hyphomicrobium* sp. was able to denitrify (Wilkinson and Hamer, 1972) and thereby not compete with the primary methane-utilizing bacterium for oxygen. This culture resulted from enrichment, rather than from reconstitution from its component bacteria. Defined, structured mixed cultures for S.C.P. production must be produced by reconstitution techniques if the exclusion of undesirable contaminants is to be guaranteed.

Several cultures that have been constructed (Harrison *et al.*, 1975) have been shown to be extremely stable in continuous culture, resistant to contamination and showing absolutely no tendency to foam; furthermore they are capable of growth at temperatures as high as 45°C. The primary bacterium in one such culture is *Methylococcus* sp. growing in association with four heterotrophic bacteria, two *Pseudomonas* spp., *Mycobacterium rhodochrous* and a *Moraxella* sp. Their role, according to Linton and Buckee (1977), is the production of extracellular enzymes to effect degradation of lysis products from *Methylococcus* sp. rather than utilization of an intermediate product of methane oxidation. This defined, structured mixed culture is capable of growth at 45°C, at growth rates in continuous culture of 0.3 h^{-1}, at cell densities of 25 g l^{-1} and yields on methane of 0.85.

The mixed-culture philosophy has also been applied, with similar advantages, to microbial methanol utilization, although some dispute exists as to whether defined, structured mixed cultures are superior to monocultures in this case (Goldberg, 1977). Snedecor and Cooney (1974) studied a thermophilic mixed culture comprising at least three morphologically distinct bacteria which, whilst being able to grow at temperatures up to 65°C, exhibited its maximum yield (0.42) at 56°C. A particularly interesting defined structured mixed culture was reconstituted by Wren *et al.* (1974) and its performance discussed in several subsequent papers (Harrison *et al.*, 1975; Harrison and Wren, 1976). This particular culture comprised the primary bacterium, an obligate methanol-utilizer, and four heterotrophs, two *Pseudomonas* spp. an *Acinetobacter* sp. and a *Curtobacterium* sp.; 42°C was the optimal temperature for growth. The obligate methanol-utilizer, when grown alone in continuous culture, exhibited a maximum growth rate of 0.16 h^{-1} and a yield of 0.30. However, when grown in association with the *Curtobacterium* sp., the maximum growth rate increased to 0.19 h^{-1} and the yield to 0.35 and, when grown in association with all four heterotrophs, the growth rate increased to 0.59 h^{-1} and the yield to 0.52. Essentially, the mixed-culture approach seeks to eliminate additional product formation, such that maximum incorporation of substrate carbon into biomass occurs.

In addition to the above studies, Crémieux *et al.* (1977) and Ballerini *et al.* (1977) have reported details concerning the isolation, identification and growth of a mixed bacterial association on methanol. Their stable culture comprised four bacteria, namely *Methylomonas methylovora*, a *Xanthomonas* sp., a *Flavobacterium* sp. and *Ps. pseudoalcaligenes*. The culture was grown at temperatures between 34°C and 41.6°C, and the maximum growth rate of the culture was 0.49 h^{-1}. The maximum yield (0.44) was achieved at 34°C and this declined to 0.36 at 41.6°C, suggesting that the performance of this culture was inferior to that described by Wren *et al.* (1974).

Irrespective of the relative merits of defined, structured mixed cultures and monocultures for optimum growth performance on either methane or methanol, it is certain that, in order to use natural gas containing even only low concentrations of gaseous alkanes, other than methane, as feedstock in the direct route for S.C.P. production, it will be necessary to employ a defined, structured mixed culture. The reason for this is the capacity of methane-utilizing bacteria to co-oxidize other gaseous alkanes such that the products of the co-oxidation accumulate in the fermentation broth to inhibitory levels unless another bacterium is present to oxidize further the products to carbon dioxide.

V. OTHER IMPORTANT PHYSIOLOGICAL FACTORS

In most studies concerned with production of S.C.P., a great deal of emphasis is placed on the carbon-energy substrate, and very little emphasis is placed on the nitrogenous substrate, even though nitrogen is a major constituent in proteins. As was suggested by Norris (1968), S.C.P. can take its place in the overall nitrogen cycle as shown in Figure 5.

Most proposed routes for S.C.P. production employ either ammonia or ammonium salts as the nitrogen source for growth. The versatility of methane-utilizing bacteria with respect to their nitrogen metabolism exceeds, by far, that of most other micro-organisms proposed for S.C.P. processes. Circumstantial evidence that natural gas-utilizing bacteria fix atmospheric nitrogen existed for many years prior to isolation of the nitrogen-fixing methane-utilizing bacterium, *Ps. methanitrificans* (Davis *et al.*, 1964) and isolation of the nitrogen-fixing *n*-butane-utilizing bacterium *Mycobacterium butanitrificans* (Coty, 1967). As methane-utilizing bacteria can also utilize nitrate as their nitrogen source, there are, conceptually, several interesting possibilities, and for growth of methane-utilizing bacteria one is no longer dependent on a single nitrogen source. Depending on the nitrogen source employed, one must

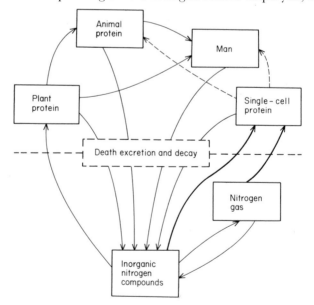

Fig. 5. . Modified nitrogen cycle showing the role of single-cell protein in overall protein synthesis. From Norris (1968).

expect variations in the yield coefficient based on the carbon-energy substrate. For *Methylococcus* sp. growing in continuous culture on methane at 45°C with ammonia as the nitrogen source, the optimized yield coefficient was found to be 0.85, whilst for dinitrogen as nitrogen source, the yield coefficient declined to 0.65 (Hamer, 1977). The value for nitrate lies between the two. The lower yield coefficient when dinitrogen is fixed is less than was generally predicted, and situations can be envisaged where it might be economically sensible to operate a direct natural gas based S.C.P.process under nitrogen-fixing conditions.

What is undoubtedly more generally important, particularly with respect to the stability of cultures, is the fact that ammonia is a competitive inhibitor for methane oxidation. The mechanisms of ammonia oxidation by methane-oxidizing bacteria have been discussed in detail (Dalton, 1977; Drozd *et al.*, 1976). The practical implication of these studies is that, although ammonia enhances the carbon substrate-based yield coefficient, it can only be used in high-productivity systems if appropriate techniques for its measurement and control of its concentration in the fermentation broth can be devised.

In industrial microbiological processes, culture stability and resistance to contamination are of particular importance. These features are directly related to the affinity of micro-organisms for their substrates. Previously, the relative merits of defined, structured mixed cultures and monocultures have been compared on the basis of their suitability for large-scale production of S.C.P.

The growth rate in carbon and energy substrate-limited continuous culture is generally considered to be regulated by uptake of the carbon-energy substrate. This uptake generally follows typical Monod kinetics with respect to substrate concentration in the liquid phase. This situation can, of course, increase in complexity when the substrate is either of a gaseous nature or is immiscible with water. In the case of potentially inhibitory substrates, such as methanol, the ability of methanol-utilizing micro-organisms to grow at high growth rates will depend on the saturation constant for growth (K_s) for the particular substrate being much less than the inhibitory concentration. Harder *et al.* (1977) reviewed the effects of selection pressures in continuous-flow fermenters with respect to substrate affinity.

Harrison (1973) examined the affinity of methane-and methanol-utilizing bacteria for their carbon substrates and was able to demonstrate that, for methanol-limited cultures, the residual methanol concentration present in the fermenter would be $< 0.005\%$ of the methanol supplied, hence invalidating any possible suggestion that residual methanol associated with biomass produced for methanol could present a toxic hazard. Further, plausible explanations concerning methanol inhibition of methane-utilizing bacteria emerged from the study, and a basis was provided for subsequent elucidation of the major interactions in a methane-utilizing mixed culture by Wilkinson and Harrison (1973). In general, for potential S.C.P. production cultures, there is no essential requirement to disrupt the metabolic regulation of the cell, as there is for production cultures in the antibiotics and enzyme industries. However, what is necessary is to ensure that the metabolic regulatory systems, which have generally evolved towards maximizing growth efficiency, are functioning optimally. Adaptation of micro-organisms to suit various process engineering solutions in S.C.P. manufacturing technology may lower growth efficiency, and ingenuity is required to circumvent such problems.

VI. PROCESS ENGINEERING ASPECTS

The process engineering problems concerned with large-scale production of S.C.P. are numerous. In most studies concerning such problems, too much emphasis has been placed on those problems that occur in the fermenter, and it is important to remember that an integrated production process will comprise some six or seven interacting unit operations.

Laine (1972) reviewed some of the bio-engineering problems occurring in production of S.C.P. from hydrocarbons. Although this discussion was oriented towards S.C.P. production from liquid n-alkanes, most of the problems discussed apply equally to processes based on natural gas and methanol. Topiwala (1974) has discussed the interactive nature of the various unit operations involved in S.C.P. production from methane and methanol.

In order to achieve economy of scale in S.C.P. production from either natural gas or methanol, it is essential that large-scale ($> 50\,000$ tonnes per annum) continuous-flow production systems, operating at minimum productivities of between 3 and 6 kg m^{-3}h^{-1}, are employed. In addition, it is also essential that both a high yield coefficient, based on the carbon feedstock, and a high conversion of the carbon feedstock are achieved,

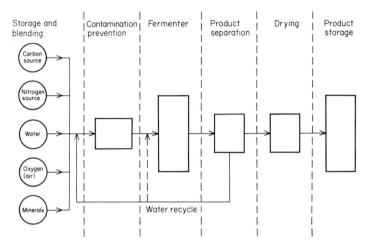

Fig. 6. Unit processes for single-cell protein production. From Topiwala (1974).

and that the process water requirements and aqueous effluent discharge are minimized.

In order to try and do justice to all the operations involved in S.C.P. manufacture from natural gas, the sequence of unit operation proposed by Topiwala (1974) and shown in Fig. 6 will be used as a basis for developing the discussion in this section. However, first it is instructive to examine the possible alternative process routes that could, theoretically, go forward to commercialization. These are shown diagrammatically in Fig. 7. Essentially, using natural gas as the carbon feedstock, one has several options available. These include using the feedstock directly, either by deploying a defined mixed bacterial culture, or a bacterial monoculture to produce S.C.P. Alternatively, one can convert the natural gas by conventional chemical processes, and perhaps in the future, by yet to be discovered biological routes (Foo and Hedén, 1976), to methanol. Using methanol as the carbon-energy substrate for the fermentation process, there are three potential choices involved in its further processing to produce S.C.P. These choices are: use of a monoculture of yeasts; use of a monoculture of bacteria; and the use of a defined mixed culture.

Even when one has produced S.C.P., one faces another three options, namely to attempt to process it further for incorporation into human food, to use it as an ingredient in compounded animal feeds, or to use it for one of several industrial uses of protein. Irrespective of the selected

M

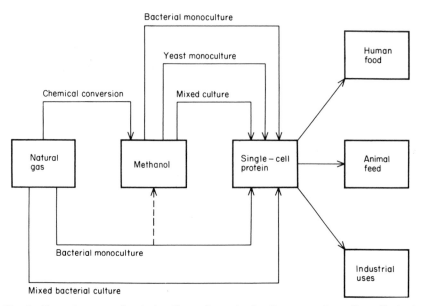

Fig. 7. Alternative routes for single-cell protein production from natural gas. From Topiwala (1974).

option, the essential unit operations in the overall process will vary remarkably little. Considering raw materials storage and blending, all routes have essentially similar inorganic nutrient requirements and, therefore, for similar scales of operation very similar facilities are required. However, with methanol-based routes, methanol storage tanks are required whilst, for the direct natural gas routes, gas delivery will, almost certainly, be by pipe-line. In the case of the more fastidious methanol-utilizing yeasts, provision must be made for vitamin addition.

The question of sterilization of the liquid medium prior to fermentation is a disputed question. Without doubt, some measure of contaminant control or prevention is essential. This can take the form of heat, chemical or filter sterilization, an essential if a food product is the ultimate objective, or varying levels of protective measures when the product is destined for either a feed or an industrial use. Such protective measures can involve use of an acidic medium, formaldehyde addition with subsequent oxidation by the methylotrophs in the fermenter, sterilization of medium components but no sterilization of the recycled medium, or the employment of defined structured mixed cultures in which all of the potential ecological niches have been filled by appropriate micro-organisms such that growth of stray contaminant micro-organisms in the reactor is effectively eliminated.

The item of the process plant that will vary the most, but not entirely because of differences in the routes, will be the fermenter. The operating volume of the fermenter or fermenters to be used for any S.C.P manufacturing venture will be controlled by both the total production required and the optimum productivity of the system, whilst total reactor volume will depend on factors such as gas hold-up and gas disengagement requirements. The maximum productivity of any particular fermenter design is dependent on its mass (gas)- and heat-transfer capacity, and the maximum productivity will vary according to the carbon feedstock used in the fermenter. The type of reactor employed will depend on the level of gaseous substrate-conversion required and on the relative capital and operating costs of mechanically agitated and sparged-induced flow systems under particular economic environments. Using the direct natural-gas route, conversion of the carbon feedstock will always be less than complete but, provided a gas-fired dryer is used, exhaust gases from the fermenter can be enriched with further natural gas to provide fuel for the drying operation.

In the design of fermenters for S.C.P. production, mass transfer, heat transfer and mixing are the most important factors, and a great deal has been written on these subjects. The problems of mass- and heat transfer in hydrocarbon-based fermentations was first discussed by Darlington (1964) and Guenther (1965), respectively, at a time when fermenter

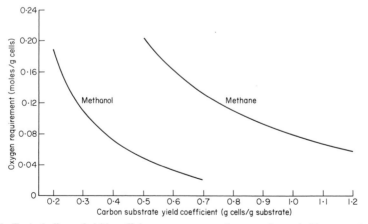

Fig. 8. Typical effects of yield coefficient on oxygen requirement when only biomass and carbon dioxide are produced. From Hamer *et al.* (1973a).

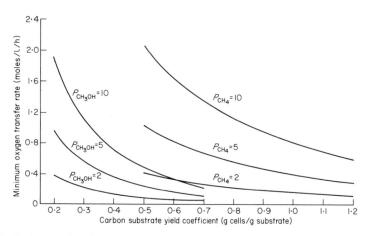

Fig. 9. Typical relationships between minimum oxygen-transfer rates and yield coefficients at various productivities (P). From Hamer *et al.* (1973a).

design was such that it seemed unlikely that substantial productivities could become a reality.

On the basis of the stoicheiometry of the overall growth reaction, and assuming conversion of the carbon feedstock to either biomass or carbon dioxide, Hamer *et al.* (1973a, b) calculated the relationships between carbon-substrate yield and oxygen requirement (Fig. 8) and between carbon-substrate yield and oxygen-transfer rate at various productivities (Fig. 9) for both methane and methanol as carbon-energy substrates. In addition, the relationship between carbon-substrate yield and heat production, using both a heat of combustion basis (Guenther, 1965) and the correlation proposed by Cooney *et al.* (1968), was calculated (Fig. 10) and, using the average value derived from the two methods, the relationship between productivity and heat production at various carbon-substrate yields was estimated (Fig. 11).

The operating costs of the fermenters, in all S.C.P. production processes, represent a considerable proportion of the total operating costs of the process, such that any improvements in the efficiency of energy utilization in the fermenters will have a significant impact on the overall process profitability. The high operating costs for the fermentation step result from a combination of gas compression and broth agitation requirements.

Transmission of energy to the fermentation broth increases the gas–liquid interfacial area and, consequently, in a system where

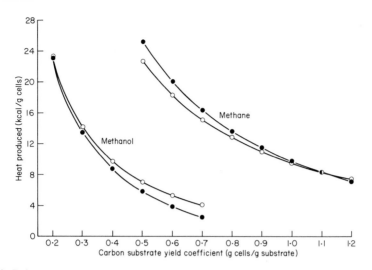

Fig. 10. Estimated relationships between heat production (O) calculated from heats of combustion and yield coefficients (●), calculated from the correlation of Cooney *et al.* (1968) in oxidation of methanol and methane by micro-organisms. From Hamer *et al.* (1973a).

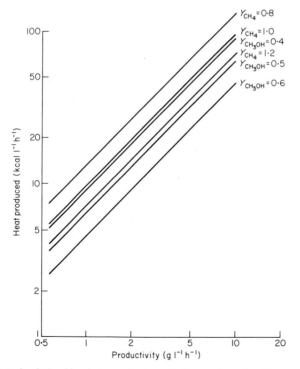

Fig. 11. Estimated relationships between heat production and productivity at various yield coefficients. From Hamer *et al.* (1973a).

gas–liquid mass transfer is the rate-determining step, increases the reaction rate. The two standard energy sources in fermenters are compressed gas expansion and mechanical agitation. The efficiency of stirring generally exceeds that for compression, suggesting that the cost to transfer a unit mass of gas is less for mechanical agitation than for gas compression and subsequent expansion. However, in these latter systems, if axial dispersion of the gas phase can be minimized, higher driving forces for mass transfer than in mechanically agitated systems where the dispersed gas phase is completely mixed, can be achieved with identical power inputs and gas compositions. In systems designed to approach plug flow of the gas phase, there may be problems of nutrient-concentration gradients in the liquid phase that will adversely affect the physiology, particularly the yield, of the culture.

It is important, when evaluating the potential performance of various fermenter designs, that they should not necessarily be compared on an equal productivity basis, but at the optimal productivity for each fermenter type. Fermenter productivity and, with the direct natural-gas route, fermenter conversion efficiency, will affect downstream operations.

Fermenter cooling is of less importance than mass-transfer efficiency for S.C.P. manufacture from natural gas, but it is still of considerable significance. When any particular process plant is built at an established site, as opposed to a greenfield site, it is an important aspect of the economics to be able to utilize existing services, e.g. cooling water from the integrated site system and the central facility for aqueous effluent treatment, provided, of course, that the new plant does not overload such services, and that the services are appropriate to the requirements of the new plant.

Heat transfer is one of the most neglected areas of fermentation technology. The optimum temperature range for growth of most micro-organisms is narrow; with many organisms it is only a few degrees. Below the optimum temperature for growth, the growth rate of any particular micro-organism increases relatively slowly with increasing temperature, whereas at temperatures in excess of the optimum for growth, there will be a very rapid decline of growth rate as the temperature is increased. Most micro-organisms used commercially are mesophilic, with temperature optima for growth in the range 20–40°C. Use of such micro-organisms for high-productivity industrial fermentations makes cooling, particularly with cooling water rather than refrigerant, a significant problem. The basis for calculating the cooling surface required for a particular design is the minimum temperature difference between the fermentation broth and the cooling water. As this difference decreases, the heat-exchange surface requirement, and therefore costs, increase. However, operating costs are directly concerned with the amount of cooling water needed, and are calculated on the basis of the average cooling-water temperature. Cooling-water temperature is dependent on a number of factors which include the origin of the cooling water, the climatic conditions and variations at the plant location, and the type of cooling-water system employed, either once-through or recycled with use of a cooling tower. Use of a saline cooling water will result in a need to use more expensive materials of construction, such as titanium, for the heat exchanger surfaces. Remarkably few data concerning heat-transfer coefficients for aerated fermentations have been published. Pollard and Topiwala (1976) have looked at the problem, but essentially ignored surface fouling, a factor cited by Hamer (1973) as being potentially a major problem with respect to cooling large S.C.P. production fermenters.

Previous sections have dealt with the process-engineering problems concerning the fermenter; finally, in this section, some brief comments on product separation and drying are necessary. In general, there is a dearth of literature concerning these aspects for S.C.P. routes based on natural gas. For any process employing a yeast as the process micro-organism, separation from the fermentation broth will be markedly easier and probably cheaper than separation of bacteria in those processes where they are used although, in the proposed routes employing yeasts, additional thermal lysis of the cells after separation and prior to drying may be necessary.

With bacterial-based routes, centrifuges are invariably proposed, but after first either flocculating or agglomerating the cells leaving the fermenter. This can be achieved by addition of either a flocculating agent or by adjusting the pH value of the fermentation broth. Care has to be exercised in the selection of flocculating agents for this use, because the agent will become associated with the final product and must, therefore, be free from any adverse toxicological properties. Provided process-water recycle is used, it is probably better to flocculate bacteria by addition of phosphoric acid, a reagent which is a necessary inorganic nutrient for the micro-organism. Centrifugal separators will be required to produce cell creams containing 20% cell dry weight as feed to the dryer. With yeast-based routes, flocculation is much less essential and a decanting operation followed by separation of cells on a drum filter is considered feasible.

With presently available equipment, asepsis during separation operations will be difficult to achieve and this will, of course, have important implications when process-water recycle is employed to achieve process-water economy. In fact, recycle of process water has a wide range of implications which have been discussed in detail by Khosrovi and Topiwala (1978). The primary dangers in operating process-water recycle are microbiological contamination of the fermentation, inhibition of the fermentation by recycling products of cell lysis and inorganic nutrient build-up unless these are very carefully balanced.

For drying S.C.P., either spray or flash dryers seem to be preferred. The economy of this step in the process is largely dependent on the water content of the cream or slurry fed to the dryer. The higher the solids content the lower will be the fuel requirements for the drying operation. The most important aspect of the drying operation is that no significant heat damage of the product occurs, as this would undoubtedly adversely affect the nutritional value of the product.

Product storage will either be in bags, which tend to be expensive, or in silos where both humidity and dust control will become important factors to which attention must be paid.

VII. PRODUCT QUALITY EVALUATION

Product-testing programmes for S.C.P. are carried out to determine, on the one hand, the nutritional value and functionality of the product and, on the other, the safety of the product. The two types of tests are essentially different, the first frequently employing the target species for which the product is ultimately intended, and the second employing those species normally used for toxicological evaluation. The former tests are designed essentially to demonstrate the quality of the product to potential customers, whilst the latter are designed to meet legislative requirements imposed by governments of the countries where the product is to be produced and marketed, although responsible producers will also wish to satisfy themselves concerning both efficacy and safety of their product.

Numerous protocols have been developed for test programmes seeking to demonstrate the two requirements. The best known guidelines concerning test procedures for S.C.P. are those proposed by the International Union of Pure and Applied Chemistry (1974), and it is their standards that most potential producers of S.C.P. are seeking to satisfy. These standards are such that virtually no currently used bulk-feed ingredient could meet them.

Nutritional aspects of S.C.P. were discussed in considerable detail by Kihlberg (1972) some years ago and, recently, the literature concerning evaluation of yeast S.C.P. for poultry feeds has been reviewed by Vananuvat (1977). In general, bacterial products are superior with respect to their contents of sulphur-containing amino acids. Relatively little information concerning detailed testing of S.C.P. derived from either the direct or the indirect production routes from natural gas have appeared, although some preliminary results have been published by D'Mello (1972, 1973).

In view of the very substantial nature of the risked capital involved in construction of a commercial scale S.C.P.-manufacturing plant, it can safely be assumed that those companies who are presently constructing manufacturing capacity have satisfied themselves, and have the appropriate information to satisfy the requirements of both their potential customers and legislation in the countries where they intend to produce and market their product.

VIII. PROCESS EVALUATION AND ECONOMICS

The market price of any particular S.C.P. will be determined by relating the product, on a relative performance basis, to the prices of conventional protein products. Compounded animal feeds are economically optimized mixtures of essential ingredients, designed to satisfy specific nutritional requirements. Complex least-cost formulation procedures are increasingly used for such optimizations. When feeds are compounded on a least-cost formulation basis, any potential substitute ingredient is unlikely to eliminate any single component from a particular feed formulation, but will almost inevitably affect the levels of several ingredients. Feed compounders utilize least-cost formulation techniques in order rapidly to switch from one feed ingredient to others as availability and prices vary, but without adversely affecting the overall performance of their products.

Opportunities for S.C.P. in compounded animal feeds must be assessed against both the complexities of the feed-compounding industry and the vagaries of agricultural product-commodity trading. It is frequently claimed that one of the major advantages to be gained from S.C.P. production will be a stabilization of protein prices, as it is commonly perceived that industrial protein production will be essentially free from unpredictable climatic changes that prove so disruptive and even disastrous for conventional agricultural production.

One of the principal problems when comparing the relative merits of agricultural and industrial protein production routes stems from the essentially different basis used for evaluation. Large-scale industrial S.C.P. production plants will be specifically designed for production from a particular carbon feedstock, and a planned plant life of some 15 years will be typical. Although a change in feedstock may, in some instances, be possible at some interim time during the plant life, it will probably result in a change from the design production capacity and adversely affect performance. A total change with respect to both feedstock and product will be out of the question for large optimized production plants. In agricultural production, where annual crops are grown on good quality land, the crop produced can be changed from season to season to meet the needs of fluctuating market predictions, hence giving the agricultural producer considerable short-term flexibility. The economic evaluation of manufacturing processes, although not identical from one organization to another, can be fitted into a general framework such that the potential economic environment in which the process will be required to function is assessed. Rudd and Watson (1968) discussed the basic terms, concepts and methodology concerned with such profitability studies.

When examining a new process for S.C.P. production, it is important to understand the technical factors that will significantly affect the overall manufacturing economics. The various process routes for S.C.P. production must be assessed relative to each other and compared with conventional protein production. Essentially, a manufacturing venture seeks to generate profit by risking capital investment.

The total investment for any manufacturing venture can be broken down into three parts according to the degree of financial risk. These three parts are the fixed investment in the immediate processing area, the investment in auxilliary services and the investment in working capital. Together, they comprise the money tied up and risked as a result of the venture. The fixed investment is the money required for the purchase and construction of all of the process equipment located within the immediate processing area, or battery limits as it is commonly known. Investment in equipment within battery limits carries the highest degree of risk since it can only be partially recovered as salvage value, should production be terminated.

Auxiliary investment refers to items such as steam plant and cooling towers, facilities commonly situated outside battery limits and which generally serve several process plants, each plant paying an appropriate portion of the costs. Such costs for a particular process plant can be charged either as an annual fee proportional to consumption or, in the case of emergency services, on a fixed-fee basis. The working capital is the capital tied up in the interests of the particular plant as ready cash to meet operating expenses and as stocks of both raw materials and product.

Manufacturing or operating costs are the costs incurred in keeping the process plant running. These costs are generally divided into three terms, one proportional to the fixed investment, one proportional to the production rate and one proportional to the labour costs. The first of these terms includes factors such as the labour and material aspects of maintenence, insurance and administration, that are independent of the rate of production and can be conveniently expressed as a percentage of the fixed investment. The second term includes raw material costs, utilities costs, the cost of chemicals, catalysts and other materials, maintenance costs related to operation, quality control, royalties and licence fees. The third term includes the direct cost of maintaining the operating labour force and includes, in addition to salaries, supervision costs and overheads for social security.

The gross profit derived from a particular plant is defined as the difference between the net income from the annual sales, after distribution, sales and promotional costs have been deducted, and the annual manufacturing costs. The net profit rate may be defined as the expected annual return on investment after deducting depreciation and taxes. Depreciation schedules are frequently complex, government-stipulated and location-dependent. Frequently, tax incentives are such as to demand optimization of the allowable depreciation schedule, thus making straight-line depreciation, i.e. depreciation inversely proportional to expected project life, a dramatic over-simplification. The cash flow for a typical manufacturing operation is summarized in Figure 12.

Topiwala (1974) has outlined the unit processes that will comprise a S.C.P. manufacturing process from either natural gas or methanol and,

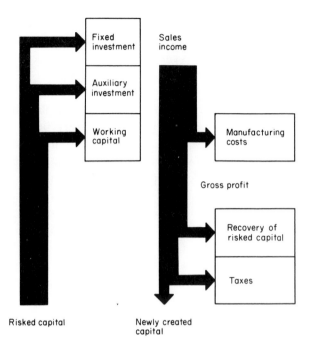

Fig. 12. Cash flow associated with a process. From Rudd and Watson (1968).

elsewhere in this chapter, both the important physiological factors and the major engineering factors contributing towards efficiency have been discussed. The physiological factors of importance are yield, growth rate, cell density, substrate affinity, culture stability, resistance to contamination and thermotolerance, whilst the engineering factors include productivity, conversion, product recovery efficiency, power input, mass transfer and heat transfer.

The yield coefficient, defined as the weight of dry product produced from unit weight of feedstock converted, is of obvious economic importance. Yield coefficient is a variable depending on cultivation conditions, and provided product quality can be maintained it is of critical importance to ensure that the yield coefficient is maximized. However, a very high yield coefficient achieved in a system where conversion, defined as the weight of dry product produced from unit weight of feedstock supplied, is poor, lacks economic attractiveness because of either feedstock wastage or feedstock recycle costs. Hence, from an economic point of view, overall conversion in the system might become more significant than the maximum yield coefficient, unless this can be achieved at consistently high conversions. High conversions of gaseous feedstocks in aerobic fermentation processes are particularly difficult to achieve. This is, of course, a particular problem for S.C.P. processes operating directly on natural gas as carbon feedstock, but is experienced in all high-productivity aerobic fermentation processes because of their essential requirement for oxygen. In fact, poor conversion has largely inhibited the introduction of tonnage oxygen in the fermentation industry. The penalty incurred for high gaseous feedstock conversion is usually an extremely high power input in order to achieve enhanced gas transfer. Yield and conversion primarily affect the operating cost term that is proportional to the production rate. In addition, in continuous fermentation processes where, in order to ensure operation under a single limiting substrate, excess dissolved mineral nutrients are supplied to the fermentation, treatment of waste process-water could result in both increased capital and increased operating costs.

From the capital-cost point of view, it is undoubtedly productivity that has the biggest single influence. In continuous fermentation processes, productivity is the product of the cell density and the growth rate (dilution rate) and defined as the weight of product produced per unit liquid volume in the fermenter per unit time. High cell densities are frequently helpful in the unit operations that follow fermentation, and high growth rates usually result in improved yield coefficients. Provided product quality can be maintained, the higher the productivity the smaller the fermenter can be, provided, of course, that the ratio of the gas hold-up and the head-space volume to the liquid volume remain unchanged. In fact, S.C.P. process productivities always tend to appear low with respect to the process industry in general because they are assessed on a dry basis, whereas growing cells comprise some 75–80 per cent water.

Substrate affinity is a factor which has relatively little direct economic impact, although its involvement with respect to resistance of the culture to contamination is important both with respect to capital costs and operating costs. In the case of the latter, a low resistance to contamination requires very high standards of engineering construction with respect to both the fermenter and ancillary equipment, resulting in high capital costs, and an effective sterilization operation which contributes both capital and operating costs.

Constantly changing environmental conditions, product and/or feedstock inhibition, and susceptibility of the culture to contamination and materials of construction, can all result in culture instability and, hence, adversely affect productivity. It is important that production cultures are sufficiently non-fastidious to permit employment of sensible engineering solutions to processing problems. Further, it is unrealistic to require an extensive use of exotic materials of construction in S.C.P. plants because of the influence that the capital cost of the plant has on the ultimate price of the product.

The thermotolerance of production cultures employed for S.C.P. manufacture is a factor having a significant economic influence, and temperature control is the only aspect of industrial protein production that is subject to climatic conditions, a subject discussed in detail in a previous section. The adverse economic implications of using mesophilic micro-organisms for S.C.P. production are obvious. It is desirable that, for a S.C.P. process to find application in a wide range of geographical locations, the production culture has a growth optimum of between $42°C$ and $48°C$, as seems possible for several natural gas- and methanol-utilizing cultures.

Remarkably few detailed economic evaluations of S.C.P. production from natural gas have been published. In general, papers concerned with the economics of S.C.P. production have tended to discuss the relative carbon feedstock costs rather than the process as a whole. The relative merits of various feedstocks have been discussed by Abbott and Clamen (1973), Harwood and Gabriel (1973) and Gaden (1974). Probably the only reasonably comprehensive evaluation concerning S.C.P. production was presented a few years ago by Brownstein and Constantinides (1975), using as their basis data contained in documentation presented at the Expert Group Meeting on the Manufacture of Protein from Hydrocarbons in Vienna (United National Industrial Development Organization, 1973). Four routes, namely gas oil (yeast), n-alkanes (yeast), methanol (bacteria) and natural gas (bacteria) were compared at the 100 000 tonnes per annum scale of production. The yield values used were, respectively: 0.94, 1.15, 0.51 and 0.75. Of these, the second is probably an overestimate by 10%, and the fourth an underestimate by 10%. The estimated fixed costs in U.S. dollars were, respectively, 106.7×10^6, 90.5×10^6, 66.0×10^6 and 72.2×10^6. The latter two are probably underestimates, but have the correct relationship with each other. Production costs, excluding raw materials, were estimated, in U.S. dollars per annum, to be: 13.2×10^6, 12.2×10^6, 9.6×10^6 and

12.0×10^6, respectively. The protein content of the four S.C.P products was assumed to be 68%, 61%, 83%, and 75%, respectively. The second value is perhaps a little low, whilst the third value is probably a little high. The outcome of the study on a 100% protein basis, for a 25% return on investment and relative to carbon feedstock price, is given in Fig. 13. The relative merits of the natural gas and methanol routes are clearly indicated, as is the advantage of the direct, as opposed to the indirect, route from natural gas, as hydrocarbon prices increase. In this latter context, it is important to recognize that hydrocarbon feedstock prices

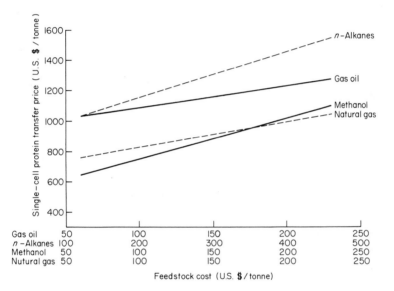

Fig. 13. Comparative economics of single-cell protein processes for a 100 000 tonnes per annum plant, calculated on a 100% protein basis and a return on investment of 25%. From Brownstein and Constantinides (1975).

can vary and are not necessarily strictly related to their production costs. The price of natural gas in major hydrocarbon-importing regions can be directly related, on an equivalent energy and quality basis, to other hydrocarbon fuel prices whilst, in major hydrocarbon-exporting regions, natural gas, because it needs to be liquefied prior to export, is correspondingly of lower relative value to liquid hydrocarbons, hence making the natural-gas routes for S.C.P. production particularly attractive for such exporting regions.

IX. MANUFACTURING VENTURES

Finally, it is important to review the status of commercialization of S.C.P.-manufacturing processes from natural gas, by either the direct route or the indirect route via methanol, using bacteria or yeasts. The major participants in research and development programmes designed to result in commercial processes have been as follows:

By the direct Route: (1) Shell Research Ltd., U.K.; (ii) B.P. Protein Ltd., U.K.; (iii) Institute of Gas Technology, U.S.A.; and (iv) Linde-Max-Planck Institute, West Germany.

By the Indirect Route: (i) Imperial Chemical Industries Ltd., Agriculture Division, U.K.; (ii) L'Institut Français du Pétrole, France; (iii) Mitsubishi Gas Chemical Co. Inc., Japan; (iv) Norprotein Group, Sweden/Norway; and (v) Hoechst AG/Uhde GmbH., West Germany. Little work performed outside these groups can be considered to have been directed towards genuine commercial objectives, although groups other than these have sought patent protection in connection with discoveries related to microbial growth on either methane or methanol.

The process research work performed by Shell, using bacteria, has been extensively documented in the scientific literature. Whilst the process-research programme performed was probably one of the most thorough of its type, the programme was cancelled prior to process development, primarily because of predicted differential inflation between natural-gas prices in Western Europe, assessed on a fuel-value equivalent basis, and the price of imported agriculturally-produced protein from either the U.S.A. or Brazil. British Petroleum Proteins have not published data concerning their investigations with bacteria, but have sought patent protection. Because of the availability of appropriate large-scale plant, it might be assumed that some development work has been undertaken. Whilst the present situation is unclear, there is no evidence that B.P. Proteins intend to commercialize the direct route for S.C.P. production from natural gas. The Institute of Gas Technology in the U.S.A. was an early participant in research concerning S.C.P.production from methane using bacteria. Their programme was discontinued about 1972, prior to any large-scale developmental work. Details of the Linde-Max-Planck Institute development are strictly limited, and there is no evidence that work has progressed beyond a relatively early stage of process research. They, of course, use bacteria. In summary, the direct route for S.C.P. manufacture, whilst having been

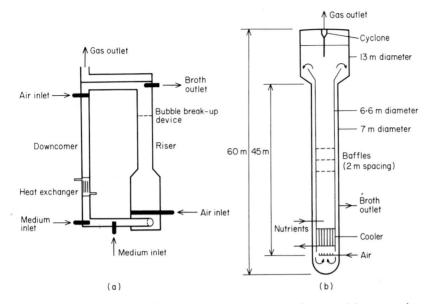

Fig. 14. Diagram of the Imperial Chemical Industries pressure-cycle fermenter. (a) as proposed; see Gow *et al.* (1973), (b) as installed.

extensively researched, still requires a major developmental programme prior to realistic evaluation and commercialization.

The indirect route has progressed further towards commercialization, but product from a commercial-scale plant, i.e. producing greater than 50 000 tonnes per annum, has not yet been marketed, although this is

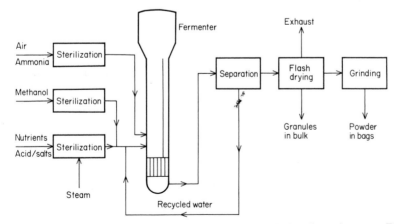

Fig. 15. Schematic diagram of the Imperial Chemical Industries single-cell protein process. From Anon. (1977).

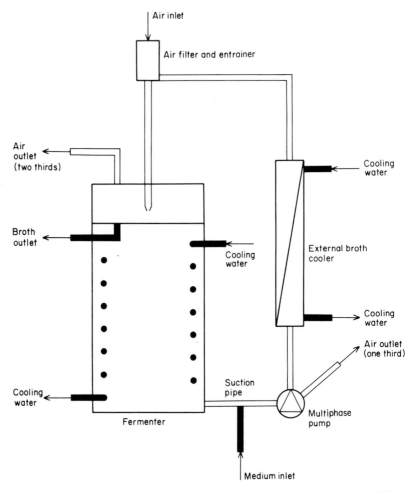

Fig. 16. Diagram of the Vogelbusch deep-jet fermenter system. From Schreier (1975).

expected to occur before 1980. Imperial Chemical Industries (I.C.I.) in Great Britain are presently building a commercial production plant with a capacity between 54 000 and 70 000 tonnes of bacterial S.C.P. per annum at Billingham, U.K., on the basis of their own technology. A demonstration plant has been operated for extended periods. Without doubt, the major technical innovation in the I.C.I. process was the design and development of their pressure-cycle fermenter (Gow *et al.*, 1973), a device that, if it meets design specification, will ensure that fermentation technology will never be quite the same again, by proving that process engineering concepts apply equally to biological as well as more conventional chemical and physical processes. In addition to the fermenter, I.C.I. have developed novel technology for bacterial separation. Some features of the I.C.I. fermenter design are shown in Fig. 14, and a process flow sheet in Fig. 15. Imperial Chemical Industries will be the first company to market S.C.P. produced on a commercial scale from natural gas by the indirect route.

L'Institut Français du Pétrole process for S.C.P. production from methanol, unlike the I.C.I. process, employs a yeast rather than a bacterium as the process micro-organism. This process, recently described by Ballerini (1978), has been operated on a pilot-plant scale in France, using an air-lift fermenter which is essentially a variation on the pressure-cycle design, and in Austria using a Vogelbusch deep-jet fermenter as shown in Fig. 16. No information exists concerning the schedule for commercialization of this process. The proposed process-flow sheet is shown in Figure 17, and the expected nominal production capacity of the first production plant is 30 000 tonnes per annum.

Mitsubishi Gas Chemical has developed a production route using methanol-utilizing yeasts (Kuraishi et al., 1977). A demonstration plant using a Hitachi-designed 20 m^3 air-lift fermenter, shown diagrammatically in Fig. 18, has been constructed. The process-flow sheet is shown in Fig. 19. No details of proposed commercialization are available, probably because of the adverse, but ill-founded, public reaction to S.C.P. that developed some years ago in Japan.

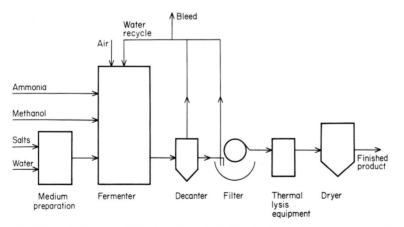

Fig. 17. Schematic diagram of l'Institut Français du Pétrole single-cell protein process. From Ballerini (1978).

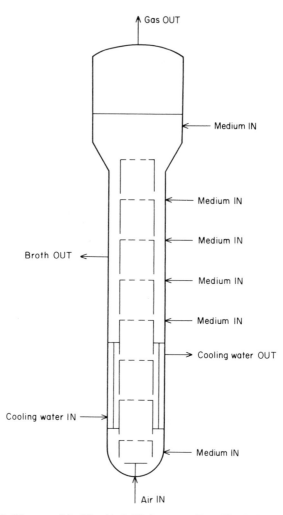

Fig. 18. Diagram of the Hitachi air-lift fermenter. From Kuraishi *et al.* (1977).

The Norprotein Group has based its research and development programme on the studies of Molin, Dostálek and Häggström, and is presumably pursuing a bacterial route. No information concerning the status of this programme is available. Finally, the last research and development programme to be established for the methanol route employing bacteria was that of the Hoechst–Uhde partnership. This programme has progressed to the point where a demonstration plant is under construction after favourable experience at the pilot-plant scale

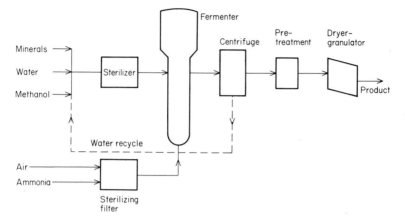

Fig. 19. Schematic diagram of the Mitsubishi Gas Chemical single-cell protein process. From Kuraishi *et al.* (1977).

(Faust *et al.*, 1977). The favoured reactor is of the internal-loop type, and the proposed process-flow sheet is shown in Fig. 20. No plans with respect to commercialization have been announced.

It must be stressed that development and commercialization of production, either directly or indirectly, from natural gas have been

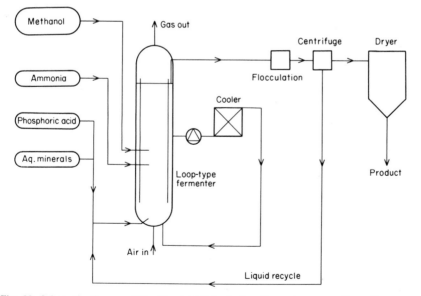

Fig. 20. Schematic diagram of the Hoechst-Uhde single-cell protein process. From Faust *et al.* (1977).

subjected to uncertainties with respect to future feedstock and energy prices on the one hand, while forecasts suggesting that the price of agriculturally-produced protein from the U.S.A. will inflate only relatively slowly have tended to cast doubt, in some sectors, on the commercial viability of S.C.P.-manufacturing ventures in Western Europe and in Japan. Obviously, confidence with respect to risking major capital investment varies. Certainly, I.C.I. have committed themselves, but no doubt expect to corner a part of the attractive European veal calf-milk replacer market where other proteins have difficulty in competing. Their confidence will be further rewarded should forecast future prices for agriculturally-produced protein be seriously in error.

However, what is perhaps more important when considering S.C.P. production from natural gas is that natural gas, in energy-abundant production areas in the Middle East and North Africa, cannot be valued on the equivalent energy basis used in essentially energy-deficient, industrially developed countries. Natural gas, unlike liquid hydrocarbons, has first to be liquefied at considerable cost prior to exportation and, secondly, requires highly sophisticated and expensive tankers for transportation. In addition, losses occur during liquefaction, during transport and during regasification. Whilst liquid hydrocarbon fractions command similar prices throughout the world, natural gas commands prices in some production areas that represent only a fraction of its price based on an equivalent energy value elsewhere. Irrespective of what happens in Western Europe and Japan, S.C.P. production from natural gas, particularly by the direct route, seems to be an appropriate technology for hydrocarbon energy-abundant countries, provided that, firstly, processes can be developed that produce a safe high-quality product; secondly, an animal feed-compounding industry can be developed to use the product; and thirdly, there is sufficient local production of eggs, poultry and meat to provide a sensible market for the compounded feeds.

REFERENCES

Abbott, B. J. and Clamen, A. (1973). *Biotechnology and Bioengineering* **15**, 117.
Aiyer, P. A. S. (1920). *Memoirs of the Department of Agriculture, India, Chemical Series* **5**, 173.
Anon. (1977). *Process Biochemistry* **12** (1), 30.
Anthony, C. (1978). *Journal of General Microbiology* **104**, 91.

Anthony, C. and Zatman, L. J. (1964). *Biochemical Journal* **92**, 609.
Asthana, H., Humphrey, A. E. and Mortiz, V. (1971). *Biotechnology and Bioengineering* **13**, 923.
Babij, T., Ralph, B. J. and Pickard, P. A. D. (1974). *Proceedings of the International Symposium on Microbial Growth on C₁ Compounds, Tokyo.* p. 213.
Ballerini, D. (1978). *Revue de l'Institut du Pétrole Française* **33**, 111.
Ballerini, D., Parlouar, D., Lepeyronnie, M. and Sri, K. (1977). *European Journal of Applied Microbiology* **4**, 11.
Barnes, L. J., Drozd, J. W., Harrison, D. E. F. and Hamer, G. (1977). *Proceedings of the Conference on Microbial Production and Utilization of Gases, Göttingen,* p. 301.
Bassalik, K. (1914). *Jahrbuch für Wissenschaftliche Botanik* **53**, 255.
Battat, E., Goldberg, I. and Mateles, R. I. (1974). *Applied Microbiology* **23**, 906.
Bauchop, T. and Elsden, S. R. (1960). *Journal of General Microbiology* **23**, 457.
Brownstein, A. M. and Constantinides, A. (1975). *Paper to the 169th Meeting of the American Chemical Society, Philadelphia.*
Byron, D. and Ousby, J. C. (1974). *Proceedings of the International Symposium on Microbial Growth on C₁ Compounds, Tokyo,* p. 23.
Chalfan, Y. and Mateles, R. I. (1972). *Applied Microbiology* **23**, 135.
Crémieux, A., Chevalier, J., Combet, M., Dumenil, G., Parlous, D. and Ballerini, D. (1977). *European Journal of Applied Microbiology* **4**, 1.
Cooney, C. L. (1974). *Proceedings of the International Symposium on Microbial Growth on C₁ Compounds, Tokyo,* p. 183.
Cooney, C. L., Wang, D. I. C. and Mateles, R. I (1968). *Biotechnology and Bioengineering* **11**, 269.
Coty, V. F. (1967). *Biotechnology and Bioengineering* **9**, 25.
Coty, V. F. (1969). *Biotechnology and Bioengineering Symposium No. 1*, 105.
Dalton, H. (1977). *Archives of Microbiology* **114**, 273.
Darlington, W. E. (1964). *Biotechnology and Bioengineering* **7**, 445.
Davis, J. R., Coty, V. F. and Stanley, J. P. (1964). *Journal of Bacteriology* **88**, 468.
van Dijken, J. P. (1976). Doctoral Dissertation: University of Gröningen.
van Dijken, J. P. and Harder, W. (1975). *Biotechnology and Bioengineering* **17**, 15.
van Dijken, J. P., Otto, R. and Harder, W. (1976). *Archives of Microbiology* **111**, 137.
D'Mello, J. P. F. (1972). *Journal of Applied Bacteriology* **35**, 145.
D'Mello, J. P. F. (1973). *British Journal of Poultry Science* **14**, 219.
den Dooren de Jong, L. E. (1927). *Zentralblatt für Bakteriologie, Parasitenkunde, Infektionskrankheiten und Hygiene, Abteilung II,* **71**, 193.
Dostálek, M. Häggström, L. and Molin, N. (1972). *Proceedings of the Fourth International Fermentation Symposium. Fermentation Technology Today, Kyoto, Japan,* p. 496.
Dostálek, M. and Molin, N. (1973). *Proceedings of the Second M.I.T. Symposium on Single Cell Protein,* p. 385.
Drozd, J. W., Bailey, M. L. and Godley, A. (1976). *Proceedings of the Society for General Microbiology* **4**, 26.
Dworkin, M. and Foster, J. W. (1956). *Journal of Bacteriology* **72**, 646.
Enebo, L. (1967). *Acta Chemica Scandinavica* **21**, 625.
Eroshin, V. K., Harwood, J. H. and Pirt, S. J. (1968). *Journal of Applied Bacteriology* **31**, 560.
Faust, V., Präve, P. and Sukatsch, D. A. (1977). *Journal of Fermentation Technology* **55**, 609.
Foo, E. L. and Hedén, C.-G. (1976). *Proceedings of the United Nations Institute for Training and Research Symposium on Microbial Energy Conversion, Göttingen,* p. 267.

Foster, J. W. and Davis, R. H. (1966). *Journal of Bacteriology* **91**, 1924.
Fukuoka, S. (1972). *Chemical Economy and Engineering Review, Japan* **4** (3), 16.
Gaden, E. L. (1974). *Proceedings of the International Symposium on Single Cell Protein, Rome,* p. 47.
Goldberg, I. (1977). *Process Biochemistry* **12** (9), 12.
Gow, J. S., Littlehailes, J. D., Smith, S. R. L. and Walter, R. B. (1973). *Proceedings of the 2nd. M.I.T. Symposium on Single Cell Protein,* p. 370.
Guenther, K. R. (1965). *Biotechnology and Bioengineering* **7**, 445.
Häggström, L. (1969). *Biotechnology and Bioengineering* **11**, 1043.
Häggström, L. and Molin, N. (1976). *Abstracts of the 5th International Fermentation Symposium, Berlin,* p. 398.
Hamer, G. (1968). *Journal of Fermentation Technology* **46**, 177.
Hamer, G. (1973). *Biotechnology and Bioengineering Symposium No. 4,* 565.
Hamer, G. (1977). *Proceedings of the Regional Seminar on Microbial Conversion Systems, Kuwait,* p. 109.
Hamer, G. and Norris, J. R. (1971). *Proceedings of the 8th World Petroleum Congress* **5**, 133.
Hamer, G., Hedén, C.-G. and Carenberg, C.-O. (1967). *Biotechnology and Bioengineering* **9**, 499.
Hamer, G., Topiwala, H. H. and Harrison, D. E. F. (1973a). *GWF–Gas/Erdgas* **114**, 531.
Hamer, G., Harrison, D. E. F., Harwood, J. H. and Topiwala, H. H. (1973b). *Proceedings of the 2nd M.I.T. Symposium on Single Cell Protein,* p. 357.
Hamer, G., Harrison, D. E. F., Topiwala, H. H. and Gabriel, A. (1976). *Institution of Chemical Engineers Symposium Series,* No. 44, 565.
Harder, W. and van Dijken, J. P. (1977). *Proceedings of the Conference on Microbial Production and Utilization of Gases, Göttingen,* p. 403.
Harder, W., Kuenen, J. C. and Martin, A. (1977). *Journal of Applied Bacteriology* **43**, 1.
Harrison, D. E. F. (1973). *Journal of Applied Bacteriology* **36**, 301.
Harrison, D. E. F. (1976). *Chemical Technology* **6**, 570.
Harrison, D. E. F. and Topiwala, H. H. (1974). *Advances in Biochemical Engineering* **3**, 167.
Harrison, D. E. F. and Wren, S. J. (1976). *Process Biochemistry* **11** (8), 30.
Harrison, D. E. F., Hamer, G. and Topiwala, H. H. (1972). *Proceedings of the 4th International Fermentation Symposium. Fermentation Technology Today, Kyoto,* p. 491.
Harrison, D. E. F., Drozd, J. W. and Khosrovi, B. (1976). *Abstracts of the 5th International Fermentation Symposium, Berlin,* p. 395.
Harrison, D. E. F., Wilkinson, T. G., Wren, S. J. and Harwood, J. H. (1975). *Proceedings of the 6th International Symposium on Microbial Physiology and Continuous Culture, Oxford,* p. 129.
Harwood, J. H. and Gabriel, A. (1973). *United Nations Industrial Development Organization Document* ID/WG 164/5.
Harwood, J. H. and Pirt, S. J. (1972). *Journal of Applied Bacteriology* **35**, 597.
Hazeu, W., de Bruyn, J. C. and Bos, P. (1972). *Archiv für Mikrobiologie* **87**, 185.
Hutton, W. E. and Zobell, C. E. (1949). *Journal of Bacteriology* **58**, 463.
International Union of Pure and Applied Chemistry (1974). *Technical Report No. 12,* 26.
Iwamoto, H. (1969). *Proceedings of the Japanese Fermentation Technology Meeting, Osaka,* p. 61.
Jensen, O. (1909). *Zentralblatt für Bakteriologie, Parasitenkunde, Infektionskrankheiten und Hygiene, Abteilung II,* **22**, 305.
Kaneda, T. and Roxburgh, J. M. (1959). *Canadian Journal of Microbiology* **5**, 87.
Kaserer, H. (1905). *Zentralblatt für Bakteriologie, Parasitenkunde, Infektionskrankheiten und Hygiene, Abteilung II,* **15**, 573.
Khosrovi, B. and Topiwala, H. H. (1978). *Biotechnology and Bioengineering* **20**, 73.
Kihlberg, R. (1972). *Annual Review of Microbiology* **26**, 427.

Klass, D. L., Iandolo, J. J. and Knabel, S. J. (1969). *Chemical Engineering Progress Symposium Series 93*, **65**, 72.

Kuraishi, M., Ohkouchi, H., Matsuda, N. and Terao, I. (1977). *Proceedings of the 2nd. International Symposium on Microbial growth on C_1 Compounds, Puschino*, p. 180.

Laine, B. (1972). *Canadian Journal of Chemical Engineering* **50**, 154.

Lawrence, A. J. and Quayle, J. R. (1970). *Journal of General Microbiology* **63**, 371.

Leadbetter, E. R. and Foster, J. W. (1958. *Archiv für Mikrobiologie* **30**, 91.

Levine, D. W. and Cooney, C. L. (1973). *Applied Microbiology* **26**, 982.

Linton, J. D. and Buckee, J. C. (1977). *Journal of General Microbiology* **101**, 219.

Linton, J. D. and Stephenson, R. J. (1978). *Federation of European Microbiological Societies Microbiology Letters* **3**, 95.

MacLennan, D. G., Gow, J. S. and Stringer, D. A. (1973). *Proceedings of the Royal Australian Chemical Institute* **40** (3), 57.

MacLennan, D. G., Ousby, J. C., Vasey, R. B. and Cotton, N. T. (1971). *Journal of General Microbiology* **69**, 395.

Mevius, W. (1953). *Archiv für Mikrobiologie* **19**, 1.

Norris, J. R. (1968). *Advancement of Science* **25** (12), 143.

Ogata, K., Nishikawa, H. and Ohsugi, M. (1969). *Agricultural and Biological Chemistry* **33**, 1519.

Ogata, K., Nishikawa, H. and Ohsugi, M. (1970). *Journal of Fermentation Technology* **48**, 389.

Patt, T. E., Cole, G. C., Bland, J. and Hanson, R. S. (1974). *Journal of Bacteriology* **120**, 955.

Patt, T. E., O'Connor, M., Cole, G. C., Day, R. and Hanson, R. S. (1977). *Proceedings of the Conference on Microbial Production and Utilization of Gases, Göttingen*, p. 317.

Payne, W. J. (1970). *Annual Review of Microbiology* **24**, 17.

Peel, D. and Quayle, J. R. (1961). *Biochemical Journal* **81**, 465.

Pilat, P. and Prokop, A. (1975). *Biotechnology and Bioengineering* **17**, 1717.

Pollard, R. and Topiwala, H. H. (1976). *Biotechnology and Bioengineering* **18**, 1517.

Quayle, J. R. (1972). *Advances in Microbial Physiology* **7**, 119.

Quayle, J. R. (1974). *Proceedings of the International Symposium on Growth on C_1 Compounds, Tokyo*, p. 59.

Reuss, M., Sahm, H. and Wagner, F. (1974). *Chemie Ingenieur Technik* **46**, 669.

Rudd, D. F. and Watson, C. C. (1968). 'Strategy of Process Engineering'. John Wiley and Sons, New York.

Sahm, H. (1977). *Advances in Biochemical Engineering* **6**, 77.

Sahm, H. and Wagner, F. (1972). *Archiv für Mikrobiologie* **84**, 29.

Sakaguchi, K., Kurane, R. and Murata, M. (1975). *Agricultural and Biological Chemistry* **39**, 1695.

Schreier, K. (1975). *Chemiker Zeitung* **99**, 328.

Sheehan, B. T. and Johnson, M. J. (1971). *Applied Microbiology* **21**, 511.

Silverman, M. P. (1964). *United States Department of the Interior Bureau of Mines Information Circular* IC 8246, 37 pp.

Skryabin, G. K., Chepigo, S. V. and Eroshin, V. K. (1975). *Proceedings of the 9th World Petroleum Congress, Tokyo* **6**, 121.

Snedecor, B. and Cooney, C. L. (1974). *Applied Microbiology* **27**, 1112.

Ström, T., Ferenci, T. and Quayle, J. R. (1974). *Biochemical Journal* **141**, 465.

Söhngen, N. L. (1905). *Zentralblatt für Bakteriologie, Parasitenkunde, Infektionskrankheiten und Hygiene, Abteilung II* **15**, 513.

Teweles, R. J., Harlow, C. V. and Stone, H. L. (1974). 'The Commodity Futures Game'. McGraw-Hill, New York.

Tonge, G. M., Drozd, J. W. and Higgins, I. J. (1977). *Journal of General Microbiology* **99**, 229.

Tonge, G. M., Harrison, D. E. F., Knowles, C. J. and Higgins, I. J. (1975). *Federation of European Biochemical Societies Letters* **58**, 293.

Tonge, G. M., Knowles, C. J., Harrison, D. E. F. and Higgins, I. J. (1974). *Federation of European Biochemical Societies Letters* **44**, 106.

Topiwala, H. H. (1974). *Proceedings of the International Symposium on Microbial Growth on C_1 Compounds, Tokyo*, p. 199.

United Nations Industrial Development Organization (1973). Report ID/128, 33.

Vananuvat, P. (1977). *Critical Reviews in Food Science and Nutrition* **9**, 325.

Vary, P. S. and Johnson, M. J. (1967). *Applied Microbiology* **15**, 1473.

Volesky, B. and Zajic, J. E. (1971). *Applied Microbiology* **21**, 614.

Whittenbury, R., Phillips, K. C. and Wilkinson, J. F. (1970). *Journal of General Microbiology* **61**, 205.

Whittenbury, R., Dalton, H., Eccleston, M. and Reed, H. L. (1974). *Proceedings of the International Symposium on Microbial Growth on C_1 Compounds, Tokyo*, p. 1.

Wilkinson, T. G. and Hamer, G. (1972). *Journal of Applied Bacteriology* **35**, 577.

Wilkinson, T. G. and Harrison, D. E. F. (1973). *Journal of Applied Bacteriology* **36**, 309.

Wilkinson, T. G., Topiwala, H. H. and Hamer, G. (1974). *Biotechnology and Bioengineering* **16**, 41.

Wolnak, B., Andreen, B. H., Chisholm, J. A. and Saadeh, M. (1967). *Biotechnology and Bioengineering* **9**, 57.

Wren, S. J., Harwood, J. H. and Harrison, D. E. F. (1974). *Proceedings of the Society for General Microbiology* **2**, 14.

Zajic, J. E., Volesky, B. and Wellman, A. (1969). *Canadian Journal of Microbiology* **15**, 1231.

Note added in Proof

With respect to manufacturing ventures, Taylor and Senior (1978, *Endeavour*, **2**, 1, 31) have confirmed that I.C.I. are constructing an S.C.P. plant of 50–70 000 tonnes per annum capacity based on their bacterial-methanol route at Billingham, U.K., with completion due in 1979, whilst Mogren (1979, *Process Biochemistry*, **14**, 3, 2) has lifted the veil of secrecy concerning the Norprotein process development. This latter process, based on the methanol-utilizing bacterium *Methylomonas methanolica*, is not dissimilar from other bacterial-methanol processes. The most interesting innovation is the use of flocculation and flotation for production harvesting. In addition to the description of the technology, interesting economic data concerning the relative breakdowns of both capital and operating expenditure for a commercial 100 000 tonnes per annum plant are provided. Because of the prevailing market situation in Scandanavia, Norprotein do not intend to commercialize their technology at present.

12. Biomass from Liquid *n*-Alkanes

J. D. LEVI[a], JEAN L. SHENNAN[b] AND G. P. EBBON[b]

aBritish Petroleum Co. Ltd., London.
bBritish Petroleum Co. Ltd., Research Centre,
Sunbury-on-Thames, England.

[a] *Present address*: Conservation of Clean Air and Water in Europe, The Hague, The Netherlands.

I. INTRODUCTION

When they wrote this review in late 1978, it seemed to the authors that they were composing a requiem for a new industrial development which had been full of promise for nearly two decades (Mateles and Tannenbaum, 1968; Gounelle de Pontanel, 1972; Davis, 1974; Tannenbaum and Wang, 1975; Litchfield, 1977). The escalation in oil prices which followed the 1973 Middle East War, and which has not been reflected in World feed prices, together with the curious inhibitions placed on the commercial development of S.C.P. production, at first in Japan and more recently in Italy, appear to have dealt economic and political death-blows to the utilization in Western countries of this new branch of fermentation technology.

But such is the adaptability and resilience of the fermentation industry, which has been 'dying' since the first synthetic chemical replaced a microbial metabolite, that certain applications of the various techniques which have been researched and developed under the title of this review will undoubtedly come to be commercialized within the next few years. Indeed, in those countries where the rules of normal Western economics do not apply, for reasons of policy connected with political ideology or economic necessity, production of S.C.P. from n-alkanes is likely to find continuing commercial application.

Ironically, most of the work on S.C.P. processes reviewed here has been with yeasts, but the sole survivor of the new wave of this technology in the West appears to be Imperial Chemical Industries' process (Slater, 1974) for production of a bacterial biomass, 'Pruteen', from methanol feedstock. A large plant (50 000 to 75 000 tonnes per annum) is due to be commissioned on Teeside during 1979 (Done, 1978) and it will be interesting to observe whether Hoechst who have also developed a bacteria-methanol S.C.P. process to the 1000 tonnes per annum scale will proceed to the construction of a full-scale plant.

The rationale behind the manufacture of S.C.P. from petroleum fractions has been explained many times during its short history and it is here reviewed (Section X, p. 409) from the perspective of former members of a group that developed what, until recently, appeared to be the process with the best chance of commercial exploitation.

Although there has been considerable work on biomass production from C_1 feedstocks, chiefly methane and methanol (see Chapter 11, p. 315), which can be utilized as carbon and energy sources by a comparatively small number of bacteria and fungi, there has been little industrial development of processes for manufacture of S.C.P. from other gaseous n-alkanes.

n-Alkanes are saturated, straight-chain hydrocarbons, commonly called paraffins. The lowest homologues of the paraffin series, which are liquids at room temperature (n-pentane, n-hexane, n-heptane and n-octane), are generally toxic, it is believed as a result of their solvent action on cell membranes or membrane-bound proteins associated with transport and oxidation (Gill and Ratledge, 1972). This property has precluded their use as fermentation feedstocks, so that the n-alkanes considered in this review are, in effect, limited to those having between 9 and 18 carbon atoms. Pure saturated liquid n-alkanes are not produced on a commercial scale, but mixtures are available in gas-oil (diesel fuel), which contains 10–30% of paraffins, and as 'cuts' of purified n-alkanes extracted from kerosine (C_{10}–C_{13}) or from gas oil (C_{14}–C_{18}).

The physical properties of the paraffins form a progression (e.g. n-pentane, b.p. 36°C; n-octadecane, b.p. 317°C, m.p. 28°C) but homologues higher than n-octadecane, although not liquids, may be present in gas-oil cuts of n-alkanes since they can be fluxed in the lower n-alkanes.

The proportion of n-alkanes varies, in different crude oils, from 0–30%. In general, many African crudes tend to be rich in saturated hydrocarbons, Middle Eastern crudes tend to contain rather less, while light crudes, e.g. some of those from the North Sea, can often be low in

Fig. 1. An aerial view of the BP-ANIC Italproteine S.C.P. production plant at Sarroch, Sardinia, Italy.

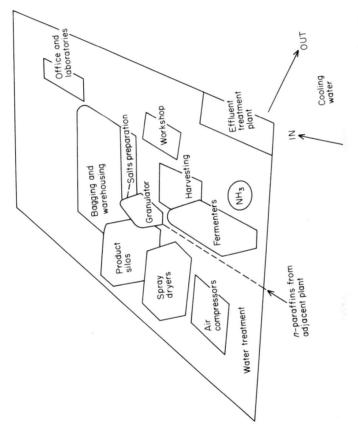

Fig. 2. Diagram of the plant layout of the BP-ANIC Italproteine S.C.P. production plant at Sarroch, Sardinia, Italy.

n-alkanes. If a conservative estimate of the *n*-alkane content of the crude oil refined in Western Europe each year is taken as 10%, then the potential feedstock for biomass production is more than 50 million tonnes per annum. Current production of *n*-alkanes is only a fraction of this figure, the principal outlets being the detergent, paint and solvent industries. The remainder, contained in kerosine and gas-oil sold as heating and automotive fuels, tends to decrease their calorific value.

Thus, there is considerable potential for adding monetary value to a proportion of current refinery throughput which has, if anything, a nuisance value. Biomass produced from hydrocarbon feedstocks in the western hemisphere would replace imported feed ingredients. In principle, this would release land in the U.S.A. and Brazil from growing soybean for the animal feed market to cultivation of soybean or other crops for human food, and release certain fisheries from the production of fish meal to the catching of fish for human consumption. However, in practice, it was expected that the steady build-up envisaged for S.C.P. production would have done no more than take up some of the forecast increase in the animal feed market that results from 'affluent man's' insatiable appetite for meat (David, 1976).

II. ORGANISMS

The capacity of certain micro-organisms to oxidize hydrocarbons has been known since the nineteenth century. An earlier review (Shennan and Levi, 1974) gives a brief historical survey of the growth of yeasts on hydrocarbons, and several other reviews cover early studies into bacterial and fungal assimilation of oils (Beerstecher, 1954; Fuhs, 1961; Zajic, 1964; van der Linden and Thijsse, 1965).

In contrast to the specialized metabolism of certain methane-utilizing bacteria, namely the obligate methylotrophs, all of the micro-organisms which have been found to assimilate liquid *n*-alkanes can grow on oxidized carbon sources. In general, their rate of growth on glucose is much higher than on paraffins. Although it is confined to certain genera and species, the capacity to oxidize hydrocarbons is widespread in each of the groups of organisms in which it has been studied.

Among the prototrophs, many bacteria and actinomycetes have been described which grow readily on *n*-alkanes. They are ubiquitous in soil and water, and play a vital part in degradation of both naturally-occurring and spilled hydrocarbons. Thus, their role in the natural cleansing of the environment as a result of the accidental activities of the oil industry and its customers, as well as in effluent treatment plants, in natural bodies of water and in the soil, is well recognized as a fundamental part of the carbon-cycle.

Heterotrophic micro-organisms with the capacity to utilize hydrocarbons have been described mainly among the yeasts and moulds, although it is clear that the appropriate enzyme systems are not limited to these groups (see Section VII. A). Reservoir levels of hydrocarbon oxidizers appear to be generally low, but cell numbers can rise rapidly in response to the presence of any mixture of oil and water, as readily to cause spoilage of cutting fluids and aviation kerosine, as to degrade oil which has been accidentally spilled.

The number of recognized species which have been reported to utilize hydrocarbons is considerable and, together with the many novel species described in the literature, would be too numerous to list in this review. Instead, the principal hydrocarbon-utilizing genera (not all species of which are hydrocarbon-positive) are listed in Table 1, and the reader is recommended to peruse some of the excellent review articles for further details (Bos and de Bruyn, 1973; Bemmann and Tröger, 1975; Bos, 1975).

The use of modern techniques (DNA base-composition affinity and serological classification, for example) causes rapid obsolescence in the most carefully prepared list of named organisms. Hence they should be interpreted with caution.

A. Examples

In part because of their long history in the brewing, baking and wine-making industries, it is the yeasts that have been most extensively studied in the development of S.C.P. processes. This historical background covers several millenia, and accounts for their general acceptability for food or feed uses. In addition, it ensures that wide industrial experience exists in all aspects of their handling. Nevertheless, much of the fundamental biochemistry of hydrocarbon oxidation has been elucidated in bacteria.

Among the asporogenous yeasts, members of the genus *Candida* are most often reported in the literature, with *Rhodotorula*, *Torulopis* and

Table 1
Some genera of micro-organisms reported to contain hydrocarbon assimilating species or strains

---

A. FILAMENTOUS FUNGI (Ainsworth *et al.*, 1973)[a]

Mucorales	*Cunninghamella*	Fungi Imperfecti	*Aspergillus*
	Mortierella		*Aureobasidium*
	Mucor		*Botrytis*
			Cephalosporium
			Cladosporium
			Fusarium
			Paecilomyces
			Penicillium
			Phialophora
			Sporotrichum
			Verticillium

B. YEASTS (Bos and de Bruyn, 1973)[a]

Ascomycetes	*Debaryomyces*	Fungi Imperfecti	*Candida*
	Endomyces		*Rhodotorula*
	Lodderomyces		*Selenotila*
	Metschnikowia		*Sporidiobolus*
	Pichia		*Sporobolomyces*
	Saccharomycopsis		*Torulopsis*
	Schwanniomyces		*Trichosporon*
	Wingea		
Basidiomycetes	*Leucosporidium*		
	Rhodosporidium		

C. BACTERIA (Buchanan and Gibbons, 1974)[a]

Phototrophs	*Rhodospirillum*		
	Rhodopseudomonas		
Gram-negative aerobic rods and cocci	*Alcaligenes*	Gram-positive cocci	*Micrococcus*
	Pseudomonas		*Sarcina*
Gram-negative facultatively anaerobic rods	*Aeromonas*	Gram-positive sporing rods	*Bacillus*
	Chromobacterium		
	Flavobacterium		
	Klebsiella		
	Vibrio		
Gram-negative cocci and coccobacilli	*Acinetobacter*		
	Moraxella		
Coryneform bacteria	*Arthrobacter*		
	Brevibacterium		
	Corynebacterium		
	Microbacterium		
Actinomycetes	*Actinomyces*		
	Mycobacterium		
	Nocardia		
	Streptomyces		

D. ACHLOROPHYLLOUS ALGAE

Prototheca

---

[a] Reference describes the classification system used.

Trichosporon species frequently mentioned. The genus *Candida* is a large 'unnatural' and diverse collection of species many of which are being reclassified into sporogenous genera, as research uncovers a diversity of life-cycles, mating systems and spore-bearing structures which permit assignment to the Ascomycetes or Basidiomycetes. One of the most familiar hydrocarbon-utilizing yeast species, *C. lipolytica*, has been renamed *Saccharomycopsis lipolytica* (von Arx, 1972) following the successful mating of two strains to produce diploid cells (Wickerham *et al.*, 1970). *Candida guilliermondii, C. intermedia, C. maltosa, C. parapsilosis, C. rugosa* and *C. tropicalis* are all popular species with research workers, while some 40 to 50 other members of this genus have been reported as hydrocarbon utilizers in the literature (Shennan and Levi, 1974).

The sporogenous yeast genus *Pichia* contains many hydrocarbon-assimilating species, and *Debaryomyces* species are also common in this group. Using the current yeast classification of Lodder (1970) with later amendments, Bos and de Bruyn (1973) concluded that, among the perfect genera, *Saccharomyces, Kluyveromyces* and *Hansenula* are completely devoid of the ability to assimilate hydrocarbons, although not, of course, *Saccharomycopsis*. For example, hydrocarbon-assimilating cultures reported by Rabinovich *et al.* (1974) to be *Saccharomyces cerevisiae* have been shown to be strains of *Candida maltosa* in taxonomic and genetic studies (Bassel *et al.*, 1978). It might be suggested, then, that the inability to assimilate hydrocarbons is a more clear-cut character at the generic level than the positive capacity for hydrocarbon oxidation.

The filamentous fungi contain many species capable of hydrocarbon oxidation but, because of their growth form, they have seldom been extensively researched in the context of S.C.P. or metabolic analysis. One exception is the 'kerosene fungus', *Cladosporium* (*Amorphotheca*) *resinae* which is important in the field of petroleum-product contamination (Walker and Cooney, 1973).

The length of the list (Table 1) of bacterial genera, which include hydrocarbon-assimilating species, testifies to their wide occurrence. Although they have not been widely exploited for S.C.P. in liquid *n*-alkanes (see Section IV, F), production of by-products from species, or mutant strains, of *Arthrobacter, Brevibacterium* and *Corynebacterium* has been extensively researched, especially in Japan (Arima, 1977; Yamada, 1977). Bacterial species have also been chosen by many workers investigating the metabolic pathways and mechanisms of substrate attack in hydrocarbon assimilation (Ratledge, 1978).

B. Ecology

The natural ecology of hydrocarbon-utilizing organisms is, despite its importance, a relatively unexplored field. Of recent years, a large volume of literature has been built up on crude-oil degradation at sea and, less often, in estuarine regions or on land (Ahearn and Meyers, 1973; Crow *et al.*, 1974; Bartha and Atlas, 1977; Malins, 1977; Wolfe, 1977), but little has yet been published on the distribution and levels of potential hydrocarbon-degrading organisms in unpolluted parts of these ecological areas (Karrick, 1977). Facultative hydrocarbon-oxidizing micro-organisms are found in relative abundance in natural oil-rich environments, such as oil seeps, but elsewhere these ubiquitous organisms remain undiscovered. Clearly, they must be members of the normal saprophytic reservoir flora of any environment where fatty, lipidic or hydrocarbon compounds from plant or animal remains are decomposed.

Unpublished work of J. L. Shennan and J. D. Levi has shown that the familiar hydrocarbon utilizer, *Saccharomycopsis lipolytica*, hitherto reported mainly from spoiled dairy products or other lipid-rich animal or plant habitats, including olive fruits (van Rij and Verona, 1949), can be isolated frequently—but at low levels—from soil-surface samples, muds and roadside dusts. It is therefore probable that the paucity of information on the broad ecological occurrence of hydrocarbon-utilizing organisms is merely due to a lack of investigation into this topic.

III. HYDROCARBONS

Zobell (1946), writing at the end of the Second World War, and basing his conclusions on the limited literature on microbial hydrocarbon oxidation then published, formulated a set of rules for the specificity of utilization. These have long since been overtaken by further discoveries, and were recently reformulated (Shennan and Levi, 1974). By a curious paradox typical of the history of industrial microbiology, Zobell's work arose through a war-time project of the American Petroleum Institute to investigate the possible synthesis of hydrocarbons by micro-organisms. No doubt, when oil prices begin to escalate seriously, at some time during the next ten years as depletion overtakes the discovery of exploitable reserves of crude oil, such studies will again assume importance.

This review is limited to liquid *n*-alkanes as substrates for S.C.P.

production. There have been several recent reviews on hydrocarbon utilization, including those of Klug and Markovetz (1971), Einsele and Fiechter (1971) and Shennan and Levi (1974). The subject of degradation of aliphatic hydrocarbons has been brought up to date by Ratledge (1978), and the reader is referred to that publication for detailed discussion of assimilation mechanisms, pathways of oxidation, and the effects of hydrocarbon metabolism on cell physiology.

Of the different types of hydrocarbons present in crude oils, aliphatic compounds are the most readily attacked by micro-organisms. A wide variety of species possess the enzymic capacity to assimilate the straight chains of unsubstituted methylene groups. For reasons of toxicity, the lower range of liquid alkanes below *n*-nonane are not assimilated by yeasts and only occasionally by bacteria, although oxidation of *n*-heptane and *n*-octane has been reported (Munk *et al.*, 1969). The higher waxy members of the homologous series, with chain lengths greater than *n*-octadecane, are assimilated provided the hydrocarbon molecule can be made available to the cell. Most reports concerning the solid *n*-alkanes describe their solubilization in inert carriers or their attack by filamentous fungi growing over the surface of the waxy substrate.

Table 2

Carbon number distribution of *n*-paraffins 'cuts' (% w/w)

Alkane	Typical 'kerosine-range'	Typical 'gas-oil range'	'Wide-range' cut from waxy crude
$n\text{-}C_{10}$	6	—	trace
$n\text{-}C_{11}$	39	trace	trace
$n\text{-}C_{12}$	42	trace	1
$n\text{-}C_{13}$	13	4	2
$n\text{-}C_{14}$	trace	28	3
$n\text{-}C_{15}$	—	29	5
$n\text{-}C_{16}$	—	22	8
$n\text{-}C_{17}$	—	12	14
$n\text{-}C_{18}$	—	4	21
$n\text{-}C_{19}$	—	1	19
$n\text{-}C_{20}$	—	—	13
$n\text{-}C_{21}$	—	—	9
$n\text{-}C_{22}$	—	—	4
$n\text{-}C_{23}$	—	—	1
$n\text{-}C_{24}$	—	—	trace

In the British Petroleum Co. Ltd. (BP) process for production of biomass from *n*-paraffins and in other similar published processes which reveal the feedstocks employed, mixtures of *n*-alkanes are used (Table 2).

Their composition depends on the crude oil, the boiling range of the cut and the n-paraffins extraction and purification process employed. Kerosine-range n-paraffins predominate in n-undecane, n-dodecane and n-tridecane, while gas-oil range cuts contain mainly n-tetradecane and higher homologues, sometimes with small quantities of n-alkanes with more than 18 carbon atoms. A wide-range n-paraffins cut extracted from a waxy North African crude may peak at C_{18} and contain 40–50% by weight of higher homologues. Such a feedstock will have a melting point close to the fermentation temperature. All of the n-alkanes in these three cuts, ranging from n-nonane to n-tetracosane and above, are likely to be assimilated by all n-paraffin-oxidizing species. However, there is evidence that, even within this range, the light n-alkanes are assimilated preferentially to the heavier compounds (Dostalek *et al.*, 1968; Ueno *et al.*, 1974b). In continuous hydrocarbon-limited fermentations, this effect is not apparent.

A. Transport

As n-alkanes are immiscible with water, the means used to effect substrate transfer into the cell is of fundamental interest in the assimilation of hydrocarbons. The mechanisms of hydrocarbon transport have been extensively studied during the past decade, and have been reviewed in several publications (e.g. Shennan and Levi, 1974; Erickson and Nakahara, 1975; Prokop and Sobotka, 1975; Ratledge, 1978). Two principal forms of transfer have been proposed: (1) through direct contact of cells with emulsified droplets of the substrate, or (2) by passage into the cell of dissolved hydrocarbons from the aqueous phase, or solubilized from hydrocarbon micelles at the cell surface and transported by some means across the cell membrane.

Hydrocarbons in fermenter broths are dispersed by the shearing action of mechanical agitation or violent aeration (as in air-lift fermenters) to a macro-emulsion containing oil droplets of 1 to 100 micrometers diameter. But when active hydrocarbon assimilation occurs, electron microscopy reveals that a micro-emulsion of very small oil droplets is formed (Yoshida *et al.*, 1973; Einsele *et al.*, 1975; Miura *et al.*, 1977; Nakahara *et al.*, 1977). These range from 0.01 to 0.5 μm in diameter, and result from the biochemical activity of the cells present. The ability to produce emulsifying surfactant compounds at the cell membrane is, in all probability, one of the determining characteristics of hydrocarbon assimilation. Various researchers have isolated lipopolysaccharides from the walls of hydrocarbon-utilizing organisms and, in some cases, these have been shown to play an important part in the process of hydrocarbon uptake (Käppeli and Fiechter, 1977). In *C. tropicalis*, this polysaccharide was found to be a mannan, containing about 4% covalently-linked fatty acids (Käppeli *et al.*, 1978).

The necessity for production of surface-active agents as a vital step in hydrocarbon assimilation has been suggested by several authors (Chiou and Chang, 1973; Barnett *et al.*, 1974; Velankar *et al.*, 1975; Aoki and Umezawa, 1976) and powerful emulsifying agents have been isolated from hydrocarbon-oxidizing organisms (Hisatsuka *et al.*, 1971; Friede, 1975; Yamaguchi *et al.*, 1976). The function of these surfactants remains to be determined precisely, but microscopic evidence indicates that contact between cells and hydrocarbon droplets is a prerequisite for growth (Ratledge, 1978). Furthermore, it seems unlikely that hydrocarbon solubilization in the aqueous phase of the broth, at a distance from the cell surface (Goma *et al.*, 1976), could provide a sufficiently rapid or effective mode of substrate transfer to actively growing cells.

In summary, the current state of knowledge indicates that adhesion is necessary between the cell and its substrate in a macro- or micro-emulsion form, and this may be mediated by production of surface-active agents. These surfactants may assist in solubilization of hydrocarbon molecules, aiding their passage across the cell membrane. The latter may be facilitated in some way by the altered chemical nature of the outer layers of the hydrocarbon-utilizing cell wall. These appear to be high in lipids or lipopolysaccharides. Recent studies into the ultrastructure of hydrocarbon-assimilating yeast cells indicate that the thickened cell walls are traversed by channels (Meisel *et al.*, 1976). Scanning electron microscopy of *n*-alkane-grown yeast cells (Osumi *et al.*, 1975) showed the cell wall to be covered with minute protrusions which may correspond with the pores or channels described elsewhere.

B. Initial Oxidation Reactions

Most *n*-alkane-degrading organisms attack a terminal methyl group although subterminal oxidation has also been reported in several species of bacteria and moulds (see Ratledge, 1978). Three mechanisms have been described for the initial attack on the *n*-alkane molecule.

1. Oxidation involving either cytochrome P450 or rubredoxin

There is evidence that molecular oxygen is incorporated to give the corresponding primary alcohol and the monocarboxylic acid of the same chain length as the substrate, this having been demonstrated in bacteria (e.g. Baptist *et al.*, 1963; Kester and Foster, 1963; Nieder and Shapiro, 1975), yeasts (e.g. Dyatlovitskaya *et al.*, 1965; Klug and Markovetz, 1966; Jones and Howe, 1968a, b, c; Souw *et al.*, 1976) and moulds (e.g. Hoffmann and Rehm, 1976a, b). Thus, it is now generally accepted that the primary alcohol is the first stable intermediate in microbial degradation of *n*-alkanes, in accordance with mechanism (1):

$$R.CH_2CH_3 + O_2 + NADPH + H^+ \rightarrow R.CH_2CH_2OH + \\ NADP^+ + H_2O \quad (1)$$

2. Hydroperoxidation

$$R.CH_2CH_3 + O_2 \rightarrow R.CH_2CH_2OOH \quad (2)$$

$$R.CH_2CH_2OOH + NADH + H^+ \rightarrow \\ R.CH_2CH_2OH + NAD^+ + H_2O \quad (3)$$

Mechanism 2 was advanced by the group led by Finnerty (Finnerty *et al.*, 1965) using the bacterium *Acinetobacter* (*Micrococcus*) *cerificans*. However, there is little evidence elsewhere in the literature to support the formation of a hydroperoxide intermediate in *n*-alkane oxidation.

3. Dehydrogenation with no involvement of oxygen

$$R.CH_2CH_3 + NAD^+ \rightarrow R.CH:CH_2 + NADH + H^+ \quad (4)$$

This scheme, postulated by Senez and Azoulay (1961), was based on evidence for the presence of alk-l-ene in reaction mixtures of *n*-alkane, an NAD^+-oxidase inhibitor (mercaptoethanol) and an enzyme system from *Pseudomonas aeruginosa* which reduced NAD^+. Alkane dehydrogenase was not detected. However, Lebeault *et al.* (1971) succeeded in isolating a typical mixed-function oxidase from *C. tropicalis* grown on *n*-tetradecane which catalysed conversion of *n*-alkane to the primary alcohol and required NADH and molecular oxygen. Later, Gallo *et al.* (1973) reported that re-examination of NAD^+ reduction by cell-free extracts of *C. tropicalis*, in the presence of *n*-decane, showed unequivocally that this reaction was due to contamination of some decane samples with an impurity and did not correspond to dehydrogenation of the *n*-alkane itself.

The presence of an anaerobic NAD^+-linked alkane dehydrogenation step would be difficult to sustain on thermodynamic grounds (Johnson, 1964). Ratledge (1978) has reviewed additional evidence against alk-l-enes as free intermediates in alkane oxidation from studies which show that different oxidation products are obtained from saturated and corresponding mono-unsaturated hydrocarbons (Klug and Markovetz, 1967a, 1968; Jones and Howe, 1968a, b, c). Bruyn (1954) showed, for example, that oxidation of n-hexadec-l-ene by *Saccharomycopsis lipolytica* gave rise to the glycol, hexadecane-1,2-diol. There is ample confirmation of formation of 1,2-diols from alk-l-enes in this species (Stewart *et al.*, 1960; Ishikura and Foster, 1961; Klug and Markovetz, 1966) and it has also been shown that the ability of a yeast to oxidize an alkane of a particular chain length does not necessarily imply utilization of the corresponding alk-l-ene (Klug and Markovetz, 1967b; Markovetz *et al.*, 1968).

Hence, the weight of experimental evidence strongly favours mechanism (1) as the primary oxidation reaction for microbial degradation of hydrocarbons. Aspects of cell physiology and the biochemistry involved have been reviewed by Ratledge (1978) who also describes subsequent metabolism of the primary oxidation products.

C. Mechanisms

Two systems have been well described for electron-transfer mechanisms underlying n-alkane oxidation, but in neither case has the biochemistry been fully worked out:

(1) Cytochrome P450

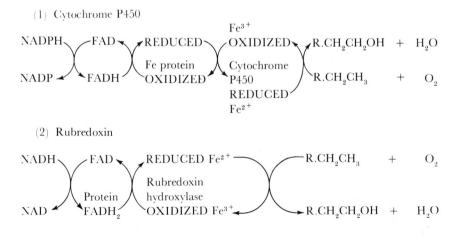

(2) Rubredoxin

The cytochrome P450 system has been found in both prokaryotic and eukaryotic cells. The system has been described, for example, in *Corynebacterium* sp. (Cardini and Jurtshuk, 1970) and *Candida tropicalis* (Gallo *et al.*, 1974). In the latter, the alkane dehydroxylase is specifically located within microsomes along with cytochrome P450 and NADPH-cytochrome *c* reductase. Phosphatidylethanolamine has been reported to be in close association with this system (Duppel *et al.*, 1973). Synthesis of cytochrome P450, the hydroxylase system (Gallo *et al.*, 1973), cytochrome *c* reductase and cytochrome b_5 is induced only by *n*-alkanes containing more than 14 carbon atoms (Gallo *et al.*, 1976), although the last two components are normal constituents of yeast microsomes at low levels. There is no evidence for association of the hydroxylase system, or its components, with microbodies in *n*-alkane-grown yeast.

Although cytochrome P450 is usually reported to be linked with NADPH, it has been shown that NADH is active with the alkane oxidase in *Saccharomycopsis lipolytica* (Rohde *et al.*, 1975). Ebbon (1970) reported that mammalian liver mixed-function oxidase could be linked to either NADPH or NADH. His conclusion that there are a number of inter-linked electron-transport chains involved, with varying sensitivities to activators and inhibitors, makes it inevitable that critical evaluation of the details will be extremely difficult. The industrially important *n*-alkane-oxidizing yeasts pose an extra problem for the biochemist, because their cells are difficult to disrupt, with the result that preparation of subcellular fractions is almost impossible.

There do not appear to be any reports of the rubredoxin system being present in organisms which are of industrial interest (Ratledge, 1978).

D. Subsequent Stages

Although discussion continues on the precise mechanism of alkane oxidation to the primary alcohol, there is general agreement about the subsequent steps, via aldehyde, to the corresponding monocarboxylic acid. Several routes exist for disposal of the fatty acids produced in this way. The best evidence, albeit indirect, is that metabolism proceeds through β-oxidation, as elucidated in animal tissues, and reviewed by Finnerty and Makula (1975).

The key role of acyl coenzyme-A synthetase in fatty acid oxidation has often been stressed (Shennan and Levi, 1974; Ratledge, 1978). Ultimate formation of shorter chain dicarboxylic acids by the β-oxidation route has been illustrated by several authors, and suggests that the alkane-degradation pathways can terminate in compounds of the tricarboxylic acid (TCA) cycle. Enzymes of the TCA cycle have been found at higher concentrations in cells grown on *n*-alkanes than in cells grown on glucose, while citric acid, isocitric acid, succinic acid, α-oxoglutaric acid and pyruvic acid have been identified in cultures of hydrocarbon-grown cells. The association of thiamin deficiency in hydrocarbon-grown yeast with accumulation of TCA-cycle acids has also been noted (Ermakova *et al.*, 1969).

According to Ratledge (1978), the evidence for α-oxidation of fatty acids is fragmentary and rests largely on indirect evidence. On the other hand, evidence for ω-oxidation is more substantial since α,ω-dioic acids as well as ω-hydroxy fatty-acids have been recovered from both bacteria and yeasts and have been considered for commerical exploitation (Jones and Howe, 1968c; Uchio and Shiio, 1972; duPont de Nemours, 1973; Phillips Petroleum, 1976).

IV. FERMENTATION

The rate of technological advance in the fermentation industry has been rapid in the last 20 years. Prior to this, the extensive use of the stirred tank reactor, developed for antibiotics production in the years immediately after the Second World War, had brought about a lull in fermenter engineering development. One line of technological advance which is characteristic of S.C.P. production is that of continuous fermentation, which, for a variety of cogent economic and technical reasons, has not yet penetrated deeply into the established fermentation industries, with the possible exception of brewing.

Several reviews of S.C.P. production over the past ten years have been published, including those edited by Mateles and Tannenbaum (1968), Gounelle de Pontanel (1972), Davis (1974) and Tannenbaum and Wang (1975). Together with the Bulletin published regularly by the Protein Advisory Group of the United Nations Organisation, these will give the reader a broad insight into the diverse nature of S.C.P. research in substrate and scale.

Since most of the teams developing commercial-scale S.C.P. processes have been associated with the oil or petrochemical industries, the emphasis on continuous unit processing is perhaps inevitable. In addition, continuous S.C.P. production should be readily capable of development since those culture changes which tend to be selected when conditions are adapted to optimum cellular protein production rates may well improve the product, the economics of the process, or both. When metabolites are being produced, the reverse is the case, particularly when highly mutated strains are employed. These frequently have low growth rates and would, in continuous culture, be rapidly overtaken by fast-growing but low-yielding revertants.

A. Theory

Fermentation theory is covered in all good microbiology texts, (e.g. Rhodes and Fletcher, 1966; Fiechter, 1975; Pirt, 1975), so only the briefest review will be given here. Monod's seminal publication (1950), extended by Herbert (1961), considers the exponential growth of a micro-organism at a rate μ_m which ultimately declines from this maximum value to the specific growth rate, as a result of a limitation in one environmental parameter. Growth rate is then proportional to the concentration of the limiting substrate. Hence, the rate of change of cellular concentration, dx, equals the specific growth rate multiplied by the cell concentration or:

$$\frac{dx}{dt} = x.$$

$$\text{Hence: } \mu = \frac{1}{x} \cdot \frac{dx}{dt} = \frac{d(\log_e x)}{dt}$$

If the cell concentration doubles (in the cell doubling time, t_d),

$$\mu = \frac{0.693}{t_d}$$

Thus, measurement of the doubling time in a batch fermentation (an easy operation requiring timed cell counts) and a simple calculation give the maximum specific growth-rate of the organism, μ_m, under the experimental conditions. This corresponds to the theoretical maximum dilution rate in continuous culture, which is the rate of replacement of medium (as a proportion of the total broth mass each hour) when the limiting substrate concentration approaches infinity. The μ_m value is the reciprocal of the mean residence time of cells in the fermenter, assuming perfect mixing.

Once 'steady state' conditions have become established in a continuous fermentation, clearly $\mu = D$, the dilution rate. If D is increased, then μ must also increase to keep the system in equilibrium, but as D approaches the value of μ_m, the limiting substrate concentration increases and the cell concentration decreases, i.e. 'wash-out' occurs.

The concentration of limiting substrate is proportional to growth rate and, over a wide range of dilution rates below μ_m, can be shown to be almost constant. However, the relationship departs rapidly from linearity as the value for μ_m is approached. The low substrate concentrations which result mean that utilization of the feedstock in continuous culture can be very efficient, the optimum dilution rate for operation being about three-quarters of the μ_m value.

B. Equipment

Most university departments of microbiology now boast at least one continuous fermenter and many different manufacturers offer equipment which varies in sophistication from batch fermenters with 'add-on' medium-supply systems, to fully instrumented and automated packages. The vessel size lies generally between one litre and ten litres, although both very small vessels and pilot-plant equipment up to a few cubic metres in working volume are also available off-the-shelf. Nearly all of this equipment is based on the standard 'Porton Pot'—a closed cylindrical baffled vessel fitted with a longitudinal shaft, bearing one to several turbines for mechanical agitation. The other basic type of fermenter—working on the 'air-lift' principle instead of mechanical agitation—is less common in small-scale applications, although a number of industrial processes are firmly based on this design.

A selection of the different types of fermenter proposed for S.C.P. production is depicted in Figure 3. These include the Porton, Waldhof, Vogelbusch, Lefrancois, Kanegafuchi and tubular fermenters and other forms.

The influence of fermenter design on oxygen transfer has been reviewed by Hatch (1975). In all schemes for production of biomass from hydrocarbon substrates, an additional factor is the requirement to

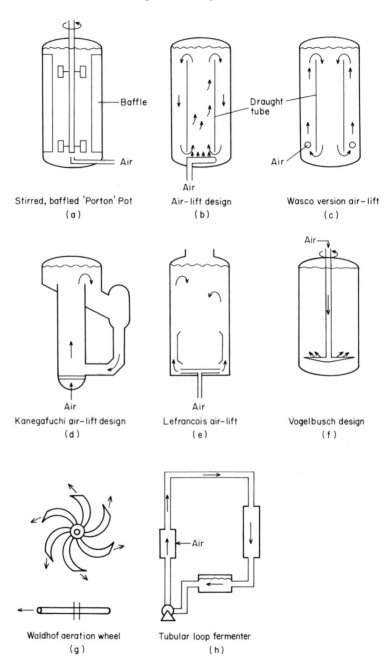

Fig. 3. A selection of fermenter designs proposed for S.C.P. production.

disperse the immiscible substrate in the aqueous phase by means of the shear forces imparted by the design and operation of the fermenter. The use of a collodial emulsion feed proposed by Yoshida and Yamane (1974) would be feasible only in laboratory-scale work.

The mechanically agitated, fully baffled fermenter with turbine mixers, in which air is introduced through a sparger, is the design used by BP for the *n*-alkane process developed in Scotland, up to the 300 m³ scale, and used in the three 1000 m³ vessels constructed in Sardinia for the Italproteine project.

For gas-oil substrates, BP adopted an air-lift system, the most important parameter in design for this substrate being the correlation between the rising speed in the draught tube and productivity (Lainé and du Chaffaut, 1975). Since the energy applied to an air-lift fermenter is supplied only with the air, it is important to obtain an optimum design for the draught tube which minimizes the energy requirement per kg of biomass produced. The same basic design (Lefrancois, 1969) has been used by l'Institut Français du Pétrole (I.F.P.) in their collaborative S.C.P. project with Elf/ERAP (Enterprise de Recherche et d'Activitiés Pétrolières). Gulf (Cooper *et al.*, 1975) also appear to have adopted a simple air-lift design in their Wasco 'petroprotein' pilot plant in California. Modified air-lift systems, in which the broth is driven by the force of the inflowing air from a large vessel through a circulatory side-arm, have been used in several projects. Examples are the Kanegafuchi design (Kanazawa, 1975) used for the Liquichimica SpA process fermenters in Italy (Masson, 1973), and the air-lift 'tubular-loop' fermenter used by a German consortium including Hoechst (Knecht *et al.*, 1977). A combined design of a turbine impellor with a draught tube, as in the fodder-yeast Waldhof fermenter, has been used in laboratory studies (Einsele and Fiechter, 1972; Prokop and Erickson, 1972; Katinger, 1973). This type of fermenter has been adopted for S.C.P. production from *n*-paraffins in Czechoslovakia (Prokop and Sobotka, 1975).

· The relative merits of mechanical agitation and the various forms of air-lift fermentation have been debated in every fermentation-development laboratory. The variety of solutions proposed by different projects show that there is no 'correct' choice of fermenter, but rather a selection of compromises. Because of its greater mechanical complexity and size limitations, less ultimate potential probably lies in the mechanically stirred fermenter but, at sizes close to the maximum usually considered to be appropriate, the higher oxygen-transfer rates obtainable with the stirred baffled design can be competitive with less complex air-lift designs.

C. Operation

A second controversy, no less intense than that over fermenter design, concerns the choice between aseptic and non-aseptic fermenter operation. The problems of designing for, and achieving and maintaining, perfect asepsis in very large pieces of equipment over long periods of time under industrial conditions are, of course, very severe. Non-aseptic operation, if it can be efficiently sustained against the onslaught of a random selection of 'wild' organisms, is, therefore, greatly to be preferred on the grounds of simplicity and cost. However, experience shows that, unless fermentation conditions are so selective that the culture organism is virtually the only one which will grow satisfactorily, contamination problems will be encountered under non-aseptic conditions. The process developed by Champagnat's team (British Petroleum Co. Ltd., 1966a) at the Lavera refinery of the Société Français des Pétroles BP in the early days of S.C.P. was entirely non-aseptic. The presence of gas-oil, which is toxic to many micro-organisms, combined with the low pH values employed, permitted successful and prolonged operation. The alternative BP process, developed by the group at Grangemouth in Scotland (British Petroleum Co. Ltd., 1963), used aseptic conditions and a pure n-paraffins feedstock. Clearly significant process advantages would follow from the development of a non-aseptic process on this substrate and no doubt various groups have sought progress in this direction.

Over the seven years of operating the Protein Demonstration Unit at Grangemouth, the practicability of maintaining an aseptic fermentation for long periods in a stirred vessel of 300 m^3 working volume was repeatedly proven. However, it was also shown that, if a mechanical fault led to the ingress of a contaminating microbe, provided that this was a bacterium (as was usual), its numbers could, if necessary, be curbed to an arbitrary working maximum of 1% weight of the total biomass by lowering the pH value of the culture broth. Because of the low pH value and selective substrate, food-poisoning bacteria have never been observed in the Grangemouth fermenters. Since growth of some wild yeasts could not be controlled, it was usual to terminate a run which became significantly contaminated, i.e. to the 1% weight level, by these organisms.

D. Raw Materials Preparation

The fermentation is the key and characteristic step in S.C.P. production. This justifies its description before turning to the preparation of the raw materials that feed it. The procurement of adequate supplies of hydrocarbon feedstock, nutrient salts, growth factors, ammonia and, not least, water of proven quality and acceptable cost, together with the quality control necessary to ensure maintenance of the required standards, are constraints that are not peculiar to the processes under discussion. Although some ingenious equipment has been devised for continuous salts solution make-up, preparation of medium solutions providing most of the ions required by the organism is more usually carried out in batch operations. Some feed ingredients, i.e. acids, alkalies and the carbon source, are liquids. For continuous non-aseptic processes, these feed streams need no further treatment, being metered into the fermenter at predetermined rates to supply adequate nutrients for the . dilution rate chosen.

Nutrient preparation for aseptic fermentation requires sterilization of all feed streams. The usual method is by heat treatment, but membrane filtration can be used for small flows. An interesting use of the self-sterilizing properties of strong acids and alkalies has been described (British Petroleum Co. Ltd., 1976a) which could provide worthwhile savings in the cost of steam for sterilization purposes. On sterilizer duty, economiser heat exchangers will be employed and the conventional equipment of plate packs, holding coils and temperature and pressure-control apparatus will be designed by the chemical engineer in materials and sizes appropriate to the duties required.

E. Conditions

Generalizations about fermentation conditions are of limited value or interest because much of the effort expended on optimizing them will be specific to the plant design and organism concerned. The advantages of low pH-value operation have been detailed, the need for efficient aeration has been stressed, and broth temperatures must be carefully controlled. The last necessitates removal of vast quantities of 'low-grade' metabolic heat and, in the Kanegafuchi process, adiabatic expansion is ingeniously employed as an inexpensive means of broth cooling (Kanazawa, 1975). The chemical requirements for growth of the organism comprise the source of carbon for energy and cellular synthesis (in this case, n-alkanes), water, mineral salts and growth factors. Phosphorus as phosphate, sulphur as sulphate and potassium, magnesium, manganese, iron, zinc and trace metals as suitable soluble salts, satisfy the mineral requirments of the micro-organism in optimized concentrations. Growth factors are supplied as pure vitamins or as mixtures in the form of yeast extract or other organic preparations. The entire process of S.C.P. production can be regarded as the catalytic conversion of ammonia into amino acids and the supply of nitrogen to the fermentation is, of course, of vital concern. Often, the rate of supply of ammonia will be fermentation-determined by a pH-value control signal, since yeast fermentations are usually acid-producing. Further details of this aspect of S.C.P. production processes can be obtained from the various reviews which have been cited. No new developments have been noted in recent years.

F. Published and Patented Processes

The breadth and depth of interest in commercial development of S.C.P. production from n-alkanes is clearly demonstrated by the large number of companies that have patented organisms or techniques in great variety. Most of these patents, which refer to pilot-plant or semi-industrial scale processes, concern yeast fermentation and, within this group, *Saccharomycopsis* (*Candida*) *lipolytica* is by far the most frequently mentioned.

The pioneering work of the two British Petroleum groups has already been described in outline. All of their n-paraffins process patents concern *Saccharomycopsis lipolytica*. These include basic process patents (British Petroleum Co. Ltd., 1963, 1966a), the yeast strains used (British Petroleum Co. Ltd., 1975a, 1976b) and various fermentation techniques including recycle of aqueous nutrient medium (British Petroleum Co. Ltd., 1966b), dispersal of the hydrocarbon substrate (British Petroleum Co. Ltd., 1967), hydrocarbon-limited operation (British Petroleum Co.

Ltd., 1971) and the use of overpressure (British Petroleum Co. Ltd., 1973).

Elsewhere in Europe, l'Institut Français du Pétrole (I.F.P.), in association with Elf/ERAP, have described a two-stage process in air-lift fermenters, probably with *Saccharomycopsis lipolytica* (Bianchi and Michaux, 1974). The second-stage fermentation is designed to eliminate residual hydrocarbons remaining in the broth from the first stage wherein biomass production took place. Lysine-enriched mutants of *Saccharomycopsis lipolytica* have also been claimed (Entreprise de Recherches et d'Activités Pétrolières, 1974).

The same species was one of several covered by a German patent (Hoechst A.G., 1973) which claimed a low cellular nucleic acid content of 3–6% dry weight, against the more usual 8–12% in other processes. *Candida guilliermondii* and *C. viswanathii* were also included. In addition, Hoechst A.G., in partnership with Uhde GmbH and Veba Chemie A.G., have described a pilot-scale process with *Saccharomycopsis lipolytica* (Birckenstaedt *et al.*, 1977) which also features in Czech patents for S.C.P. production from hydrocarbons (e.g. Stross *et al.*, 1974).

With their wide experience in the fermentation field, the Japanese have been active in patenting S.C.P. processes, although these are greatly outnumbered by projects for recovery of metabolites from batch processes. Kyowa Hakko Kogyo Co. Ltd. has patented S.C.P. biomass processes based on *Saccharomycopsis lipolytica*, in addition to *C. tropicalis* and *Torulopsis famata* (Kyowa Hakko Kogyo Co. Ltd., 1972, 1973). The Morinaga Milk Industry Ltd. has published patents (e.g. Morinaga Milk Industry Ltd., 1973) involving *Saccharomycopsis lipolytica* or *C. guilliermondii* biomass which is claimed to have a high digestibility by virtue of the use of mutants with lowered cell-wall integrity (Harada *et al.*, 1972). Mitsui Petrochemical Industries Ltd. have patented a process for production of *Saccharomycopsis lipolytica* (Mitsui Petrochemical Industries Ltd., 1974) on *n*-alkanes. Miura (1973) has described an interesting technique of two-stage continuous culture in which the first vessel contains a bacterium and the second a yeast (*Saccharomycopsis lipolytica* or *C. tropicalis*), with recycle of broth to the first vessel. Finally, the Mobil Oil Corporation is one of the relatively few American groups to show evidence of work in this field through a patent on *Saccharomycopsis lipolytica* (Mobil Oil Corporation, 1973a) and another concerning a *Pichia* strain (Mobil Oil Corporation, 1973b).

Candida guilliermondii is another popular organism for patent cover. It has been mentioned above (Hoechst A.G., 1973; Morinaga Milk Industries Ltd., 1973), while both the Russians (Chepigo, 1972; All Union Scientific Research Institute for Protein Biosynthesis, 1972a, 1974) and the Bulgarians (Rodionova *et al.*, 1970) have described non-aseptic processes using this species. Texaco Ltd. in Belguim have a patent (Texaco Ltd., 1968) on growth of *C. guilliermondii* or *C. rugosa* on *n*-paraffins, while in Taiwan (Ting *et al.*, 1971) *C. guilliermondii* is the organism used on a semi-industrial scale (150 m³) batch process.

Candida tropicalis, in addition to its mention in British Petroleum gas-oil patents (e.g. British Petroleum Co. Ltd., 1966a, 1976b) and those already cited (Kyowa Hakko Kogyo Co. Ltd., 1972, 1973; Miura, 1973), is the culture organism of an I.F.P. process using a Vogelbusch 'liquid jet' fermenter (Ballerini, 1978). Gulf Research and Development Co. have patented continuous culture of this species (Gulf Research and Development Co., 1973), for which Silver and Cooper (1972) claimed that the biomass was rich in sulphur-containing amino acids. In fermentations up to 50 m³ in size, they demonstrated increased total nitrogen, lysine and methionine + cystine contents by use of a two-stage fermentation. Thiosulphate was added to the second stage, wherein the residual hydrocarbon content was also lowered.

Candida zeylanoides is the last of the established yeasts to be exploited for an S.C.P. patent, granted to Imada and Hirotani (1972).

Three companies, all Japanese, have described processes based on novel yeast species. The first of these, which was licensed to Liquichimica SpA for their plant of 100 000 tonnes per annum built in Calabria, was developed by the Kanegafuchi Chemical Co. Ltd. Takata (1969) summarized the early stages of this company's work which led to a process (Kanegafuchi Chemical Industry Co. Ltd., 1973) using a species taxonomically close to *C. tropicalis*, which was named *C. novellus*. This has since been shown to be a strain of *C. maltosa* (Meyer *et al.*, 1975). Carnitine and its precursors, crotonobetaine and butyrobetaine, gave increased yields of this species when added to fermentations (Kanegafuchi Chemical Industry Co. Ltd., 1975a). The same company have also patented a mixed fermentation which is inoculated with a hydrocarbon-utilizing species and one which does not oxidize hydrocarbons but enhances the yield, presumably by growing on oxidation products (Kanegafuchi Chemical Industry Co. Ltd., 1970, 1972, 1975b).

The Mitsui group have described an S.C.P. process based on *C. kofuensis* in a fermentation carried out at 37°C, but this was tested in batch conditions at a scale of only 20 litres (Ueno *et al.*, 1974a). Lastly in this sector, Dainippon Ink and Chemicals Inc. have claimed a high-protein S.C.P based on *C. paraffinica* (Dainippon Ink and Chemicals Inc., 1975). This is presumed to relate to the process licensed to the Romaniam Ministry of Chemical Industry. The Roniprot plant (60 000 tonnes per annum) is reported (Anon, 1978b) to be complete, with commissioning due to start in December 1978.

Other S.C.P. enterprises which have been reported in the scientific press include that of the Regional Research Laboratory, Jorhat, Assam, in association with I.F.P., using *Saccharomycopsis lipolytica* (Lonsane *et al.*, 1973), where a 15 000–20 000 tonnes per annum plant is planned for 1980 (Anon, 1978c) and the Korean Institute of Science in Seoul, where a mixed culture of *C. tropicalis* and *Trichosporon cutaneum* has been employed (Pyun *et al.*, 1977).

We have found few references to processes in which bacteria are used to produce S.C.P from liquid *n*-alkanes, in addition to the first stage of the two-stage process described by Miura (1973) and already cited. One version of the Esso-Nestlé product, Protecel, is bacterial biomass (*Acinetobacter cerificans*) grown on *n*-paraffins (Mauron and Wuhrmann, 1971), but about which we have seen no recent publications. Pilot-scale studies (6 m³ batch fermentation) were reported from Taiwan in the 1960s using species of *Achromobacter* (Ko *et al.*, 1964) and *Pseudomonas* (Ko and Yu 1968), while the Indian workers, Kasturi and Tamhane (1974), described produciton of biomass from *Ps. aeruginosa* grown on diesel oil.

V. PRODUCT HARVESTING

Removal of water from fermenter broth containing only a few grams of microbial cells per litre to achieve the end product—substantially dry biomass—is a complex and expensive procedure. Hence it has attracted a great deal of development work.

Although the fermentation may have been operated aseptically, it is not practicable to harvest biomass in the same way. If the first stage of dewatering is a settling or flotation procedure, then the equipment may be closed and capable of aseptic design, but a centrifugation step is usually included and large centrifuges are not available which can be operated aseptically. Therefore, the entire liquid-side harvesting train will generally be designed for 'hygienic' operation. This may involve duplication of some or all of the equipment to allow for periodic cleaning. The precautions employed in handling liquid milk, for example, offer a reasonable analogy, although it must be remembered that S.C.P. plants will run round the clock for months at a time, rather than in shift-long campaigns. This will influence design and may lead to the adoption of automated cleaning-in-place equipment. Two factors pose extra problems which are peculiar to the handling of biomass once it leaves the fermenter. Firstly, it is warm (fermentation temperature is likely to be at least 30°C in a yeast process and from 37°C to 60°C or even higher for bacteria) and, secondly, even if the pH value of the broth is low, continued respiration of the live cells will rapidly deplete the dissolved oxygen and the pH value will rise in the harvested broth. Thus, micro-aerophilic and anaerobic bacteria will be provided with ideal conditions and a plentiful source of substrate—the biomass—on which to proliferate. It may therefore prove necessary to cool the broth which emerges from the fermenter, to pass forward to the harvesting train, below the temperatures at which rapid bacterial growth will occur. Pasteurization of the broth may also be required at some stage of the harvesting sequence, to keep contaminants down to acceptable levels and (depending on the mode of final drying) the pasteurization step may also be essential to kill the biomass. In some processes it is, in effect, a cooking procedure, rendering the cell walls of the biomass more readily digestible by the consuming animal.

The aqueous phase from the dewatering stages is high in Biological Oxygen Demand and requires extensive effluent treatment if it is to be disposed of. In large plants, it is more usual to recycle these streams and it may even be possible to re-use all of the water in the fermenter feed with appropriate adjustment of nutrient levels and resterilization in an aseptic process. This provides important economies, not only in water usage but also in salts requirements, because significant residual concentrations remain in the spent medium. Clearly, however, recycling of metabolites may pose problems to the culture organism and hence to the fermentation technologist developing the process.

Final dryers may be spray-dryers, flash-dryers or other types of equipment that will handle a fluid feed and provide a dry product which may be in powder or granular form. Some type of storage silo will be needed, together with equipment for sack-filling and bulk-loading to suitable transport.

Table 3

Amino-acid composition of yeasts and other selected foodstuffs

Content (g/16 g nitrogen) of amino acid in

Amino acid	Feed yeasts					Traditional feed proteins		Food proteins	
	Toprina G[1]	U.S.S.R.[2]	Kanegafuchi-Liquichimica[2]	Pekilo[3]	Saccharomyces cerevisiae[4]	Fish-meal[1]	Extracted soybean[1]	Egg[4]	F.A.O. reference[5]
Isoleucine	5.1	5.9	4.6	4.8	4.6	4.6	5.4	6.7	4.2
Leucine	7.4	7.2	7.1	7.4	7.0	7.3	7.7	8.9	4.8
Phenylalanine	4.3	4.4	4.3	4.0	4.1	4.0	5.1	5.8	2.8
Tyrosine	3.6	4.0	3.8	3.5	—	2.9	2.7	4.2	2.8
Threonine	4.9	5.2	5.4	4.1	4.8	4.2	4.0	5.1	2.8
Tryptophan	1.4	1.1	1.0	1.1	1.0	1.2	1.5	1.6	1.4
Valine	5.9	6.3	5.5	5.1	5.3	5.2	5.0	7.3	4.2
Arginine	5.1	5.0	5.1	5.3	—	5.0	7.7	—	—
Histidine	2.1	2.4	2.4	1.9	—	2.3	2.4	—	—
Lysine	7.4	7.6	7.0	6.5	7.7	7.0	6.5	6.5	4.2
Cystine	1.1	0.9	1.5	1.0	—	1.0	1.4	2.4	2.0
Methionine	1.8	1.6	1.4	1.9	1.7	2.6	1.4	5.1	2.8
Total sulphur-containing acids	2.9	2.5	2.9	2.9	—	3.6	2.8	7.5	4.8

[1] Shacklady and Gatumel (1972).
[2] British Petroleum (unpublished data).
[3] Anon (1973).
[4] Kihlberg (1972).
[5] Senez (1972).

VI. PROPERTIES

A. Composition

Inevitably, most of the comparative compositional data on S.C.P. concern yeast biomass, as the *n*-alkanes processes brought to large-scale operation have usually employed these organisms. A good quality sample of S.C.P. can contain at least 60% crude protein, about 20% carbohydrate and about 10% lipids.

1. Amino-acids

Some 90% of the crude protein content (%N × 6.25) of a typical S.C.P. sample contains amino acids composing the cellular protein. Most of the remaining cellular nitrogen resides in nucleic acids.

Amino-acid analyses of a number of different types of biomass grown on *n*-alkanes are shown in Table 3 together with typical values for some competing protein sources (fish-meal and solvent-extracted soybean meal) and the F.A.O. reference protein and whole egg for comparison. S.C.P. products are compared in animal-feeding studies with soybean meal and fish-meal, which are highly regarded as feed proteins.

The yeast amino-acid analyses in Table 3 were carried out on an auto-analyser according to the method of Moore and Stein (1951) with separate determination of tryptophan. Comparison of the spectra shows that, apart from the somewhat higher methionine content in fish-meal and the enhanced lysine content in the yeast samples, there is a considerable similarity in the pattern in all cases. Thus, the nutritive value of yeast proteins should be at least as high as that of soybean meal and should equal that of fish-meal when supplemented with methionine. That this is the case is discussed later (see Table 5).

No qualitative changes occurred in the amino-acid composition of *Saccharomycopsis lipolytica* when hydrocarbon replaced glucose as the carbon source (Rosenfeld and Disler, 1971), but the amounts of free amino-acids increased when the concentration of thiamin in the medium was lowered. *Saccharomycopsis lipolytica* contained more glutamate decarboxylase than *C. tropicalis*, and the former also accumulated γ-aminobutyric acid in the amino-acid pool (Rosenfeld *et al.*, 1971). The principal deamination reactions in this yeast were glutamic to pyruvic and oxaloacetic acid, and alanine to α-oxoglutaric acid. Muntyan (1971)

observed a decline in the thiamin requirement of *C. tropicalis* growing on
n-hexadecane, and suggested that this was linked to the decreased role of
α-oxoglutaric acid decarboxylase on this substrate compared with
glucose.

Similar amino-acid compositions were observed in a number of
different strains of *C. tropicalis* grown on hydrocarbon, but one strain
grown on different feedstocks showed a range of amino-acid spectra
(Kvasnikov *et al.*, 1969).

2. Lipids

Yeast grown on carbohydrates, although not free of fatty acids with odd
numbers of carbon atoms, contains predominantly even-numbered fatty
acids. However, biomass grown on hydrocarbons generally contains
significant amounts of fatty acids with the same chain length as the
substrate. In the case of bacterial biomass, the situation is obscure but a
number of reports on yeasts (Lowery, 1968; Lideman and Yudintseva,
1969; Hug and Fiechter, 1973) show that, with n-hexadecane and n-
octadecane as substrates, C_{16} (palmitic) and C_{18} (stearic) acids
predominate among the saturates and $C_{16:1}$ (palmitoleic) and $C_{18:1}$
(oleic) acids among the unsaturates (Mizuno *et al.*, 1966; Pelechova *et al.*,
1971). Both chain lengthening and chain shortening occur (Jwanny,
1975). In yeast grown on n-tetradecane, C_{16} and C_{18} fatty acids are
present in saturated and unsaturated forms (Klug and Markovetz,
1967a; Hornei *et al.*, 1972), as well as saturated C_8 and C_{10} acids
(Dyatlovitskaya *et al.*, 1965).

A mixture of odd and even numbered fatty acids is obtained when
yeast is grown on odd numbered n-alkanes (e.g. Zhukov *et al.*, 1970;
Mishina *et al.*, 1977b), suggesting that both α-oxidation and β-oxidation in
addition to C_2 chain-elongation reactions are present.

It has been reported (Nyns *et al.*, 1968; Zaikina, 1971; Chenouda and
Jwanny, 1972; Mishina *et al.*, 1977a) that hydrocarbon-grown yeast can
contain three times as much cellular lipid as carbohydrate-grown cells.
The lipid content appears to depend on the yeast strain and on the
fermentation conditions, but the qualitative composition of the fatty
acids is said to be independent of the conditions used (Pelechova *et al.*,
1971).

Phospholipids can comprise 40–50% of the total lipids in n-alkane
grown yeast (Mizuno *et al.*, 1966; Greshnykh *et al.*, 1968; Chenouda and

Jwanny, 1972). Thorpe and Ratledge (1972) suggest that these high phospholipid concentrations are required to maintain the integrity of the deeply invaginated cytoplasmic membranes found in yeasts which are assimilating hydrocarbons.

Low triglyceride levels have been reported in *C. petrophilum* (Mizuno *et al.*, 1966) but high concentrations were found in *Saccharomycopsis lipolytica* (26% of the total lipids; Chenouda and Jwanny, 1972). Growth on hydrocarbons also enhances the cellular ergosterol content (Shirota *et al.*, 1970).

The lipids of *n*-alkane-grown bacteria have been studied by Finnerty's group. In a cogent review on the biochemistry of microbial alkane oxidation, Finnerty (1977) also discusses quantitative and qualitative changes in the lipid composition of *n*-hexadecane-grown *Acinetobacter* sp. (Makula *et al.*, 1975).

3. Carbohydrates

Yeast cell carbohydrate content may be lower in hydrocarbon-grown biomass (e.g. Rudenko, 1971) but the monosaccharide composition may be the same as in glucose-grown cells (Iizuka *et al.*, 1965; Rudenko *et al.*, 1970). The ratio of mannan to glucan varies with the species (Nabeshima *et al.*, 1970). *Saccharomycopsis lipolytica* contains a higher and *C. tropicalis* a lower mannan:glucan ratio when grown on *n*-alkanes compared with growth on glucose.

4. General

Yeasts growing on hydrocarbons show an enhanced rate of synthesis of the B-group of vitamins (Kvasnikov *et al.*, 1967; Popova, 1968). Cell composition depends to some extent on cultural conditions, especially on the ratio of carbon to nitrogen in the culture medium, and the dilution rate in continuous culture (or the time of harvesting in batch fermentation). A well-known method for increasing the lipid content in S.C.P. grown on both carbohydrates and hydrocarbons is to restrict nitrogen supply, i.e. increasing the C:N ratio. This has been employed with *Rhodotorula* sp. (Woodbine, 1959), and Ratledge (1968) used a nitrogen-deficient medium to obtain a yield of 21 g fatty acids/litre from a *Candida* sp. grown on *n*-alkanes. On the other hand, Alent'eva *et al.* (1969) have observed that low nitrogen contents decreased the content of mono-

and diglycerides, sterol and fatty acids of *Candida*, while Lideman and Yudintseva (1969) reported that excess nitrogenous nutrients gave high yields of cellular lipids; the level increased from 7.5% to 15% of dry weight. Altering the nitrogen source can change the cellular lipid content and its composition in *C. tropicalis* (Greshnykh *et al.*, 1968), while increasing the concentration of phosphate in the medium can increase the content of cellular phospholipids and decrease the triglyceride fraction in *Candida* sp. grown on *n*-alkanes.

An increase of the fermentation pH value from 4.0 to 4.5 was found to increase the free fatty acids produced by *Saccharomycopsis lipolytica* (Dyatlovitskaya *et al.*, 1969), with a concomitant decrease in mono-, di- and triglycerides. In addition, polyglycerophosphatides increased sharply, while lecithin and phosphatidylserine both declined in content. However, it was also shown that the fatty-acid composition of *C. guilliermondii* was unaffected by the same change in pH value. Overall, therefore, the primary determinant of cellular composition appears to be cultural conditions rather than carbon source.

B. Product Specifications

The following aspects of product quality must be considered: chemical composition, bacteriological content and nutritional suitability of the S.C.P. in relation to its intended use.

Nutritional and toxicological testing have often been considered together, and a discussion of them will be found in the next section of this review. The setting of a bacteriological quality-control standard for S.C.P. is also described in Section VII, but it should be noted that the attainment of high final product quality is the end result of careful quality control throughout the production process.

The same is true of chemical quality control. Rigorous checking and control of large quantities of water and inorganic raw materials required as salts, acids and alkalies for medium make-up and cleaning is essential. This ensures that suppliers maintain their quality standards and that, for example, the very low levels of heavy metals essential to ensure maintenance of satisfactory product quality are not exceeded.

Regular analysis of prepared medium salts mixes gives assurance that adequate supplies pass forward to the fermenters where further checks are carried out to ensure that the appropriate ions at required concentrations are being fed to the culture. If spent medium is recycled, analysis is required at this stage so that additional nutrient salts can be added as necessary. Obviously, regular process-control analysis of the product for key parameters will be a matter of daily routine. In addition, a representative sample of each final product batch will receive complete analysis to ensure compliance with the established quality-control protocols. The most important fermenter feedstock—the *n*-paraffins— will probably be produced at an adjacent site and piped into the S.C.P. plant. Very stringent quality control of this ingredient is clearly of fundamental importance.

Bacteriological quality control is also applied at each key stage in the production process. Just which stages are 'key' is learnt only by operational experience with a particular process. As a result of several years development of the BP *n*-alkanes process, it was found that certain checks could be dropped from the chemical and bacteriological test schedules initially adopted, once potential trouble spots had been identified. Other stages, at which defined parameters required control within narrow limits, were pinpointed and, in some cases, automatic on-line chemical analysis was developed to serve these functions.

The microbiological standard shown in Table 4 is comparable with the

Table 4

Microbiological standards for a single-cell protein product (Toprina)

	Maximum counts of contaminating organisms per gram		
	For general use	For milk-replacer use	UNICEF specification for dried weaning formulae for humans
	Toprina G		
Lancefield Group D streptococci	10 000[a]	1000[a]	100
Total clostridia	3000	1000	—
Clostridium perfringens	100	10	10
Species of Enterobacteriaceae	10	1	1
Total aerobic bacteria	100 000	100 000	10 000
Total aerobic spores	100 000	100 000	10 000
Yeasts[b] and moulds	100	100	100
Staphylococcus aureus	<1	<1	<1
Bacillus cereus	1	1	100
Salmonella spp.	absent from 50 g		absent from 25 g

[a] Pressor amines being absent.
[b] Not culture yeast.

UNICEF standard for infant foods also given. Experience during production from the 300 m³ fermenter at Grangemouth indicates that careful control can result in regular attainment of a product with excellent microbiological quality.

C. Shelf-Life

The keeping quality of some high-protein feed ingredients is poor; fish-meal tends to go rancid and soybeans are difficult to store in bulk for long periods. In contrast Toprina has been shown to have a virtually unlimited shelf-life. A three-year storage trial was completed without incident; after an initial rise in moisture content, no measured parameter changed significantly. The water relations of Toprina have been studied (Pixton and Warburton, 1977) and found to be similar to those of high-lipid flours. No spoilage occurs below 12% moisture content (J. L. Shennan and J. D. Levi, unpublished observations) which is attained at 75% relative humidity under equilibrium conditions at 25°C. However, as Toprina was produced at 6% moisture maximum, this resulted in a high safety margin. The first colonizers of moist Toprina are, as expected, xerophilic moulds.

D. Functional Properties

Only one application has so far been revealed in which use is made of the functional properties of S.C.P. This is the incorporation of biomass in liquid feeds for calves, as a partial replacement of skimmed milk (British Petroleum Co. Ltd., 1978). This opens up the potentiality of intensive feeding of pre-ruminant animals on S.C.P. in parts of the world where milk is scarce. It relies on the fat-binding properties of the biomass, because, in the intensive rearing of these animals, high fat contents are employed in order to maximize the metabolizable energy content of the diet. Yeast grown on *n*-alkanes can be blended with high proportions of fat of various types suitable for calf diets but with retention of the free-flowing powder characteristics essential in bulk handling (British Petroleum Co. Ltd., 1977).

Much effort has been expended on attempts to fractionate S.C.P. in order to obtain high-value products which could compete with conventional foods by virtue of functional properties deriving from proteins or carbohydrates. Despite numerous patents, it appears that these efforts have enjoyed limited success (Litchfield, 1977; Anon, 1978a).

VII. PRODUCT TESTING

Chemical and bacteriological testing of every batch of product made is an essential part of quality control, and has here been described. However, the product will find purchasers only if it is both economically priced and, in addition, has been shown to be wholesome, reliable and free from harmful properties. In the case of S.C.P., the test programmes that have been pursued during its development have been by far the most comprehensive ever devised (Anon, 1976; Truhaut and Ferrando, 1976).

A. Toxicity Testing

Predictions that S.C.P. from liquid *n*-alkanes would partially replace conventional high-protein constituents of animal feeds, i.e. fish-meal and soybean meal (Takata, 1969; Shacklady and Gatumel, 1972), now seem unlikely to be fulfilled in the near future, as a result of the refusal of the Italian authorities to allow the two 100 000 tonnes per annum S.C.P. plants built by Italproteine and Liquichimica in the middle 1970s to operate. However, the proposed Italproteine product, Toprina, had been marketed by the British Petroleum Co. Ltd. from its semi-industrial development plant at Grangemouth, Scotland at about 1500 tonnes per annum from 1971 to 1978 inclusive.

Shacklady and Gatumel (1972) have published an extensive range of toxicological and nutritional test results on this product, and on the sister product made at Lavera, France, on gas-oil. Although new food products require to be tested and their safety established, no such requirement exists in the United Kingdon for animal feeds. Nevertheless, Toprina has been subjected to a long series of tests, the design of which has been described by Engel (1972), who also shows that no deleterious results were obtained. The test protocols included acute toxicity (6-week rat feeding test with 40% of the diet as Toprina), subchronic toxicity (three months) and chronic toxicity (two years, equal to the life span) and multigeneration tests (15 generations in rats; 23 generations in quails) (de Groot *et al.*, 1975; Shacklady, 1975).

1. Carcinogens

Since some polycyclic aromatic compounds are recognized carcinogenic agents and are present in petroleum, the content of these compounds in S.C.P. grown on crude oil derivatives is of considerable interest. However, yeasts grown on mineral oil and pure *n*-alkanes were found by Grimmer (1974) to contain minimal levels of polycyclic aromatic compounds as were dietary yeasts. These differed from baker's yeasts which contained 6- to 10-fold more of these compounds (Grimmer and Wilhelm, 1969; Truhaut and Ferrando, 1976). On the other hand, McGinnis and Norris (1975) reported levels of polycyclic aromatic compounds in *n*-paraffin-grown yeast of 1 to 11 p.p.b. whereas the occurrence in glucose-grown yeast was less than 1 p.p.b. Polycyclic aromatic compounds could not be found in the *n*-paraffin feedstock and, since their presence in the yeast could not therefore be traced to that source, extraneous contamination must be suspected.

In processes based on purified *n*-alkanes, it is common to find that food-grade paraffins are used, which contain less than 40 p.p.m. total aromatic compounds. British Petroleum feedstock has been shown to have less than 1 p.p.b. (i.e. the lower limit of detection) of 3,4-benzpyrene, 1,2,5,6-dibenzanthracene or methylcholanthrene (British Petroleum Co. Ltd., unpublished observations). Nor could these three compounds be detected in S.C.P. or its derivatives grown in this substrate (unpublished work, cited in Shacklady, 1975). It was concluded that careful control of the feedstock purity (in the case of *n*-alkanes) or of the post-fermentation solvent-extraction procedure (in the case of gas-oil) was essential to ensure freedom from polycyclic aromatic compounds in the S.C.P.

2. Hydrocarbons

Mammalian utilization of *n*-alkanes has been known for at least 30 years. Nevertheless, interest has been expressed in the fate of the residual hydrocarbon which many single-cell proteins contain, after ingestion by the target animal. Shacklady (1977a) has reported comparisons between calves, pigs and poultry fed on diets containing Toprina S.C.P., and on yeast-free diets. Muscle, fat, brain, liver, kidney, heart, spleen and, in the case of birds, eggs were analysed. The largest differences in *n*-alkane content of the tissues were found in the perirenal fat of calves and pigs, namely about 50 p.p.m., and in the dorsal fat of pigs (about 37 p.p.m.). It

is, however, interesting to note that the control animals also contained
n-alkanes in their tissues. The tendency of hydrocarbons to migrate to the
lipidic tissues confirms other work (e.g. Tulliez and Bories, 1975a, b) in
which broadly similar retentions were noted. Unpublished work on
broiler carcasses carried out at the Food Science Unit of the Ministry of
Agriculture, Fisheries and Food, Norwich, Great Britain (D. McWeeny,
unpublished observations) shows that the n-alkanes derived from
Toprina in the feed, which are detected in the carcasses, are turned over
in the animal.

The significance of these enhanced levels of hydrocarbons in the tissues
of food animals and their products has been considered by Lester (1979).
He arrived at several conclusions: n-alkanes are ubiquitous in living
tissues, so that they are therefore normal components of the diets of
humans and higher animals at a concentration not usually exceeding
0.1% of the weight. This has been confirmed by analysis. Moreover,
n-alkanes are subject to metabolic turnover, slow oxidation to fatty acids
having been demonstrated in many studies. Also, they can accumulate in
the reticulo-endothelial system of man, up to a concentration sufficient
for a separate oil phase to exist in the tissues. This condition (follicular
lipidosis) arises only when very large quantities of paraffins are ingested,
but is not associated with any overt pathology. Finally, n-alkanes are
absorbed by the fatty tissues but no morphological or functional changes
are observable.

The biological significance of odd-numbered fatty acids, derived from
hydrocarbon-grown yeast, has been investigated by Bizzi et al. (1976)
using the Liquichimica SpA product 'Liquipron'. They concluded that
the presence of odd-numbered fatty acids in the body at limited
concentrations, not exceeding 10% of the total fatty acids, is compatible
with normal physiological functions since they can be metabolized.

Long-term and multigeneration studies on animals fed Toprina-
containing diets have been reported (Engel, 1972; de Groot, 1974; de
Groot et al., 1975). The yeast contained up to 1500 p.p.m of residual
n-alkanes, and was present at up to 30% weight of the total diet. In no case
was any treatment-related pathology observed in the yeast-fed animals.
More than 20 successive generations of quail, a 'sensitive' species, have
been raised on 30% yeast diets, but no untoward effects have been
detectable and reproductive performance was normal.

Rats have been fed on diets containing meat, fat and eggs from animals
which had been fed Toprina. No difference could be found between these

animals and control rats fed on protein from yeast-free animals, after tests of six weeks to three months duration. Furthermore, Shacklady (1977b) notes that levels of n-alkanes several times higher than the levels in the animals described in de Groot's tests have been found in the lipids of grazing cattle slaughtered in Italy (the latter had not, of course, been fed on yeast-containing diets). In addition, analysis of random samples of butter, margarine, sunflower-oil, olive-oil and cereal products, such as grissini and rice, has shown that n-alkanes are present in concentrations equal to or greater than those in the tissues of Toprina-fed animals. Chain lengths of the fatty acids are similar in the two cases. It is concluded that, having regard to the relative rates of consumption of some of the staple foods analysed and of the meat and egg products that would be derived from yeast-fed animals, introduction of Toprina to the animal-feed market would have no significant qualitative or quantitative effect on human diets. Very similar conclusions have been published on the results of safety evaluation studies on other hydrocarbon-grown yeast S.C.P., for example in Japan (Yoshida, 1972), Italy (Giolitti, 1975), and Russia (Gradova et al., 1977).

B. Microbiological Quality

Recommendations have been published for microbiological standards of new or unconventional proteins for use in animal feeds (International Union of Pure and Applied Chemistry, 1974; Protein Advisory Group, 1974). It is clear from the survey reported by Mossel et al. (1973) that mixed animal feeds and milk replacers purchased on the British and Dutch markets sometimes contain significant numbers of species of Enterobacteriaceae although Salmonella spp. were not detected in any of the 100 samples examined. The experience of BP over a seven-year period indicates that this group of bacteria does not readily invade or proliferate in the S.C.P. produced by the process they employed. A working protocol for microbiological quality control of the Toprina process and product range has been detailed (Section VI. B).

Maximum numbers of organisms in each class were comparable with, or lower than, those in the market samples analysed (Mossel et al., 1973), a position confirmed by regular analyses (J. D. Levi, unpublished observations) of animal feed samples in the laboratories of the Central Institute for Nutrition and Food Research (C.I.V.O.) in the

Netherlands. Indirect evidence for the microbiological safety of S.C.P. is provided by the satisfactory results (de Groot *et al.*, 1970a, b, 1971) of the early feeding trials carried out on Toprina produced in the Grangemouth pilot plant before the 300 m³ demonstration plant was built. These samples contained a qualitatively similar microflora but the counts were higher.

C. Nutritional Testing

Toxicological test regimes are not intended to reflect desirable commerical practice, since concentrations of the test substance in the former are many times those which will be used in practical animal husbandry. Thus, it is necessary to evaluate the performance of S.C.P. in rations formulated on commercial lines. Again, the pioneers in this approach were Shacklady and his collaborators. Detailed results have been published in full (van Weerden *et al.*, 1970, 1972) and also reviewed (Engel, 1972; Shacklady and Gatumel, 1972; Shacklady, 1974; Shacklady and Walker, 1974) and only a brief survey will be given here. Pigs, even if early-weaned, thrive on diets containing Toprina replacing fish or milk proteins completely. Equally, soy- or fish-meal can be fully substituted by S.C.P. in poultry feeds. Several other research groups have reported successful use of Toprina in pig and poultry nutrition (Kneale, 1972; Shannon and McNab, 1972, 1973; Woodham and Deans, 1973; Beck and Gropp, 1974; Nielsen *et al.*, 1974; Tiens *et al.*, 1974; Shannon *et al.*, 1976; Mordenti *et al.*, 1977). Many more reports have been published on yeast S.C.P. from other identified or unidentified hydrocarbon-based processes (e.g. Pillai *et al.*, 1972; D'Mello, 1973; Seno *et al.*, 1973; Yarov *et al.*, 1973; Chierici, 1974; Tkachev and Grigorov, 1975; Russo *et al.*, 1976). The potential for S.C.P. in milk replacers for young animals, particularly calves, has been discussed in Section VI. C.

Mixed feeds for intensively reared animals, which provide much of the meat and eggs consumed in developed countries are, in the main, formulated by animal-feed compounders according to computer programmes based on the nutritional requirements of the target animal: requirements which are more precisely known than those for man. The formulation arrived at in any given case will depend on availability and current price of a wide range of ingredients. Agricultural products can fluctuate widely in price. For example, the ban placed by the United

Table 5

Net protein utilization, true digestibility and biological value of various materials with and without methionine supplementation. From Shacklady and Gatumel (1972).

	Net protein utilization	True digestibility	Biological value
BP Yeast Toprina G	59	96	61
BP Yeast Toprina G with 0.3% DL-methionine	88	96	91
Yeast, food grade	41	88	46
Yeast, food grade with 0.3% DL-methionine	87	90	96
Commercial soybean protein	42	100	42
Commercial soybean protein with 0.3% DL-methionine	64	99	65
Dried whole egg	90	100	90
Dried whole egg with 0.3% DL-methionine	97	100	97

States of America some years ago on soybean exports and the non-arrival of the Peruvian anchovy, which normally satisfies much of the world fish-meal market, during one recent fishing season, had very severe effects on prices in the animal feed market. Therefore, availability of an industrial product—S.C.P.—as an animal-feed ingredient could have an important stabilizing effect on the market, since its price is independent of weather, pestilence or crop failure.

However, it will not attract purchasers unless its composition is suitable for the intended range of uses. Comparative amino-acid analyses are given in Table 3 (p. 389) for S.C.P., conventional high-protein feed ingredients and reference proteins. Single-cell protein produced using n-alkanes is rich in lysine, often the first limiting amino-acid of the cereal grains which compose the bulk of animal-feed formulations, but it is low in sulphur-containing amino-acids. The latter are relatively abundant in cereals but, if supplementation is required, cheap synthetic methionine is available as a normal product of commerce. In tests for Biological Value and Net Protein Utilization of Toprina, it has been usual to supplement with methionine to produce the amino-acid balance that would normally, in a rich commercial diet, be provided by suitable admixture of S.C.P. and methionine-rich cereals (Table 5).

VIII. GENETIC MANIPULATION

A. Mutation

The improvement of protein content is the most obvious property of S.C.P. which the industrial microbiologist might seek to improve, since the value of the product is directly dependent on the 'crude protein content'. Clearly it is not easy to devise techniques for the selection of mutants with enhanced protein content, nor can it be assumed that such mutants will appear among the survivors of mutation treatments similar to those used to generate high-yielding strains of antibiotic-producing organisms. In the latter case, factors which inhibit growth may assist in diverting metabolic activity towards synthesis of secondary metabolites, whereas cellular protein synthesis is such a fundamental feature of the organism's economy that it may be doubted whether much improvement is possible.

The experience of one group of workers suggests that it is indeed possible, as reported in a patent (British Petroleum Co. Ltd., 1975a). They describe a technique for improving the protein content of *Saccharomycopsis lipolytica* in a continuous *n*-alkane fermentation. It is not clear from the specification what was the inducing agent for the presumed mutation, which is claimed to occur in a variety of strains of the species, but ingenious use was made of the continuous fermentation conditions to select the desired high-protein mutant. This, by virtue of its higher protein content reflecting a generally more efficient phenotype, outgrew the wild type. Although the increase in specific growth rate was apparently too small to measure, it appeared to be sufficient to ensure that the mutant predominated within a few hundred hours after inoculating a fermentation with a pure culture of the wild type. Fortunately, the mutant was recognizable in the microbiology laboratory because it had lost the capacity to produce pseudomycelium possessed by the wild-type culture. As there was no change in physiological characteristics, this illustrates yet another example of the artificiality of the generic barrier between *Candida* and *Torulopsis*, the non-pseudomycelial genus into which this mutant strain could have been classified had it been isolated *de novo* from nature (Yarrow and Meyer, 1978). This change in the organism at the level of cell structure cannot, however, be directly linked to the increase in intracellular protein-rich components as, under the normal highly-aerated continuous fermentation conditions maintained with n-alkanes as substrate, the wild type was observed to be free from pseudomycelium. Furthermore, any increase in average cell size, sufficient to account for the 10% increase in total nitrogen content obtained, was too small to be detectable.

Conveniently, loss of pseudomycelium is readily detected by examination of giant colonies of the yeast on streak plates. On certain commercially available sugar-based agar-containing media, the 'smooth' colony character of the mutant, non-pseudomycelial, strain contrasts sharply with the 'rough' pseudomycelial colony of the wild-type. Thus, in this instance, it was possible to define, control and patent 'smooth' cultures of *Saccharomycopsis lipolytica* which were subsequently in use in continuous fermentations (usually 5000 hours) for ten years following their original isolation, without any further detectable change in characteristics. This remarkable testimony to the rapid and efficient selective properties of chemostat culture, resulting so quickly in the dominance of a high-protein mutant, may illustrate the difficulties likely to be encountered by microbiologists seeking development of continuous-culture techniques for properties which are dissociated from (and may be counter to) rapid cell growth.

More specific means of enhancing the commercially important properties of S.C.P. cultures have been sought. A group of BP investigators (Jenkins and Raboin, 1978) obtained mutants of *Saccharomycopsis lipolytica* with decreased glycogen content by means of mutagen treatment followed by plating of survivors. Low-glycogen mutants so obtained were found to contain more protein than the corresponding wild-type cells.

Outside Great Britain, other groups have sought to improve process strains by mutational techniques (see Section IV. D). In Russia, significant increases in crude protein content or maximum specific growth rate were reported by Gradova *et al.* (1977) for *C. guilliermondii*, but they were unable to define precisely the mechanism responsible for the observed changes.

B. Breeding Techniques

In the field of baker's yeast, conventional cross-breeding techniques have been employed with most satisfactory results for commercially important properties of yeast used in the bakery trade, namely 'baking strength' (speed of carbon dioxide production in dough) and 'keeping quality' (i.e. percent baking quality retained after one week's storage). It may be presumed that the yield of yeast on substrate has also been improved. The techniques used have been described by Fowell (1969) and some of the results by Burrows (1970). When elucidation of sexuality in *Saccharomycopsis lipolytica* by Wickerham *et al.* (1970) opened up the possibility of cross-breeding in this species, the potentialities were perhaps overestimated. The range and diversity of distinct genetic lines of *Saccharomyces cereviseae* is extremely rich; in effect, every brewery, winery and baker's yeast factory in the world was likely to have evolved, over the years and under plant-specific conditions imposed, its own characteristic range of yeast races. No such industrial evolution, save the example described above, of *Saccharomycopsis lipolytica* could have occurred, since it has only been in use for fermentation for a few years in a limited number of locations. However, it was clearly worthwhile to assess the potentialities of cross-breeding in this yeast, having first established that Wickerham's work could be repeated. Some results from the BP team which pursued this course have been published (Forbes, 1978). Because of the limited genetic variation available from breeding stocks, attention was first focused on the production of diploids. Tests showed that, as expected, these had a larger cell size than normal haploids (in yeasts, cell volume is roughly proportional to ploidy), and the consequent higher protein content in the biomass was confirmed by analysis. Stability of the diploid cultures in the chemostat over prolonged periods of operation was likely to be a problem, and this point was checked carefully, but it was claimed that good stability was in fact obtained. No special techniques for ensuring this have been described. Obviously, monitoring is easy because of the greater cell size (about 30%) in diploid cultures.

In contrast to the situation in *Sacch. cerevisiae*, it is necessary to cross auxotrophs of *Saccharomycopsis lipolytica*, and a further restriction to the scope of any breeding programme is the need to find haploids of both mating types in nature, the haplophase being the 'natural' form of the species. In practice, most wild-type strains are of mating type B and the original breakthrough by Wickerman *et al.* (1970) arose from their discovery of the rare mating type A. It is clear that the I.F.P./Elf-E.R.A.P. group in France have been working on similar lines since they have patented diploid strains of *Saccharomycopsis lipolytica* which gave a high yield of α-oxoglutaric acid (Maldonado *et al.*, 1974).

Current research in other techniques of genetic modification includes protoplast fusion, which offers the possibility of conferring properties not previously found in the species by transfer of cellular elements from other organisms. Plasmid transfer may be of particular importance to the future of hydrocarbon microbiology, since the hydrocarbon substrate range of bacteria has already been extended by the use of this technique (Friello *et al.*, 1976). A recent review (Williams, 1978) gives details on this subject.

IX. BY-PRODUCTS

Although not strictly relevant to production of biomass, a discussion of metabolic by-products can hardly be divorced from it, because so much of the work published, especially from Japan (Arima, 1977; Yamada, 1976) shows that 'two-product' schemes are extremely popular. These are process routes in which biomass is produced, but a metabolite is coproduced or post-produced and harvested separately or extracted at some stage of the harvesting procedure. In some developments, the valuable metabolic product has become the *raison d'être* of the process, the biomass being harvested as the by-product.

Three reviews (Abbott and Gledhill, 1971; Fukui and Tanaka, 1971; Markovetz, 1978) consider the extracellular accumulation of metabolic products by hydrocarbon-utilizing micro-organisms. It now seems that interest in *n*-alkanes as substrates for this function is waning on economic grounds, methanol currently occupying the limelight.

The plant built by the Italian petrochemical group, Liquichimica, in Calabria, Italy, was designed to produce not only biomass but also citric acid and fatty acids (Venturini, 1977, 1978a). This plant, like that of Italproteine in Sardinia, has not been allowed to operate by the Italian authorities and the reported precarious financial position of the parent company (Venturini, 1978a,b) places the plant's future in some doubt. In addition, Italproteine was in the process of liquidation during 1978 (Done, 1978).

A. Lipids

Although microbial fats are not thought likely soon to become economic, several workers have considered the possibilities of exploiting the increase in cellular lipids of yeasts grown on n-alkanes, compared with glucose. Harries and Ratledge (1969) analysed the lipids of *Candida* sp. 107 grown on n-paraffins and found that the triglycerides possessed fatty acids which were typical of plant oils. In addition, about 75% of the triglycerides were of the symmetrical disaturated type: such a composition is known only in the valuable cocoa and illipé butters which show 'snapping'—the property characteristic of chocolate.

A cocoa-butter substitute derived from a microbiological source could be an attractive commercial prospect but, unfortunatly, insufficient quantities were synthesized by *Candida* 107 and *C. tropicalis* to make development worthwhile (Thorpe and Ratledge, 1972).

There may be more future in the biosynthesis of specific fatty acids, for example, those with the same chain length as the n-alkane substrate. Neither C_{12} nor C_{14} fatty acids, nor those with odd numbers of carbon atoms, are accessible from other sources. Even in this case, however, despite high conversion ratios (Whitworth and Ratledge, 1974), an economic process is not envisaged. Patents have been issued for the production of sebacic acid (1,10-decanedioic acid) from n-decane by *Pichia polymorpha* (Arima and Shigeo, 1970) and for lipids from C_{11} to C_{23} n-alkane mixtures by *C. guilliermondii, C. tropicalis* and *Saccharomycopsis lipolytica* (All Union Scientific Research Institute for Protein Biosynthesis, 1969, 1972b, 1973).

A Japanese group obtained yields of 5.8 mg of ergosterol/g dry weight from *C. tropicalis* and other species cultivated on n-alkane mixtures (Tanaka *et al.*, 1971). A patent exists for extraction of this compound from *C. tropicalis* (British Petroleum Co. Ltd., 1975b).

B. Carbohydrates

The increased yields of carbohydrates observed in some hydrocarbon fermentations have been exploited in the production of D-mannitol by *Saccharomycopsis lipolytica* (Dezeeuw and Tynan, 1972) and of unspecified sugars from a number of micro-organisms including the same species (Kyowa Hakko Kogyo Co. Ltd., 1969a).

Other reports concern conversion of citric acid fermentations to polyol fermentations in *Saccharomycopsis lipolytica* (Tabuchi and Hara, 1973) and *C. zeylanoides* (Hattori and Suzuki, 1974d), production of erythritol by *C. zeylandoides* (Hattori and Suzuki, 1974a) and conversion of this process to recover D-mannitol (Hattori and Suzuki, 1974b). The same group also report the production of D-arabitol from *C. tropicalis* (Hattori and Suzuki, 1974c).

C. Acids

Acids of the tricarboxylic acid cycle are among the degradation products of hydrocarbon oxidation, and special techniques can be employed to increase their production in fermentations. Much Japanese work has been published in this field, particularly on citric acid synthesis. Since the review of Shennan and Levi (1974), many more publications on citric acid synthesis by *n*-alkane-grown yeasts have appeared. *Saccharomycopsis lipolytica*, for instance, is cited as the organism in processes by Monsanto (Gledhill *et al.*, 1973); I.F.P. (Marchal *et al.*, 1977a, b) and the Tianjin Institute for Industrial Microbiology (1976) in Taiwan.

A team of workers at the University of Tokyo have used *C. brumptii* for production of succinic acid (Sato *et al.*, 1972) and, more recently, malic acid (Sato *et al.*, 1977). Production of α-oxoglutaric acid from various yeasts has been reviewed (Shennan and Levi, 1974). The use by I.F.P. of diploid strains of *Saccharomycopsis lipolytica* for α-oxoglutaric acid production has already been noted (Maldonado *et al.*, 1974).

D. Amino-Acids

Several Japanese companies have amino-acids in commercial production by fermentation, but to what extent they are using *n*-alkanes as substrates is by no means clear. Presumably, choice is determined by economic considerations. The Kyowa Hakko Kogyo Co. Ltd. were pioneers in this application of *n*-alkane fermentation techniques, with patents for L-tryptophan from *Saccharomycopsis lipolytica* (Kyowa Hakko Kogyo Co. Ltd., 1970a) and *C. tropicalis* (Kyowa Hakko Kogyo Co. Ltd., 1970b) and isoleucine from several organisms including *Saccharomycopsis lipolytica* (Kyowa Hakko Kogyo Co. Ltd., 1969b). A general amino-acid production patent for *Saccharomycopsis lipolytica* has been published (Ajinomoto Co. Ltd., 1962), while L-phenylalanine (Ajinomoto Co. Ltd., 1971), L-tyrosine (Ajinomoto Co. Ltd., 1972) and L-lysine (Gaillardin *et al.*, 1975) can be obtained from mutants of *Candida* species.

Several species produce L-glutamic acid. Miura *et al.* (1976), for example, have a patent for its production by a *Brevibacterium* strain on spent medium from *n*-alkane fermentations with *Saccharomycopsis lipolytica*.

E. Vitamins and Cofactors

Overproduction of certain B-group vitamins can be induced in yeasts and bacteria growing on *n*-alkanes. *Candida albicans, C. intermedia* and *C. tropicalis* all made large quantities of pyridoxin in the hands of Tanaka and Fukui (1967). In cultures containing glucose, *C. albicans* released pyridoxin into the broth but, on *n*-hexadecane, a very high intracellular yield of the vitamin was obtained (Tanaka *et al.*, 1967). Vitamin B_{12} production from photosynthetic bacteria, species of *Rhodopseudomonas* or *Rhodospirillum*, grown on *n*-alkanes, has been patented (Uehisa, 1972).

Certain strains of *C. krusei, C. parapsilosis, C. tropicalis* and *Pichia farinosa* (Teranishi *et al.*, 1971) and of *P. guilliermondii* (Nishio and Kamikubo, 1970, 1971) produce riboflavin: more that 250 mg/l being obtained from the latter species under optimized conditions.

Under defined conditions, including the addition of *p*-hydroxy-benzoic acid as a precursor, coenzyme Q was produced by yeasts in attractive quantities (Shimizu *et al.*, 1970; Ajinomoto Co. Ltd., 1973). Using an *n*-undecane-rich mixture of alkanes, a yield of 4.52 mg of the coenzyme/g was reported after only 43 hours of batch growth (Takeda Chemical Industries Ltd., 1970). Selected bacterial strains grown on *n*-paraffins produced coenzyme Q (Takeda *et al.*, 1968) and coenzyme A (Kuno *et al.*, 1973).

F. Other Products

Earlier reports of production of the 5'-purines, DNA, RNA, ribonuclease, crude protease and hydrolases from yeasts grown on *n*-alkanes have been reviewed (Shennan and Levi, 1974). More recently, *C. tropicalis* grown on *n*-alkanes has been cited as a potential source of catalase (Tanaka *et al.*, 1976), D-amino acid oxidase (Kawamoto *et al.*, 1977) and uricase (Tanaka *et al.*, 1977). A mutant strain of *Fusarium* sp. giving a high yield of alkaline protease is reported by Suzuki *et al.* (1976).

G. Co-oxidation

Most published examples of hydrocarbon co-oxidation by non-hydrocarbon-utilizing species are among the true bacteria and actinomycetes (Abbott and Gledhill, 1971; Raymond *et al.*, 1971). There are also a few examples among the yeasts; Tulloch *et al.* (1971) found that *Torulopsis magnoliae* could oxidize even-numbered *n*-alkanes from C_{16} to C_{24} to hydroxy fatty-acid sophorosides when growing on glucose. Similarly, Jones and Howe (1968a, b, c, d) reported almost complete conversion of *n*-alkanes, alk-1-enes and branched alkanes to glycolipids by *T. gropengeisseri*.

Conversion of *n*-hexadecane to *n*-hexadecanoic acid (palmitic acid) by *n*-alkane-grown cells of *Saccharomycopsis lipolytica* immobilized in polyacrylamide has been reported (British Petroleum Co. Ltd., 1976c).

X. THE MARKET FOR SINGLE-CELL PROTEIN

There can be no doubt that the delay to production of S.C.P. from *n*-alkanes imposed by the Japanese authorities some years ago, on the pretext of consumer pressure (Katoh, 1974; Yamada, 1976), coupled with the more recent difficulties placed in the way of operating S.C.P. plants actually built with official approval in Italy (Done, 1978), must raise grave doubts about the future of this technology for the time being. Only in the United Kingdom (I.C.I.'s methanol process) and India (Anon, 1978c) is there clear evidence of the intention to embark on full-scale production in Western-style economies. However, the continuing development of these projects and the existence of a number of S.C.P. plants based on liquid *n*-alkanes in the Soviet Union, possibly in Czechoslovakia, and one using a Japanese process in Romania (see Section IV. F) suggests that the setbacks described may only be temporary.

Certainly it would indeed be tragic if all of the careful development and testing lavished on this promising new technology were to be entirely wasted. It is to be hoped that those projects now in abeyance will be placed 'on-the-shelf' for possible future use and not abandoned completely. Clearly, quite minor changes in the economic climate, or its meteorological counterpart, resulting in a series of poor harvests, could transform both the profitability of S.C.P. production and the need for it in world terms. At the present time, however, it cannot be doubted that the 'protein gap' which was predicted in the early days of the S.C.P. story has not yet emerged as a problem sufficiently pressing to the world as a whole to give large-scale exploitation of the now proven technology for S.C.P. production from *n*-alkanes a high priority. It must also be admitted that the prospect for continuing escalation of oil prices is strong, while world agriculture has been able to cope with demand more effectively than had been predicted. Therefore, the prospect for profitable S.C.P. production seems unlikley to improve in the present circumstances, and the technique will thrive only where special considerations make it worthwhile.

REFERENCES

Abbott, B. J. and Gledhill, W. E. (1971). *Advances in Applied Microbiology* **14**, 249.

Ahearn, D. G. and Meyer, S. P., eds. (1973). 'Microbial Degradation of Oil Pollutants'. Centre for Wetland Resources, Louisiana State University, Baton Rouge, U.S.A.

Ainsworth, G. C., Sparrow, F. K. and Sussman, A. S. (1973). *In* 'The Fungi' (G. C. Ainsworth and A. S. Sussmann, eds.), Vol. IVa. Academic Press, London.

Ajinomoto Co., Ltd. (1962). Japanese Patent 29 206.

Ajinomoto Co., Ltd. (1971). British Patent 1 258 848.

Ajinomoto Co., Ltd. (1972). British Patent 1 297 989.

Ajinomoto Co., Ltd. (1973). United States Patent 3 769 170.

Alent'eva, E. S., Zhdannikova, E. N. and Kozlova, L. I. (1969). *In* 'Continuous Cultivation of Micro-organisms. Proceedings 4th Symposium', (I. Malek, ed.), p. 573. Czechoslovak Academy of Sciences.

All Union Scientific Research Institute for Protein Biosynthesis (1969). British Patent 1 173 484.

All Union Scientific Research Institute for Protein Biosynthesis (1972a). United States Patent 3 654 084.

All Union Scientific Research Institute for Protein Biosynthesis (1972b). French Patent 2 115 524.

All Union Scientific Research Institute for Protein Biosynthesis (1973). British Patent 1 3 9 114.

All Union Scientific Research Institute for Protein Biosynthesis (1974). British Patent 1 382 329.

Anon. (1973) *Process Biochemistry* **8** (11), 24.

Anon. (1976). *Protein Advisory Group Bulletin* **6** (2), 1.

Anon. (1978a). *Farmer's Weekly*, July 28th, p. 61.

Anon. (1978b). *Chemical Age*, 27th October, p. 14.

Anon. (1978c). *Trends in Biological Sciences*, July, p. N.160.

Aoki, H. and Umezawa, M. (1976). *Journal of Food Science and Technology, Tokyo*, **23**, 32.

Arima, K. (1977). *Developments in Industrial Microbiology* **18**, 79.

Arima, K. and Shigeo, H. (1970). Japanese Patent 24 392.

von Arx, J. A. (1972). *Antonie van Leeuwenhoek* **38**, 289.

Ballerini, D. (1978). *Revue de l'Institut Français du Pétrole*, **33**, 101.
Baptist, J. N., Gholson, R. K. and Coon, M. J. (1963). *Biochimica et Biophysica Acta* **69**, 40.
Barnett, S. M., Velankar, S. K. and Houston, C. W. (1974). *Biotechnology and Bioengineering* **16**, 863.
Bartha, R. and Atlas, R. M. (1977). *Advances in Applied Microbiology* **22**, 225.
Bassel, J., Phaff, H. J., Mortimer, R. K. and Miranda, M. (1978). *International Journal of Systematic Bacteriology* **28** (3), 427.
Beck, H. and Gropp, J. (1974). *Zeitschrift Tierphysiologie, Tierenährung und Futtermittelkunde* **33**, 158.
Beerstecher, E., Jr. (1954). 'Petroleum Microbiology'. Elsevier, New York.
Bemmann, W. and Tröger, R. (1975). *Zentralblatt für Bakteriologie, Parasitenkunde, Infektionskrankheiten und Hygiene. II Abteilung* **129**, 742.
Bianchi, J. P. and Michaux, J. P. (1974). 'Microbiologie Pétrolière', Companie Français des Pétroles.
Birckenstaedt, J. W., Faust, U. and Sambeth, W. (1977). *Process Biochemistry* **12** (9), 7.
Bizzi, A., Tacconi, M. T., Veneroni, E., Jori, A., Salmona, M., de Gaetano, G., Paglialunga, S. and Garattini, S. (1976). *Protein Advisory Group Bulletin,* **6**, 24.
Bos, P. (1975). PhD. Thesis: University of Delft.
Bos, P. and de Bruyn, J. C. (1973). *Antonie van Leeuwenhoek* **39**, 99.
British Petroleum Co., Ltd. (1963). British Patent 914 568.
British Petroleum Co., Ltd. (1966a). British Patent 1 017 584.
British Petroleum Co., Ltd. (1966b). British Patent 1 021 697.
British Petroleum Co., Ltd. (1967). British Patent 1 059 888.
British Petroleum Co., Ltd. (1971). British Patent 1 206 466.
British Petroleum Co., Ltd. (1973). British Patent 1 307 836.
British Petroleum Co., Ltd. (1975a). British Patent 1 401 277.
British Petroleum Co., Ltd. (1975b). French Patent 2 240 234.
British Petroleum Co., Ltd. (1976a). British Patent 1 419 953.
British Petroleum Co., Ltd. (1976b). British Patent 1 421 155.
British Petroleum Co., Ltd. (1976c). Belgian Patent 841 057.
British Petroleum Co., Ltd. (1977). British Patent 1 459 297.
British Petroleum Co., Ltd. (1978). British Patent 1 504 877.
Bruyn, J. (1954). *Koninke Nederlansch Akademie van Wetenschappen, Proceedings Series C* **57**, 41.
Buchanan, R. E. and Gibbons, N. E., eds. (1974). 'Bergey's Manual of Determinative Bacteriology', 8th Edition. Williams and Wilkins Co., Baltimore.
Burrows, S. (1970). *In* 'The Yeasts', (A. H. Rose and J. S. Harrison, eds.), Vol. 3, pp. 349–420. Academic Press, London.
Cardini, G. and Jurtshuk, P. (1970). *Journal of Biological Chemistry* **245** (11), 2789.
Chenouda, M. S. and Jwanny, E. W. (1972). *Journal of General and Applied Microbiology, Tokyo* **18** (3), 181.
Chepigo, S. V. (1972). *Journal 'D. I. Mendeleev' All-Union Chemical Society* **17**, 504.
Chierici, L. (1974). *Oli, Grassi Derivati* **10** (2), 11.
Chiou, C. J. and Chang, S. T. (1973). *Chung Kuo Nung Yeh Hua Hsueh Hui Chih* **11** (1–2), 21.
Cooper, P. G., Silver, R. S. and Boyle, J. P. (1975). *In* 'Single Cell Protein II', (S. R. Tannenbaum and D. I. C. Wang, eds.), p. 454. M.I.T. Press, Cambridge Massachusetts.
Crow, S. A., Meyers, S. P. and Ahearn, D. G. (1974). *La Mer* **12** (2), 95.
Dainippon Ink and Chemicals Inc. (1975). Japanese Patent 52 047 979.

David, R. (1976). *Financial Times*, October 5th.

Davis, P., ed. (1974). 'Single Cell Protein'. Academic Press, London.

Dezeeuw, J. R. and Tynan, E. J. (1972). German Patent 2 203 467.

Done, K. (1978). *Financial Times*, July 20th.

Dostalek, M., Munk, V., Volfova, O. and Fencl, Z. (1968). *Biotechnology and Bioengineering* **10**, 865.

Duppel, W., Lebeault, J. M. and Coon, M. J. (1973). *European Journal of Biochemistry* **36**, 583.

Dyatlovitskaya, E. V., Greshnykh, K. P. and Bergelson, L. D. (1965). *Prikladnaya Biochimiya Mikrobiologiya* **1**, 473.

Dyatlovitskaya, E. V., Greshnykh, K. P., Zhdannikova, E. N., Kozlova, L. I. and Bergelson, L. D. (1969). *Prikladnaya Biochimiya Mikrobiologiya* **5**, 511.

Ebbon, G. P. (1970). PhD. Thesis: University of London.

Einsele, A. and Fiechter, A. (1971). *Advances in Biochemical Engineering* **1**, 169.

Einsele, A. and Fiechter, A. (1972). *Pathologia et Microbiologia* **34**, 149.

Einsele, A., Schneider, H. and Fiechter, A. (1975). *Journal of Fermentation Technology* **53** (4), 241.

Engel, C. (1972). *In* 'Proteins from Hydrocarbons', (H. Gounelle de Pontanel, ed.), pp. 53-81. Academic Press, london.

Enterprise de Recherche et d'Activités Pétrolières (1974). British Patent 1 500 340.

Erikson, L. E. and Nakahara, T. (1975). *Process Biochemistry* **10** (5), 9.

Ermakova, I. T., Rosenfeld, S. M., Novakovskaya, N. S., Neklyudova, L. V. and Disler, E. N. (1969). *Prikladnaya Biokhimiya Mikrobiologiya* **5**, 252.

Fiechter, A. (1975). *In* 'Methods in Cell Biology', (D. M. Prescott, ed.), Vol. 11. pp. 97-130. Academic Press, New York.

Finnerty, W. R. (1977). *Trends in Biological Sciences*, April, p. 73.

Finnerty, W. R. Den, H., Jensen, D. and Voss, E. (1965). *Abhandlungen der Deutschen Akademie der Wissenschaften zu Berlin* (2), 143.

Finnerty, W. R. and Makula, R. A. (1975). *Critical Reviews of Microbiology* **4**, 1.

Forbes, A. D. (1978). *Proceedings of the First European Congress of Biotechnology*, Interlaken, Switzerland. (in press).

Fowell, R. R. (1969). *In* 'The Yeasts', (A. H. Rose and J. S. Harrison, eds.), Vol. I, pp. 303-383. Academic Press, London.

Friede, J. D. (1975). *Office of Naval Research, Contract N00014-73-C-0186, Annual Report No. 1*, U.S. Department of Commerce, Virginia, U.S.A.

Friello, D. A., Mylroie, J. R. and Chackrabarty, A. M. (1976). *Proceedings of the Third International Biodeterioration Symposium, 1975*, p. 205.

Fuhs, W. G. (1961). *Archiv für Mikrobiologie* **39**, 374.

Fukui, S. and Tanaka, A. (1971). *Proceedings of the Eighth World Petroleum Congress*, PD 21 (5).

Gaillardin, C., Heslot, H., Maldonado, P. and Sylvestre, G. (1975). French Patent 2 254 638.

Gallo, M., Bertrand, J. C., Roche, B. and Azoulay, E. (1973). *Biochimica et Biophysica Acta* **296**, 624.

Gallo, M., Roche-Penverne, B. and Azoulay, E. (1974). *Federation of European Biochemical Societies Letters* **46** (1), 78.

Gallo, M., Roche, B. and Azoulay, E. (1976). *Biochimia et Biophysica Acta* **419**, 425.

Gill, C. O. and Ratledge, C. (1972). *Journal of General Microbiology* **72**, 165.

Giolitti, G. (1975). *Annali della Accademia Agricoltura, Torino*, **117**, 79.

Gledhill, W. E., Hill, I. D. and Hodson, P. H. (1973). *Biotechnology and Bioengineering* **15**, 963.

Goma, G., Ani, D. A. and Pareilleux, A. (1976). *Proceedings of the Fifth International Fermentation Symposium, Berlin.* Abstracts, p. 131.

Gounelle de Pontanel, H., ed. (1972). 'Proteins from Hydrocarbons'. Academic Press, London.

Gradova, N. B., Rodionova, G. C., Zaikina, A. I., Bravicheva, R. N., Osipova, V. G., Kuranova, N. F. and Tretyakova, V. N. (1977). *Proceedings of the Fifth International Specialised Symposium on Yeasts,* Keszthely, Hungary. Part II, p. 65.

Greshnykh, K. P., Grigoryan, A. N., Dikanskaya, E. M., Dyatlovitskaya, E. V. and Bergelson, L. D. (1968). *Mikrobiologiya* **37** (2), 251.

Grimmer, G. (1974). *Deutsche Lebensmittel Rundschau* **70** (11), 394.

Grimmer, G. and Wilhem, G. (1969). *Deutsche Lebensmittel Rundschau* **65,** 229.

de Groot, A. P. (1974). *In* 'Single Cell Protein', (P. Davis, ed.), pp. 75–92. Academic Press, London.

de Groot, A. P., Dreff van der Meulen, H. C., Til, H. P. and Feron, V. J. (1975). *Food and Cosmetics Toxicology* **13,** 619.

de Groot, A. P., Til, H. P. and Feron, V. J. (1970a). *Food and Cosmetics Toxicology* **8,** 267.

de Groot, A. P., Til, H. P. and Feron, V. J. (1970b). *Food and Cosmetics Toxicology* **8,** 499.

de Groot, A. P., Til, H. P. and Feron, V. J. (1971). *Food and Cosmetics Toxicology* **9,** 787.

Gulf Research and Development Co. (1973). United States Patent 3 721 604.

Harada, Y., Ono, J. and Nagasawa, T. (1972). *Proceedings of the Fourth International Fermentation Symposium,* Kyoto, Japan, p. 479.

Harries, P. C. and Ratledge, C. (1969). *Chemistry and Industry,* p. 582.

Hatch, R. T. (1975). *In* 'Single Cell Protein II', (S. R. Tannenbaum and D. I. C. Wang, eds.), pp. 46–68. M.I.T. Press, Cambridge, Massachusetts.

Hattori, K. and Suzuki, T. (1974a). *Agricultural and Biological Chemistry* **38,** 581.

Hattori, K. and Suzuki, T. (1974b). *Agricultural and Biological Chemistry* **38,** 1203.

Hattori, K. and Suzuki, T. (1974c). *Agricultural and Biological Chemistry* **38,** 1875.

Hattori, K. and Suzuki, T. (1974d). *Agricultural and Biological Chemistry* **38,** 2419.

Herbert, D. (1961). *Society of Chemistry and Industry, Monographs* **12,** 21.

Hisatsuka, K., Nakahara, T., Sano, N. and Yamada, K. (1971). *Agricultural and Biological Chemistry* **35,** 686.

Hoechst AG (1973). German Patent 2 348 753.

Hoffmann, B. and Rehm, H. J. (1976a). *European Journal of Applied Microbiology* **3,** 19.

Hoffmann, B. and Rehm, H. J. (1976b). *European Journal of Applied Microbiology* **3,** 31.

Hornei, S., Kohler, M. and Weide, H. (1972). *Zeitschrift für Allgemeine Mikrobiologie* **12,** 19.

Hug, H. and Fiechter, A. (1973). *Archives of Microbiology* **88,** 87.

Iizuka, H., Suekane, M. and Nakajima, Y. (1965). *Journal of General and Applied Microbiology, Tokyo* **11,** 153.

Imada, O. and Hirotani, S. (1972). Japanese Patent 13 107.

Ishikura, I. and Foster, J. W. (1961). *Nature, London* **192,** 892.

International Union of Pure and Applied Chemistry (1974). Technical Bulletin No. 12.

Jenkins, P. G. and Raboin, D. (1978). *Journal of Applied Bacteriology* **44,** 279.

Johnson, M. J. (1964). *Chemistry and Industry,* p. 1532.

Jones, D. F. and Howe, R. (1968a). *International Union of Pure and Applied Chemistry. Fifth International Symposium on the Chemistry of Natural Products,* C63, p. 167.

Jones, D. F. and Howe, R. (1968b). *International Union of Pure and Applied Chemistry. Fifth International Symposium on the Chemistry of Natural Products,* C69, p. 174.

Jones, D. F. and Howe, R. (1968c). *Journal of the Chemical Society, Section C,* 2801.

Jones, D. F. and Howe, R. (1968d). *Journal of the Chemical Society, Section C,* 2809.

Jwanny, E. W. (1975). *Zeitschrift für Allgemeine Mikrobiologie* **15**, 423.

Kanazawa, M. (1975). *In* 'Single Cell Protein II'. (S. R. Tannenbaum and D. I. C. Wang, eds.), pp. 438–453. M.I.T. Press, Cambridge, Massachusetts.

Kanegafuchi Chemical Co., Ltd. (1970). British Patent 1 288 666.

Kanegafuchi Chemical Co., Ltd. (1972). British Patent 1 294 810.

Kanegafuchi Chemical Co., Ltd. (1973). British Patent 1 307 434.

Kanegafuchi Chemical Co., Ltd. (1975a). German Patent 2 454 048.

Kanegafuchi Chemical Co., Ltd. (1975b). Japanese Patent 16 430.

Käppeli, O. and Fiechter, A. (1977). *Journal of Bacteriology* **131**, 917.

Käppeli, O., Mueller, M. and Fiechter, A. (1978). *Journal of Bacteriology* **133,** 1952.

Karrick, N. L. (1977). *In* 'Effects of Petroleum on Arctic and Subarctic Marine Environments and Organisms', (D. C. Malins, ed.), Vol. 1, pp. 225–299. Academic Press, London.

Kasturi, K. and Tamhane, D. V. (1974). *Indian Journal of Experimental Biology* **12**, 101.

Katinger, H. (1973). *In* 'Advances in Microbial Engineering', (B. Sikyta, A. Prokop and M. Novak, eds.), pp. 485–505. Wiley London.

Katoh, K. (1974). *In* 'Single Cell Protein', (P. Davis, ed.), pp. 223–230, Academic Press, London.

Kawamoto, S., Kobayashi, M., Tanaka, A. and Fukui, S. (1977). *Journal of Fermentation Technology* **55**, 13.

Kester, A. S. and Foster, J. W. (1963). *Journal of Bacteriology* **85,** 859.

Kihlberg, R. (1972). *Annual Review of Microbiology* **26**, 427.

Klug, M. J. and Markovetz, A. J. (1966). *Bacteriological Proceedings*, Abst A39, p. 7.

Klug, M. J. and Markovetz, A. J. (1967a). *Journal of Bacteriology* **93,** 1847.

Klug, M. J. and Markovetz, A. J. (1967b). *Nature, London* **215,** 1082.

Klug, M. J. and Markovetz, A. J. (1968). *Journal of Bacteriology* **96,** 1115.

Klug, M. J. and Markovetz, A. J. (1971). *Advances in Microbial Physiology* **5,** 1.

Kneale, W. A. (1972). *Experimental Husbandry* **22,** 55.

Knecht, R., Prave, P., Seipenbusch, R, and Sukatsch, D. A. (1977). *Process Biochemistry* **12**, (4), 11.

Ko, P. C. and Yu, Y. (1968). *In* 'Single Cell Protein', (R. I. Mateles and S. R. Tannenbaum, eds.), pp. 255–262. M.I.T. Press, Cambridge, Massachusetts.

Ko, P. C., Yu, Y. and Li, C. S. (1964). 'Protein from Petroleum by Fermentation Process'. Chinese Petroleum Corporation, Taiwan.

Kuno, M., Kikuchi, M. and Nakao, Y. (1973). *Agricultural and Biological Chemistry* **37,** 313.

Kvasnikov, E. I., Isakova, D. M. and Vaskivnykh, V. T. (1967). *Mikrobiologiya* **36,** 932.

Kvasnikov, E. I., Masumyan, V. Y., Osadchaya, A. I., Shchelokova, I. F., Kuberskaya, S. L., Semenov, V. F. and Ostapchenko, T. P. (1969). *Mikrobiologiya Zhurnal, Kiev* **31**, 431.

Kyowa Hakko Kogyo Co., Ltd. (1969a). British Patent 1 141 107.

Kyowa Hakko Kogyo Co., Ltd. (1969b). British Patent 1 213 087.

Kyowa Hakko Kogyo Co., Ltd. (1970a). British Patent 1 186 952.

Kyowa Hakko Kogyo Co., Ltd. (1970b). British Patent 1 196 391.

Kyowa Hakko Kogyo Co., Ltd. (1972). Japanese Patent 38 183.

Kyowa Hakko Kogyo Co., Ltd. (1973). Japanese Patent 75 789.

Lainé, B. M. and du Chaffaut, J. (1975). *In* 'Single Cell Protein II', (S. R. Tannenbaum and D. I. C. Wang, eds.), pp. 424–437. M.I.T. Press, Cambridge, Massachusetts.

Lebeault, J. M., Lode, E. T. and Coon, M. J. (1971). *Biochemical and Biophysical Research Communications* **42,** 413.

Lefrancois, L. (1969). *Chimie et Industrie, France* **8,** 1038.

Lester, D. E. (1979). *Progress in Food and Nutrition Science* **3**, 1.
Lideman, L. F. and Yudintseva, I. M. (1969). *Prikladnaya Biochimiya Mikrobiologiya* **5**, 20.
van der Linden, A. C. and Thijsse, G. J. E. (1965). *Advances in Enzymology* **27**, 469.
Litchfield, J. H. (1977). *Food Technology*, May, p. 175.
Lodder, J., ed. (1970). 'The Yeasts. A Taxonomic Study'. North-Holland, Amsterdam.
Lonsane, B. K., Vadalkar, K., Nigam, J. N., Singh, H. D., Baruah, J. N. and Iyengar, M. S. (1973). *Indian Journal of Experimental Biology* **11**, 413.
Lowery, C. E. (1968). *Dissertation Abstracts* 461 B.
McGinnis, E, L. and Norris, M. S. (1975). *Preprints, Division of Petroleum Chemistry, American Chemical Society* **20**, 829.
Makula, R. A., Lockwood, P. A. and Finnerty, W. R. (1975). *Journal of Bacteriology* **121**, 250.
Maldonado, P., Gaillardin, C., Desmarquest, J. P. and Daniel, B. (1974). German Patent 2 358 312.
Malins, D. C., ed. (1977). 'Effects of Petroleum on Arctic and Subarctic Marine Environments and Organisms'. Academic Press, New York.
Marchal, R., Chaude, O. and Metche, M. (1977a). *European Journal of Microbiology* **4**, 111.
Marchal, R., Vandecasteele, J. P. and Metche, M. (1977b). *Archives of Microbiology* **113**, 99.
Markovetz, A. J. (1978). *Journal of the American Oil Chemists' Society* **55**, 430.
Markovetz, A. J., Cazin, J. and Allen, J. E. (1968). *Applied Microbiology* **16**, 487.
Masson, J. C. (1973). *In* 'Proceedings of the UNIDO Expert Group on Manufacture of Proteins from Hydrocarbons', Vienna, Austria.
Mateles, R. I. and Tannenbaum, S. R., eds. (1968). 'Single Cell Protein'. M.I.T. Press, Cambridge, Massachusetts.
Mauron, J. and Wuhrmann, J. J. (1971). FAO/WHO/UNICEF Protein Advisory Committee. Single-Cell Protein Meeting, Moscow. Item 5.
Meisel, M. N., Medvedeva, G. A. and Kozlova, T. M. (1976). *Mikrobiologiya* **45**, 844.
d'Mello, J. P. F. (1973). *Nutrition Reports International* **8**, 105.
Meyer, S. A., Anderson, K., Brown, R. E., Smith, M. T., Yarrow, D. Mitchell, G. and Ahearn, D. G. (1975). *Archives of Microbiology* **104**, 225.
Mishina, M., Isurugi, M., Tanaka, A. and Fukui, S. (1977a). *Agricultural and Biological Chemistry* **41**, 517.
Mishina, M., Isurugi, M., Tanaka, A. and Fukui, S. (1977b). *Agricultural and Biological Chemistry* **41**, 635.
Mitsui Petrochemical Industries Ltd. (1974). Japanese Patent 69 884.
Miura, Y. (1973). United States Patent 3 767 534.
Miura, Y., Okazaki, M. and Honda, T. (1976). Japanese Patent 35 488.
Miura, Y., Okazaki, M., Hamada, S., Murakawa, S. and Yugen, R. (1977). *Biotechnology and Bioengineering* **19**, 701.
Mizuno, M., Shimojima, Y. Iguchi, T., Takeda, I. and Senoh, S. (1966). *Agricultural and Biological Chemistry* **30**, 506.
Mobil Oil Corporation (1973a). United States Patent 3 767 527.
Mobil Oil Corporation (1973b). German Patent 2 152 549.
Monod, J. (1950). *Annales de l'Institut Pasteur, Paris* **79**, 390.
Moore, S. and Stein, W. H. (1951). *Journal of Biological Chemistry* **192**, 663.
Mordenti, A., Cenni, B., Tocchini, M., Monetti, P. G. and Parisini, P. (1977). *Zootechnica e Nutritzione Animale* **3** (2), 99.
Morinaga Milk Industry Ltd. (1973). British Patent 1 312 181.

Mossel, D. A. A., Shennan, J. L. and Vega, C. (1973). *Journal of the Science of Food and Agriculture* **24**, 499.

Munk, V., Volfova, O., Dostalek, M., Mostecky, J. and Pecka, K. (1969). *Folia Microbiologica* **14**, 334.

Muntyan, L. N. (1971). *Microbiologiya* **40**, 1005.

Nabeshima, S., Tanaka, A. and Fukui, S. (1970). *Journal of Fermentation Technology* **48**, 556.

Nakahara, T., Erikson, L. E. and Gutierrez, J. R. (1977). *Biotechnology and Bioengineering* **19**, 9.

Nieder, M. and Shapiro, J. (1975). *Journal of Bacteriology* **122**, 93.

Nielsen, H. E., Sriwaranard, P., Danielsen, V. and Eggum, B. O. (1974). *Zeitschrift für Tierphysiologie, Tierernährung und Futtermittelkunde* **33**, 151.

Nishio, N. and Kamikubo, T. (1970). *Journal of Fermentation Technology* **48**, 1.

Nishio, N. and Kamikubo, T. (1971). *Agricultural and Biological Chemistry* **35**, 485.

Nyns, E. J., Chiang, N. and Wiaux, A. L. (1968). *Antonie van Leeuwenhoek* **34**, 197.

Osumi, M., Fukuzumi, F., Yamada, N., Nagatani, T., Terranishi, Y., Tanaka, A. and Fukui, S. (1975). *Journal of Fermentation Technology* **53**, 244.

Protein Advisory Group (1974). Guideline No. 15.

Pelechova, J., Krumphanzl, V., Uher, J. and Dyr, J. (1971). *Folia Microbiologica* **16**, 103.

Phillips Petroleum Co. (1976). United States Patent 3 963 571.

Pillai, K. R., Singh, H. D., Barua, B., Baruah, J. N. and Iyengar, M. S. (1972). *Nutritional Reports International* **6**, 209.

Pirt, S. J. (1975). 'Principles of Microbe and Cell Cultivation'. Blackwell Scientific . Publications London.

Pixton, S. W. and Warburton, S. (1977). *Journal of Stored Product Research* **13**, 35.

duPont de Nemours (1973). United States Patent 3 773 621.

Popova, T. E. (1968). *Prikladnaya Biochimiya Mikrobiologiya* **4**, 103.

Prokop, A. and Erickson, L. E. (1972). *Biotechnology and Bioengineering* **14**, 533.

Prokop, A. and Sobotka, M. (1975). *In* 'Single Cell Protein II', (S. R. Tannenbaum and D. I. C. Wang, eds.), pp. 127–157. M.I.T. Press, Cambridge, Massachusetts.

Pyun, Y. R., Kwon, T. W. and Yu, J. H. (1977). *Hanguk Sitp'um Kwahakhoe Chi* **9**, 306.

Rabinovich, E. G., Yegorova, V. N. and Inge-Vechtomov, S. G. (1974). *Genetika* **10**, 92.

Ratledge, C. (1968). *Biotechnology and Bioengineering* **10**, 511.

Ratledge, C. (1978). *In* 'Developments in Biodegradation of Hydrocarbons', (R. J. Watkinson, ed.), Vol. I, pp. 1–46. Applied Science Publications.

Raymond, R. L., Jamison, V. W. and Hudson, J. O. (1971). *Lipids* **6**, 453.

Rhodes, A. and Fletcher, D. L. (1966) 'Principles of Industrial Microbiology'. Pergamon Press, London.

van Rij, N. J. W. and Verona, O. (1949). *Atti dell'Accademia Nazionale dei Lincei* **7**, 249.

Rodionova, G. S., Bravicheva, R. N. and Osipova, V. G. (1970). Russian Patent 1 310 671.

Rohde, H. G., Schröder, S., Schirpke, B. and Weide, H. (1975). *Zeitschrift für Allgemeine Mikrobiologie* **15**, 195.

Rosenfeld, S. M. and Disler, E. N. (1971). *Mikrobiologiya* **40**, 218.

Rosenfeld, S. M., Disler, E. N., Adanin, V. M., Zyakun, A. M. and Grishchenko, V. M. (1971). *Referativnyi Zhurnal Biologiya Khimiya* no. 17F 674.

Rudenko, V. I. (1971). *Mikrobiologiya Zhurnal , Kiev* **33**, 358.

Rudenko, V. I., Kvasnikov, E. I., and Shchelokova, I. F. (1970). *Mikrobiologiya Zhurnal, Kiev* **32**, 715.

Russo, V., Catalano, A., Mariani, P. and del Monte, P. (1976). *Animal Feed Science Technology* **1**, 25.

Sato, S., Nakahara, T. and Minoda, Y. (1977). *Agricultural and Biological Chemistry* **41,** 967.
Sato, M., Nakahara, T. and Yamada, K, (1972). *Agricultural and Biological Chemistry* **36,** 1969.
Senez, J. C. (1972). *In* 'Proteins from Hydrocarbons', (H. Gounelle de Pontanel, ed.), pp. 1–26. Academic Press, London.
Senez, J. C. and Azoulay, E. (1961). *Biochimica et Biophysica Acta* **47,** 1307.
Seno, F., Tada, M., Iwamoto, T., Murata, T. and Kawasaki, Λ. (1973). *Nippon Kakin Gakkaishi* **10,** 189.
Shacklady, C. A. (1974). *Process Biochemistry* **9** (10), 9.
Shacklady, C. A. (1975), *In* 'Single Cell Protein II', (S. R. Tannenbaum and D. I. C. Wang, eds.), pp. 489–504. M.I.T. Press, Cambridge, Massachusetts.
Shacklady, C. A. (1977a). Protein Advisory Group Symposium 'Investigations on Single Cell Protein'. Milan, Italy; Proceedings.
Shacklady, C. A. (1977b). *Proceedings of the Symposium on Zootechnology*, Milan, Italy.
Shacklady, C. A. and Gatumel, E. (1972). *In* 'Proteins from Hydrocarbons', (H. Gounelle de Pontanel, ed.), pp. 27–52. Academic Press, London.
Shacklady, C. A. and Walker, T. (1974). *Proceedings of the Fourth International Congress of Food Science and Technology* **1,** 715.
Shannon, D. W. F. and McNab, J. M. (1972). *British Poultry Science* **13,** 267.
Shannon, D. W. F. and McNab, J. M. (1973). *Journal of the Science of Food and Agriculture* **24,** 27.
Shannon, D. W. F., McNab, J. M. and Anderson, G. B. (1976). *Journal of the Science of Food Agriculture* **27,** 471.
Shennan, J. L. and Levi, J. D. (1974). *Progress in Industrial Microbiology* **13,** 1.
Shimizu, S., Tanaka, A. and Fukui, S. (1970). *Journal of Fermentation Technology* **48,** 549.
Shirota, S., Hayakawa, S., Watanabe, S., Iguchi, T. and Takeda, J. (1970). *Journal of Fermentation Technology* **48,** 747.
Silver, R. S. and Cooper, P. (1972). *American Chemical Society, Division of Petroleum Chemistry, Preprints* **17,** E 61.
Slater, L. E. (1974). *Food Engineering*, July, p. 68.
Souw, P., Reiff, I. and Rehm, H. J. (1976). *European Journal of Applied Microbiology* **3,** 43.
Stewart, J. E., Finnerty, W. R., Kallio, R. E. and Stevenson, D. P. (1960). *Science, New York* **132,** 1254.
Stross, F., Prokop, A., Hauser, K., Svojer, M. and Ademek, L. (1974). British Patent 1 348 074.
Suzuki, M., Kuno, M. and Nakao, Y. (1976). *Agricultural and Biological Chemistry* **40,** 365.
Tabuchi, T. and Hara, S. (1973). *Journal of the Agricultural Chemistry Society of Japan* **47,** 485.
Takata, T. (1969). *Hydrocarbon Processing*, March. p. 99.
Takeda Chemical Industries Ltd. (1970). British Patent 1 200 646.
Takeda, T., Tsuchimoto, N. and Shoichi, M. (1968). *Applied Microbiology* **16,** 1806.
Tanaka, A. and Fukui, S. (1967). *Journal of Fermentation Technology* **45,** 611.
Tanaka, A., Ohishi, N. and Fukui, S. (1967). *Journal of Fermentation Technology* **45,** 617.
Tanaka, A., Takahashi, R., Kawamoto, S. and Fukui, S. (1976). *Journal of Fermentation Technology* **54,** 850.
Tanaka, A., Yamada, R., Shimizu, S. and Fukui, S. (1971). *Hakko Kogaku Zasshi* **49,** 792.
Tanaka, A., Yamamura, M., Kawamoto, S. and Fukui, S. (1977). *Applied and Environmental Microbiology* **34,** 342.
Tannenbaum, S. R. and Wang, D. I. C., eds. (1975). 'Single Cell Protein II'. M.I.T. Press, Cambridge, Massachusetts.

Teranishi, Y., Shimizu, S., Tanaka, A. and Fukui, S. (1971). *Journal of Fermentation Technology* **49**, 213.

Texaco, Ltd. (1968). British Patent 1 204 646.

Thorpe, R. F. and Ratledge, C. (1972). *Journal of General Microbiology* **72**, 151.

Tianjin Institute for Industrial Microbiology (1976). *Wei Sheng Wu Hsueh Pao* **16**, 214.

Tiens, J., Gropp, J., Schulz, H., Erbersdobler, A. and Beck, H. (1974). *Zeitschrift für Tierphysiologie, Tierernährung und Futtermittelkunde* **34**, 86.

Ting, S., Li, C. and Lee, C. (1971). *Proceedings of the Symposium on Use of Radiation and Radioisotopes in Industrial Micro-organisms*, p. 259.

Tkachev, I. F. and Grigorov, V. V. (1975). *Internationale Zeitschrift der Landwirtschaft* **30**, 329.

Truhaut, R. and Ferrando, R. (1976). *Protein Advisory Group Bulletin* **6**, 15.

Tulliez, J. and Bories, G. (1975a). *Annales de la Nutrition et de l'Alimentation* **29**, 201.

Tulliez, J. and Bories, G. (1975b). *Annales de la Nutrition et de l'Alimentation* **29**, 213.

Tulloch, A. P., Hill, A. and Spencer, J. F. T. (1971). *Chemical Communications*, p. 584.

Uchio, R. and Shiio, I. (1972). *Agricultural and Biological Chemistry* **36**, 1389.

Uehisa, Y. (1972). Japanese Patent 50 397.

Ueno, K., Asai, Y., Shimada, M. and Kametani, T. (1974a). *Journal of Fermentation Technology* **52**, 867.

Ueno, K., Asai, Y., Yonemura, H. and Kametani, T. (1974b). *Journal of Fermentation Technology* **52**, 873.

Velankar, S. K., Barnett, S. M., Houston, C. W. and Thompson, A. R. (1975). *Biotechnology and Bioengineering* **17**, 241.

Venturini, M. (1977). *Chemical Age*, 4th February, p. 2.

Venturini, M. (1978a). *Chemical Age*, 28th April, p. 1.

Venturini, M. (1978b). *Chemical Age*, 29th September, p. 6.

Walker, J. D. and Cooney, J. J. (1973). *Canadian Journal of Microbiology* **19**, 1325.

van Weerden, E. J., Shacklady, C. A. and van der Wal, P. (1970). *British Poultry Science* **11**, 189.

van Weerden, E. J., Shacklady, C. A. and van der Wal, P. (1972). *Archiv für Geflügelkunde* **36**, 1.

Whitworth, D. A. and Ratledge, C. (1974). *Process Biochemistry* **9** (9), 14.

Wickerman, L. J., Kurtzman, C. P. and Herman, A. I. (1970). *Science, New York* **167**, 1141.

Williams, P. A. (1978). In 'Developments in Biodegradation of Hydrocarbons', (R. J. Watkinson, ed.), Vol. I. pp. 135–164. Applied Science Publishers.

Wolfe, D. A., ed. (1977). 'Fate and Effects of Petroleum Hydrocarbons in Marine Ecosystems and Organisms'. Pergamon Press, Oxford.

Woodbine, M. (1959). *Progress in Industrial Microbiology* **1**, 179.

Woodham, A. A. and Deans, P. S. (1973). *British Poultry Science* **13**, 569.

Yamada, K. (1976). *Protein Advisory Group Bulletin* **6**, 10.

Yamada, K. (1977). *Biotechnology and Bioengineering* **19**, 1563.

Yamaguchi, M., Sato, A. and Yukayama, A. (1976). *Chemistry and Industry*, p. 741.

Yarov, I. I., Basagin, N. N. and Shcherbak, L. I. (1973), *Vestnik Sel'skokhozyaistvennoi Nauki, Moscow* (2), 50.

Yarrow, D. and Meyer, S. A. (1978). *International Journal of Systematic Bacteriology* **28**, 611.

Yoshida, F. and Yamane, T. (1974). *Biotechnology and Bioengineering* **16**, 635.

Yoshida, F., Yamane, T. and Nakamoto, K. (1973). *Biotechnology and Bioengineering* **15**, 257.

Yoshida, M. (1972). *Sekiyu To Sekiyu Kagaku* **16**, 137.

Zaikina, A. I. (1971). *Mikrobiologiya* **40**, 236.

Zajic, J. E. (1964). *Developments in Industrial Microbiology* **6,** 16.
Zhukov, A. V., Davydova, I. M. and Vereshchagin, A. G. (1970). *Prikladnaya Biochimiya Microbiologiya* **6,** 627.
Zobell, C. E. (1946). *Bacteriological Reviews* **10,** 1.

AUTHOR INDEX

Numbers in italic are those pages on which the references are listed

Curley, R. L., 76, *88*
Currie, J. A., 10, *27*
Cysewski, G. R., 275, 280, *288*

D

Dabes, J. N., 184, *203*
Dainippon Ink and Chemicals Inc., 387, *411*
Dalton, H., 325, 333, *357, 360*
Dam, R., 198, 200, *203, 204*
Daniel, B., 404, 407, *415*
Danielson, V., 400, *416*
Darlington, W. E., 337, *357*
Dart, P. J., 68, 78, 81, 83, 84, 85, 86, *88*
Darwis, A., 116, *138*
Das, K., 279, *288*
Date, R. A., 67, 76, *88*
David, R., 366, *412*
Davidov, R. B., 218, *268*
Davidson, L. I., 95, *112*
Davis, J. B., 23, *27*
Davis, J. G., 223, *269*
Davis, J. R., 320, 332, *357*
Davis, P., 14, *27*, 362, 377, *412*
Davis, R. H., 320, *357*
Davydova, I. M., 391, *419*
Dawes, I. W., 22, *27*
Dawson, P. S. S., 19, *27*
Day, J., 85, *88*
Day, J. M., 68, 69, 78, 81, 82, 83, 84, 85, 86, *88*
Day, R., 321, *359*
Deans, P. S., 400, *418*
De Cesari, L., 48, *62*
Delaney, A. M., 259, *268*
Delbrück, M., 5, *27*
Delcaire, J. R., 144, *175*
Demmler, G., 215, 220, 232, 251, *268*
Den, H., 374, *412*
Denham, T. G., 149, *175*
Dennan, K. L., 184, *204*
Dennert, G., 168, *175*
Deschodt, C. C., 71, *89*
Desmarquest, J. P., 404, 407, *415*
Dezeeuw, J. R., 406, *412*
Di Fiore, L., 16, *27*
Dijken, J. P., van, 323, 324, 325, 327, 329, *357, 358*

Dikansky, S., 215, 219, 227, 233, 251, 258, *269*, 302, *312*
Dikauskaya, E. M., 391, 393, *413*
Diokno-Palo, N., 119, *138*
Disler, E. N., 377, 390, *412, 416*
D'Mello, J. P. F., 343, *357*
Dobereiner, J., 68, 69, 78, 81, 82, 83, 84, *88*
Dommergues, Y., 85, *89*
Donath, W. P., 118, 133, *139*
Done, K., 13, 26, *27*, 363, 405, 409, *412*
den Dooren de Jong, L. E., 321, *357*
Dostálek, M., 329, *357*, 371, 372, *412, 416*
Drake, B. B., 100, *113*
Dreff van der Meulen, H. C., 396, 398, *413*
Drews, B., 40, 41, *60*
Drews, S. M., 208, 209, 264, *268*
Droop, M. R., 185, *203*
Drozd, J. W., 327, 333, *357, 358, 360*
D'Silva, E., 71, *88*
Dube, J. N., 71, *88*
Dubrow, H., 215, *268*
Dugan, G. L., 192, *203*
Dulmage, H. T., 101, 102, 103, 104, 105, *113*
Dunlap, C. E., 274, 280, 281, 282, *286, 288*
Dumenil, G., 331, *357*
Dunn, I. J., 41, *60*
Dunnill, P., 10, *27*
Dupont, A., 119, *138*
Duppel, W., 376, *412*
Durand-Chastel, H., 193, 195, *203, 204*
Duthie, I. F., 298, 306, *311*
Dutky, S. R., 95, 106, 107, *113, 114*
Dworkin, M., 319, *357*
Dyatlovitskaya, E. V., 374, 391, 393, *412, 413*
Dyr, J., 391, *416*

E

Ebbon, G. P., 376, *412*
Ebbutt, L. I. K., 47, *60*
Eccleston, M., 325, *360*
Edani, Y., 133, *139*
Edelsten, D., 212, 218, *268*
Edwards, R. L., 146, *175*
Eger, G., 151, 171, *175*
Eggum, B. O., 400, *416*
Einsele, A., 371, 372, 381, *412*

P

AUTHOR INDEX   433

Raymond, R. L., 409, *416*
Read, W. H., 159, *175*
Reade, A. E., 306, *311, 312*
Recheigl, M., 7, *28*
Reed, G. R., 32, 33, 41, *62*
Reed, H. L., 325, *360*
Reese, E. T., 279, *288*
Rehm, H. J., 374, *413, 417*
Rehnborg, C., 102, *113*
Reiff, I., 374, *417*
Reilly, H. C., 168, *176*
Reilly, P. J., 276, *288*
Rennie, S. D., 46, *62*
Reuss, M., 329, *359*
Reusser, F., 39, *62*
Reuter, H., 219, 220, 225, 226, 229, 230, 232, 238, 239, 243, 245, 251, 258, 264, *268*
Rhodes, A., 300, *312*, 378, *416*
Rhodes, L. J., 106, *113*
Rhodes, R. A., 94, 95, 101, 105, 106, 107, 109, *112, 113, 114*
Rice, W. A., 85, *89*
Richards, M., 39, *62*
Ricica, J., 21, *28*
Ridge, E. H., 86, *89*
Righelato, R. C., 290, 292, 293, 294, 295, 309, *311, 312*
Rij, N. J. W., van, 370, *416*
Rinaudo, G., 85, *89*
Rixford, C. E., 192, *203*
Robbins, D. J., 127, 133, 134, *140*, 260, *268*
Robbins, E. A., 10, *28*
Robe, K., 216, 227, 256, 258, *269*, 302, *312*
Roberts, C., 51, *63*
Robinson, R. F., 303, *312*
Roche, B., 374, 376, *412*
Roche-Penverne, B., 376, *412*
Rodgers, N. E., 219, 236, *268*
Rodionova, G. C., 399, 403, *413*
Rodionova, G. S., 386, *416*
Roedjito, S. W., 121, *139*
Roelofsen, P. A., 118, 120, 133, *139*
Roels, J. A., 297, *312*
Rogoff, M., 105, *112, 113*
Rohde, H. G., 376, *416*
Rohr, M., 279, *286*
Rolle, I., 191, *205*
Rolz, C., 308, *312*
Romantschuk, H., 17, *28*

Rose, A. H., 11, 22, *28*, 294, *311*
Rosen, K., 33, *62*
Rosenfeld, S. M., 390, *416*
Roughley, R. J., 76, *88*
Rovira, A. D., 86, *89*
Roxburgh, J. M., 321, *358*
Rozenfeld, S. M., 377, *412*
Rubenchik, L. L., 76, 78, 80, *89*
Rudd, D. F., 344, 346, *359*
Rudenko, V. I., 392, *416*
Rusmin, S., 122, 125, 127, *139*
Russell, N. A., 85, *87*
Russell, R. M., 33, *62*
Russo, V., 400, *416*
Ryder, D. N., 294, *312*
Ryther, J. H., 182, *204*

S

Saadeh, M., 320, *360*
Sahm, H., 321, 325, 326, 329, *359*
Sahni, V. P., 74, *89*
St. Julian, G., 94, 95, 101, 106, 107, *112, 113*
Sak, S., 4, *28*, 33, *62*
Sakaguchi, K., 321, *359*
Salmona, M., 398, *411*
Salwin, H., 48, *62*
Sambeth, W., 385, *411*
Sandbank, E., 192, 201, *204*
Sanke, Y., 133, *139*
Sano, N., 373, *413*
Santillen, C., 193, *204*
Saono, S., 116, *139*
Sarkäny, J., 216, 219, 227, 239, 251, 258, *269*
Sasaki, T., 168, *175*
Sato, A., 373, *418*
Sato, H., 168, *176*
Sato, M., 407, *417*
Sato, S., 407, *417*
Sato, T., 44, *62*
Sauer, K. H., 184, *203*
Saunders, R. M., 43, 44, *62*
Scammell, G. W., 295, 299, 306, *311*
Scargill, I., 3, *28*
Schaarschmidt, B., 52, *61*
Schaefer, G., 118, 119, 123, 131, 133, *140*
Schank, S. C., 83, 84, *89*

SUBJECT INDEX

A

Accelerating growth phase, 20
Acceptability
  algal biomass, 178
  micro-algal single cell protein, 200
Achlya bisexualis, growth rate, 295
Achromobacter spp.
  for azofication, 85
  hydrocarbon fermentation by, 387
Acidification, soybeans in tempe manufacture, 125
Acids, production in hydrocarbon fermentation, 407
Acinetobacter sp.
  growth on methanol, efficiency, 330, 331
  hydrocarbon assimilation, 368
  lipids in, 392
Acinetobacter calcoaceticus, growth on ethanol, 16
Acinetobacter cerificans
  hydrocarbon fermentation by, 387
  hydrocarbon utilization, hydroperoxidation in, 374
Actinomyces spp., hydrocarbon assimilation, 368
Actinomycetes in single cell protein manufacture from cellulose, 280
Active dried yeast, 32, 45-48
Additives
  in active dried yeast preparation, 46
  in biomass production from whey, 239
4,9,-Adenylbutyric acid, 2(*R*),3(*R*)-dihydroxy- — *See* Eritadenine

Aeration
  in *Agaricus bisporus* manufacture, 152
  in baker's yeast production, 32, 35-37
  in biomass production from whey, 255
  in hydrocarbon fermentation, 384
Aerobacter spp., for azofication, 85
Aerobacter aerogenes
  acceptability, 9
  as insecticide, 92
  cultivation on whey, 214
Aeromonas spp., hydrocarbon assimilation, 368
Aflatoxin in tempe manufacture, 133
Agglomeration, single cell protein from methane, 341
Agitation
  in hydrocarbon fermentation, 381
  in methane fermentation, 340
Agaricus spp., cultivation, 2
Agaricus bisporus
  competitors, 157
  consumption, 146
  cultivation, 142, 170
  germination, 147
  life cycle, 142
  manufacture pests and diseases, 156-159
  morphology, 142, 143
  production, 142
    procedures, 147-160
  world production, 144, 146
Agaracus campestris, morphology, 142
Agricultural wastes, single cell protein production on, 17
Agritrol, 98, 99

439

X

Y

Z